Integrated Circuits and Systems

T0191719

Series Editor

A. Chandrakasan
Massachusetts Institute of Technology
Cambridge, Massachusetts

For further volumes:
http://www.springer.com/series/7236

Stephen W. Keckler · Kunle Olukotun ·
H. Peter Hofstee

Editors

Multicore Processors
and Systems

 Springer

Editors
Stephen W. Keckler
The University of Texas at Austin
1 University Station, C0500
Austin, TX 78712-0233
USA
skeckler@cs.utexas.edu

H. Peter Hofstee
IBM Systems and Technology Group
11500 Burnet Rd.
Austin, TX 78758
USA
hofstee@us.ibm.com

Kunle Olukotun
Department of Electrical Engineering and
    Computer Science
Stanford University
Stanford, CA 94305-9030
USA
kunle@stanford.edu

ISSN 1558-9412
ISBN 978-1-4614-2450-5   e-ISBN 978-1-4419-0263-4
DOI 10.1007/978-1-4419-0263-4
Springer New York Dordrecht Heidelberg London

Springer is part of Springer Science+Business Media (www.springer.com)

Preface

While the last 30 years has seen the computer industry driven primarily by faster and faster uniprocessors, those days have come to a close. Emerging in their place are microprocessors containing multiple processor cores that are expected to exploit parallelism. The computer industry has not abandoned uniprocessor performance as the key systems performance driver by choice, but it has been forced in this direction by technology limitations, primarily microprocessor power. Limits on transistor scaling have led to a distinct slowdown in the rate at which transistor switching energies improve. Coupled with clear indications that customers' tolerance level for power per chip has been reached, successive generations of microprocessors can now no longer significantly increase frequency. While architectural enhancements to a single-threaded core cannot pick up the slack, multicore processors promise historical performance growth rates, albeit only on sufficiently parallel applications and at the expense of additional programming effort.

Multicore design has its challenges as well. Replication of cores communicating primarily with off-chip memory and I/O leads to off-chip bandwidth demands inversely proportional to the square of the feature size, while off-chip bandwidths tend to be limited by the (more or less fixed) perimeter from which signals must escape. This trend leads to significant pressure on off-chip signaling rates, which are beginning to exceed the clock frequency of the processor core. High-bandwidth packaging such as that provided by 3D-integration may soon be required to keep up with multicore bandwidth demands.

Compared to a uniprocessor, a multicore processor is somewhat less generic, as uniprocessors are generally more efficient at emulating a parallel processor than vice versa. In the space of emerging applications with inherent concurrency, an ideal multicore processor would become part of a general-purpose system with broad enough markets to sustain the required investments in technology. However, this reality will only come to pass if the programming tools are developed and programmers are trained to deliver the intrinsic performance of these processors to the end applications.

This book is intended to provide an overview of the first generation of multicore processors and systems that are now entering the marketplace. The contributed chapters cover many topics affecting, and associated with, multicore systems including technology trends, architecture innovations, and software and programming issues. Chapters also cover case studies of state-of-the-art commercial and research multicore systems. The case studies examine multicore implementations across different application domains, including general purpose, server, signal processing, and media/broadband. Cross-cutting themes of the book are the challenges associated with scaling multicore systems to hundreds and ultimately thousands of cores. We expect that researchers, practitioners, and students will find this book useful in their educational, research, and development endeavors.

We organize the chapters around three major themes: (1) multicore design considerations, (2) programmability innovations, and (3) case studies. While each of these areas are broader than this book can cover, together the chapters introduce the fundamentals of modern and future multicore design. While the third section of this book is dedicated to case studies, the first section also discusses prototype multicore architectures to exemplify the design considerations. In all, this book describes commercial or research implementations of seven different multicore systems or families of systems.

Multicore Design Considerations

Chapter 1 introduces tiled multicore designs, first embodied in the Raw processor chip. Tiling is both a natural and necessary method of constructing a complex system from multiple copies of a simple processor tile. Furthermore, because tiled architectures are inherently partitioned, they are less vulnerable to the deleterious effects of decreasing wire speed relative to transistor speed. Each tile also includes a network router and ports to its neighbors so that scaling a tiled design can be accomplished merely by adding more tiles, logically snapping them together at the interfaces. The perimeter of the tiled array can be used for connections to memory or to connect multiple chips together. The Raw prototype chip consists of a 4×4 array of tiles, with each tile comprising a simple processing pipeline, instruction memory, and data memory. The same design principles apply to Tilera's TILE64 chip, a commercial venture that is descended from Raw.

The key feature of the Raw system is the tight integration of the network into the processor pipeline. Raw exposes inter-tile interaction through register-mapped network ports; reading or writing the designated register name injects an operand into or extracts an operand from the network. Raw also employs a statically programmable router which executes a sequence of routing instructions to route operand packets in concert with the computation instructions executed by the core pipeline. The authors of the chapter show that applications with different characteristics can be executed by the Raw tiled substrate because the underlying computation and communication mechanisms are exposed to software. For example, exposing the operand communication enables serial programs to be partitioned across the tiles with producer/consumer instructions interacting through the operand network. The network also enables fine-grained parallel streaming computation in which small data records are processed in a pipelined fashion across multiple tiles. The Raw chip makes a compelling case for distributed architectures and for exposing communication to software as a first-class object. These themes recur in different ways throughout the chapters of this book.

Chapter 2 focuses on the design and implementation of on-chip interconnects (networks-on-chip or NOCs), as a critical element of multicore chips. When a chip contains only a few processors and memory units, simple interconnects such as buses and small-scale crossbars suffice. However, as the demand for on-chip communication increases, so does the need for more sophisticated on-chip networks. For example, the IBM Cell chip employs a bidirectional ring network to connect its eight data processors, its Power core, and its off-chip interfaces. The Tilera chip employs a collection of mesh networks to interconnect 64 on-chip processors. This chapter first examines the fundamentals of interconnection network design, including an examination of topologies, routing, flow control, and network interfaces in the context of NOCs. The chapter also discusses the trade-offs for router microarchitecture designed to meet the area, speed, and power constraints for multicore chips.

The chapter also includes two case studies of NOCs deployed in different ways in two recent prototype chips. The TRIPS processor employs a routed network, rather than a conventional broadcast bus, to deliver operands from producing instructions to consuming instructions in a distributed uniprocessor microarchitecture and to connect the processing cores to the level-1 caches. The operand network has 25 terminals and optimizes the routers for latency with a total of one clock cycle per hop. The network is connected directly to the inputs and the outputs of the ALUs and supports fixed length, operand-sized packets. By contrast, the Intel TeraFLOPS chip connects 80 on-chip cores with a high-speed, clock-phase-skew tolerant, inter-router protocol. The TeraFLOPS routers employ a five-stage pipeline and run at 5 GHz in 65 nm. Message injection and reception is explicit with send and receive instructions that use local registers as interfaces to the network. TRIPS and the TeraFLOPS chip make different choices about routing algorithms and flow control based on the demands of their particular deployments. One message from this chapter is that NOCs are not "one-size-fits-all" and should be tailored to the needs of the system to obtain the fastest and the most efficient design. The area of NOC architecture is extremely active at present, with new innovations being published on topologies, routing algorithms, and arbitration, among other topics.

Chapter 3 examines the granularity of cores within a multicore system. A survey of commercial multicore chips shows that core size and number of cores per chip vary by almost two orders of magnitude. At one end of the spectrum are *bulldozer* processors which are large, wide issue, and currently are deployed with only 2–4 cores per chip. At the other end of the spectrum are the *termites* which are small and numerous, with as many as 128 cores per chip in 2008 technology. The trade-off between these two ends of the spectrum are driven by the application domain and the market targeted by the design. Bulldozers are the natural extensions of yesterday's high-performance uniprocessors and target the general-purpose market. Termites typically target a narrower market in which the applications have inherent concurrency that can be more easily exploited by the more numerous yet simpler cores.

This chapter points out that the diversity in multicore granularity is an indicator of greater hardware and software specialization that steps further away from the general-purpose computing platforms developed over the last 30 years. To address this challenge, the chapter describes a general class of composable core architectures in which a chip consists of many simple cores that can be dynamically aggregated to form larger and more powerful processors. The authors describe a particular design consisting of 32 termite-sized cores which can be configured as 32 parallel uniprocessors, one 32-wide, single-threaded processor, or any number and size of processors in between. The chapter describes a set of inherently scalable microarchitectural mechanisms (instruction fetch, memory access, etc.) that are necessary to achieve this degree of flexibility and that have been prototyped and evaluated in a recent test chip. Such a flexible design would enable a single implementation to match the parallelism granularity of a range of applications and even adapt to varying granularity within a single program across its phases. Further, this class of flexible architectures may also be able to feasibly provide greater single-thread performance, a factor that is being sacrificed by many in the multicore space.

Programming for Multicores

While there are a number of challenges to the design of multicore architectures, arguably the most challenging aspect of the transition to multicore architectures is enabling mainstream application developers to make effective use of the multiple processors. To address this challenge we consider in this section the techniques of thread-level speculation (TLS) that can be used to automatically parallelize sequential applications and transactional memory (TM) which can simplify the task of writing parallel programs. As is evident from the chapters that describe these techniques, their efficient implementation requires the close interaction between systems software and hardware support.

Chapter 4 introduces speculative multithreaded or thread-level speculation architectures. The authors first provide motivation for these architectures as a solution to the problems of limited performance and implementation scalability associated with exploiting ILP using dynamic superscalar architectures and the parallel programming problems associated with traditional CMPs.

Speculatively multithreaded architectures eliminate the need for manual parallel programming by using the compiler or hardware to automatically partition a sequential program into parallel tasks. Even though these speculatively parallel tasks may have data-flow or control-flow dependencies, the hardware executes the tasks in parallel but prevents the dependencies from generating incorrect results. The authors describe in detail the approach taken to speculative multithreading in the Multiscalar architecture which was the earliest concrete example of speculative multithreading. A key element of the Multiscalar approach is the use of software and hardware to their best advantage in the implementation of speculative multithreaded execution. Software is used for task partitioning and register data-dependency tracking between tasks, which is possible with static information. Hardware sequences the tasks at runtime and tracks inter-task memory dependencies, both of which require dynamic information. The hardware support for memory dependency tracking and memory renaming required to support speculative multithreading in the Multiscalar architecture is called the speculative version cache (SVC) and adds considerable complexity to the design of a CMP. As the authors describe, this added complexity can buy significant performance improvements for applications that cannot be parallelized with traditional auto-parallelizing technology. Despite the hardware complexity of speculative-multithreading architectures, the desire to reduce the software disruption caused by CMPs spurs continued interest in these architectures especially in conjunction with dynamic compiler technology.

Chapter 5 presents the design of Transactional Memory (TM) systems. The authors begin by describing how TM can be used to simplify parallel programming by providing a new concurrency construct that eliminates the pitfalls of using explicit locks for synchronization. While this construct is new to the mainstream parallel programming community it has been used and proven for decades in the database community. The authors describe how programming with TM can be done using the *atomic* keyword. With the high-level atomic construct, programmers have the potential to achieve the performance of a highly optimized lock-based implementation with a much simpler programming model.

The authors explain that the key implementation capabilities of a TM system include keeping multiple versions of memory and detecting memory conflicts between different transactions. These capabilities can be implemented in a software TM (STM) or in a hardware TM (HTM) using mechanisms similar to those required for speculative multithreading or a combination of both software and hardware in a hybrid-TM. Because one of the major challenges of STM implementations is performance, the chapter goes into some detail in describing compiler-based optimizations used to improve the performance of an STM by reducing the software overheads of STM operations. The key to efficient optimization is a close coupling between the compiler and the STM algorithm. While STMs support TM on current CMPs, significant performance improvements are possible by adding hardware support for TM. The chapter explains that this support can range from hardware-enhanced STM that reduces software overheads to a pure HTM that completely eliminates all software overheads and achieves the highest performance. The authors conclude that TM deployment will include the use of STMs for existing CMP hardware, but must encompass language, compiler, and runtime support to achieve adequate performance. Ultimately, hardware acceleration will be required to achieve good TM performance.

Case Studies

The last set of chapters comprise four case studies of multicore systems. In each case study, the authors describe not only the hardware design, but also discuss the demands on the systems and application software needed as a part of the multicore ecosystem. We selected these chapters as they represent different points in the design space of multicore systems and each target different applications or multicore programming models.

Chapter 6 presents the architecture and system design of the AMD Opteron family of general-purpose multicore systems. The authors start by describing the trends and challenges for such general-purpose systems, including power consumption, memory latency, memory bandwidth, and design complexity. While multicore systems can help with some aspects, such as design complexity through replication, others are much more challenging. Power consumption must be shared by multiple processors and with strict power envelopes, not all of the processors can simultaneously operate at their peak performance. Likewise, external bandwidth (memory and interconnect) becomes a greatly constrained resource that must be shared. The challenges are no less severe on the software side, as parallel programming environments for general-purpose applications are in their infancy and system software, such as operating systems, are not yet prepared for increasing levels of concurrency of emerging multicore systems.

The authors then present the Opteron family of multicore systems, which were designed from the outset to be scaled both internally (more processors per core) and externally (more chips per system). These processor cores fit into the category of bulldozers, as they are physically large and can support high single-thread performance. The authors emphasize that because all but small-scale computer systems will be composed of many multicore chips, design for system-level scalability is critical. The chapter describes the details of the system-level architecture that is implemented on the processor chips, including multiple DRAM interfaces, interchip interconnection (Hypertransport), and the system-level cache hierarchy. The lessons from this chapter are that general-purpose multicore systems require a balanced design but that substantial challenges in programmability, system software, and energy efficiency still remain.

Chapter 7 details Sun Microsystems' Niagara and Niagara 2 chips, which are designed primarily for the server space in which job throughput is more important than the execution latency of any one job. The authors make the case that simple processors that lack the speculation and out-of-order execution of high-end uniprocessors are a better match for throughput-oriented workloads. Such processors with shallower pipelines and slower clock rates achieve better power efficiency and can employ multithreading as a latency tolerance technique since independent threads are abundant in these workloads. Because these processors are smaller and more power efficient, more of them can be packed onto a single chip, increasing overall throughput. These processors can be classified as *chainsaws*, as they are smaller than the bulldozers, yet larger and more full-featured than termites.

In their case studies, the authors describe in detail the microarchitecture of Sun's line of multicore chips. The first generation Niagara chip employs eight four-way multithreaded cores, for a total of 32 simultaneously executing threads. The processors share a four-banked level-2 cache, with each bank connected to an independent DRAM memory controller. The second-generation Niagara 2 doubles the number of threads per core (for a total of 64 simultaneously executing threads) and doubles the number of L2 banks to increase on-chip memory-level parallelism. While Niagara had a single floating-point that is shared across all of the cores, each Niagara 2 core has its own floating-point and graphics unit. Both Niagara systems employ cache coherence that spans the cores on the chip and multiple chips in a system. The chapter includes experimental results that show nearly a factor of 2 boost in power efficiency (performance/Watt) over more complex core systems. Like the authors of Chapter 6, the authors of this chapter indicate that the expected increase in cores and threads will demand more system support, such as operating system virtualization.

Chapter 8 describes a family of stream processors, which are multicore systems oriented around a data-streaming execution model. Initially amenable to multimedia and scientific applications, stream processing exploits parallelism by simultaneously operating on different elements of a stream of data in a fashion similar to vector processors. By making data streams first-class objects that are explicitly moved through levels of the memory hierarchy, stream processors eliminate power and area inefficient caches. SIMD-style computation reduces the amount of control logic required for stream processors. The authors claim that these factors provide a 10–30 times improvement in power efficiency over conventional multicore architectures.

The chapter first details a stream program model which partitions programs into computation kernels and data streams. Kernels can be thought of as data filters in which an input stream or streams is transformed into an output stream. A stream program is represented as a stream flow graph which has nodes that correspond to kernel computations and edges that correspond to the streams. The program itself includes explicit operations on streams to move them through the memory hierarchy and instantiations of kernels between stream transfers. High-level programming languages such as Brook and Sequoia have been developed to target stream processing execution models.

The chapter then describes a general stream processor microarchitecture as well as three case studies: (1) Imagine stream processor, (2) Stream Processors Inc. Storm I processor, and (3) Merrimac streaming supercomputer architecture. Each of these designs shares several common design principles including an explicitly managed memory hierarchy, hardware for bulk data transfer operations, and simple lightweight arithmetic cores controlled through a hybrid SIMD/VLIW execution model. These simple processing elements fall into the category of termites, since they have simple control logic and little memory per core. For example, the Storm I processor has a total of 80 ALUs organized into 16 lanes that execute in parallel. Stream processors exemplify performance and power efficiency that can be obtained when the programming model and application domain allows for explicit and organized concurrency.

Chapter 9 describes the Cell Broadband Engine architecture and its first implementations. This family of processors, jointly developed by IBM, Toshiba, and Sony falls at the upper end of the chainsaw category. The Cell Broadband Engine (Cell B.E) is a heterogeneous multiprocessor with two types of programmable cores integrated on a single chip. The Power processor element (PPE) is a Power architecture compliant core that runs the operating system and orchestrates the eight synergistic processor elements (SPEs). Per-core performance of the SPEs on compute-intensive applications is comparable to that of bulldozer cores, but each SPE requires substantially less power, less area, and fewer transistors to implement, allowing a nine-core Cell B.E. in a chip the size of a typical dual-core bulldozer.

Like the stream processors described in Chapter 8, the SPEs achieve their efficiency by explicitly managing memory. The SPEs manage one additional level of memory than typical processors, a 256 kB local store included in each SPE. SPE DMA operations are the equivalents of the PPE's load and store operations, and access coherent shared memory on the processor in exactly the same way as the PPE's load and stores. Instead of targeting the SPE's register file, however, the DMA operations place code and data in the local store (or copy it back to main memory). The SIMD-RISC execution core of the SPE operates asynchronously on this local store. Each SPE is an autonomous single-context processor with its own program counter capable of fetching its own code and data by sending commands to the DMA unit.

The management of the local store is an added burden on the software. Initial versions of the software development environment for the Cell Broadband Engine required the application programmer to partition code and data and control their movement into and out of the local store memory. A more recent version removes the burden of code partitioning from the programmer. Following the pioneering efforts of a number of research compilers for the Cell B.E. and other multicore processors, standards-based approaches to programming this processor are now available. These programming languages and frameworks abstract the notions of locality and concurrency and can be targeted at a wide variety of multicore CPUs and (GP)GPUs allowing portable approaches to developing high-performance software and providing a path towards bringing heterogeneous computing into the mainstream.

Current and Future Directions

While the case studies in this book describe a range of architectures for different types of applications, we recognize that the design and application space is much broader. For example, we do not discuss the multicore designs of existing and emerging graphics processing units (GPUs), such as those from NVidia, Intel, and AMD, a subject which merits its own book. The graphics application domain has traditionally exposed substantial pixel and object-level parallelism, a good match to the growing number of on-chip processing elements. Interestingly, there is substantial diversity among GPUs in the number and complexity of the processing elements. The NVidia GeForce 8800 employs 128 simple cores that can operate as independent MIMD processors with non-coherent memory. Intel's Larrabee system employs fewer but larger x86-based cores with local cache and a coherent memory. AMD's Radeon HD 2900 employs four parallel stream processing units, each with 80 ALUs controlled using a hybrid of VLIW and SIMD. Successive generations of GPUs have become more programmable and currently require domain-tailored program systems such as NVidia's CUDA and AMD's CAL to achieve both performance and programmability. The differences in approaches among the architectures targeted at the graphics domain share similarities with the distinctions between some of the architectures described in the chapters of this book.

Looking forward, Moore's law of doubling integrated circuit transistor counts every 18–24 months appears likely to continue for at least the next decade. The implication for multicore systems is the expected doubling of core count per technology generation. Given the starting point of 4–64 processors in today's commercial systems, we expect to see chips that have the capacity to contain hundreds or even thousands of cores within 10 years. However, the challenges that are described throughout this book will remain. Keeping power consumption within thermal limits will not become any easier with increasing transistor density and core counts. Programming these parallel systems is far from a solved problem, despite recent advances in parallel programming languages and tools. Developing solutions to these problems will be critical to the continued growth and performance of computer systems.

As technology matures further, eventually performance improvements will be possible only with gains in efficiency. Besides increased concurrency, the only significant means to gain efficiency appears to be increased specialization. In the context of multicore systems, cores would be specialized through configuration at run-time, at manufacture, or through unique design. Such differentiated (hybrid) multicore chips will likely be even more difficult to build and program than conventional multicore processors. However, if maturing silicon technologies lead to a lengthening of semiconductor product cycles, manufacturers will be able to afford the cost of specialization and other innovations in computer design, perhaps signaling a new golden age of computer architecture.

Acknowledgments

We would like to thank all of the contributing authors for their hard work in producing a set of excellent chapters. They all responded well to editorial feedback intended to provide a level of cohesion to the independently authored chapters. We would also like to thank Anantha Chandrakasan, Chief Editor of Springer's Series on Integrated Circuits and Systems, for recruiting us to assemble this book. Thanks go to the editorial staff at Springer, in particular Katelyn Stanne and Carl Harris for their patience and support of the production of this book.

Texas, USA   Stephen W. Keckler
California, USA   Kunle Olukotun
Texas, USA   H. Peter Hofstee

Contents

1   **Tiled Multicore Processors** 1
Michael B. Taylor, Walter Lee, Jason E. Miller, David Wentzlaff,
Ian Bratt, Ben Greenwald, Henry Hoffmann, Paul R. Johnson,
Jason S. Kim, James Psota, Arvind Saraf, Nathan Shnidman,
Volker Strumpen, Matthew I. Frank, Saman Amarasinghe, and
Anant Agarwal

2   **On-Chip Networks for Multicore Systems** 35
Li-Shiuan Peh, Stephen W. Keckler, and Sriram Vangal

3   **Composable Multicore Chips** 73
Doug Burger, Stephen W. Keckler, and Simha Sethumadhavan

4   **Speculatively Multithreaded Architectures** 111
Gurindar S. Sohi and T.N. Vijaykumar

5   **Optimizing Memory Transactions for Multicore Systems** 145
Ali-Reza Adl-Tabatabai, Christos Kozyrakis, and Bratin Saha

6   **General-Purpose Multi-core Processors** 173
Chuck Moore and Pat Conway

7   **Throughput-Oriented Multicore Processors** 205
James Laudon, Robert Golla, and Greg Grohoski

8   **Stream Processors** 231
Mattan Erez and William J. Dally

9   **Heterogeneous Multi-core Processors: The Cell
Broadband Engine** 271
H. Peter Hofstee

Index 297

Contributors

Ali-Reza Adl-Tabatabai Intel Corporation, Hillsboro, OR USA

Anant Agarwal MIT CSAIL, Cambridge, MA USA

Saman Amarasinghe MIT CSAIL, Cambridge, MA USA

Ian Bratt Tilera Corporation, Westborough, MA USA

Doug Burger The University of Texas at Austin, Austin, TX USA

Pat Conway Advanced Micro Devices Inc., Sunnyvale, CA USA

William J. Dally Stanford University, Stanford, CA USA

Mattan Erez The University of Texas at Austin, Austin, TX USA

Matthew I. Frank University of Illinois at Urbana – Champaign, Urbana, IL USA

Robert Golla Sun Microsystems, Austin, TX USA

Ben Greenwald Veracode, Burlington, MA USA

Greg Grohoski Sun Microsystems, Austin, TX USA

Henry Hoffmann MIT CSAIL, Cambridge, MA USA

H. Peter Hofstee IBM Systems and Technology Group, Austin, TX USA

Paul R. Johnson MIT CSAIL, Cambridge, MA USA

Stephen W. Keckler The University of Texas at Austin, Austin, TX USA

Jason S. Kim MIT CSAIL, Cambridge, MA USA

Christos Kozyrakis Stanford University, Stanford, CA USA

James Laudon Google, Madison, WI USA

Walter Lee Tilera Corporation, Westborough, MA USA

Jason E. Miller MIT CSAIL, Cambridge, MA USA

Chuck Moore Advanced Micro Devices Inc., Sunnyvale, CA USA

Li-Shiuan Peh Princeton University, Princeton, New Jersey, USA

James Psota MIT CSAIL, Cambridge, MA USA

Bratin Saha Intel Corporation, Hillsboro, OR USA

Arvind Saraf Swasth Foundation, Bangalore India

Simha Sethumadhavan Columbia University, New York, NY USA

Nathan Shnidman The MITRE Corporation, Bedford, MA USA

Gurindar S. Sohi University of Wisconsin, Madison, WI USA

Volker Strumpen IBM Austin Research Laboratory, Austin, TX USA

Michael B. Taylor University of California, San Diego, La Jolla, CA USA

Sriram Vangal Intel Corporation, Hillsboro, OR USA

T.N. Vijaykumar Purdue University, West Lafayette, IN USA

David Wentzlaff MIT CSAIL, Cambridge, MA USA

Chapter 1
Tiled Multicore Processors

Michael B. Taylor, Walter Lee, Jason E. Miller, David Wentzlaff, Ian Bratt, Ben Greenwald, Henry Hoffmann, Paul R. Johnson, Jason S. Kim, James Psota, Arvind Saraf, Nathan Shnidman, Volker Strumpen, Matthew I. Frank, Saman Amarasinghe, and Anant Agarwal

Abstract For the last few decades Moore's Law has continually provided exponential growth in the number of transistors on a single chip. This chapter describes a class of architectures, called *tiled multicore* architectures, that are designed to exploit massive quantities of on-chip resources in an efficient, scalable manner. Tiled multicore architectures combine each processor core with a switch to create a modular element called a tile. Tiles are replicated on a chip as needed to create multicores with any number of tiles. The Raw processor, a pioneering example of a tiled multicore processor, is examined in detail to explain the philosophy, design, and strengths of such architectures. Raw addresses the challenge of building a general-purpose architecture that performs well on a larger class of stream and embedded computing applications than existing microprocessors, while still running existing ILP-based sequential programs with reasonable performance. Central to achieving this goal is Raw's ability to exploit all forms of parallelism, including ILP, DLP, TLP, and Stream parallelism. Raw approaches this challenge by implementing plenty of on-chip resources – including logic, wires, and pins – in a tiled arrangement, and *exposing* them through a new ISA, so that the software can take advantage of these resources for parallel applications. Compared to a traditional superscalar processor, Raw performs within a factor of 2x for sequential applications with a very low degree of ILP, about 2x–9x better for higher levels of ILP, and 10x–100x better when highly parallel applications are coded in a stream language or optimized by hand.

J.E. Miller (✉)
MIT CSAIL 32 Vassar St, Cambridge, MA 02139, USA
e-mail: jasonm@alum.mit.edu

Based on "Evaluation of the Raw Microprocessor: An Exposed-Wire-Delay Architecture for ILP and Streams", by M.B. Taylor, W. Lee, J.E. Miller, et al. which appeared in The 31st Annual International Symposium on Computer Architecture (ISCA). ©2004 IEEE. [46]

1.1 Introduction

For the last few decades, Moore's Law has continually provided exponential growth in the number of transistors on a single chip. The challenge for computer architects is to find a way to use these additional transistors to increase application performance. This chapter describes a class of architectures, called *tiled multicore* architectures, that are designed to exploit massive quantities of on-chip resources in an efficient and scalable manner. Tiled multicore architectures combine each processor core with a switch to create a modular element called a tile. Tiles are replicated on a chip as needed to create multicores with a few or large numbers of tiles.

The Raw processor, a pioneering example of a tiled multicore processor, will be examined in detail to explain the philosophy, design, and strengths of such architectures. Raw addresses the challenge of building a general-purpose architecture that performs well on a larger class of stream and embedded computing applications than existing microprocessors, while still running existing ILP-based sequential programs with reasonable performance. Central to achieving this goal is Raw's ability to exploit all forms of parallelism, including instruction-level parallelism (ILP), data-level parallelism (DLP), task or thread-level parallelism (TLP), and stream parallelism.

1.1.1 The End of Monolithic Processors

Over the past few decades, general-purpose processor designs have evolved by attempting to automatically find and exploit increasing amounts of parallelism in sequential programs: first came pipelined single-issue processors, then in-order superscalars, and finally out-of-order superscalars. Each generation has employed larger and more complex circuits (e.g., highly ported register files, huge bypass networks, reorder buffers, complex cache hierarchies, and load/store queues) to extract additional parallelism from a simple single-threaded program.

As clock frequencies have increased, wire delay within these large centralized structures has begun to limit scalability [9, 28, 2]. With a higher clock frequency, the fraction of a chip that a signal can reach in a single clock period becomes smaller. This makes it very difficult and costly to scale centralized structures to the sizes needed by large monolithic processors. As an example, the Itanium II processor [32] spends over half of its critical path in the bypass paths of the ALUs (which form a large centralized interconnect). Techniques like resource partitioning and super-pipelining attempt to hide the realities of wire delay from the programmer, but create other inefficiencies that result in diminishing performance returns.

Besides the performance implications, many of these large centralized structures are power-hungry and very costly to design and verify. Structures such as bypass networks and multiported register files grow with the square or cube of the issue-width in monolithic superscalars. Because power relates to both area and frequency in CMOS VLSI design, the power consumed by these complex monolithic processors is approaching VLSI limits. Increased complexity also results in increased design and verification costs. For large superscalar processors, verification can account for as much as 70% of total development cost [10].

Due to these limited performance improvements, skyrocketing energy consumption, and run-away design costs, it has become clear that large monolithic processor architectures will not scale into the billion-transistor era [35, 2, 16, 36, 44].

1.1.2 Tiled Multicore Architectures

Tiled multicore architectures avoid the inefficiencies of large monolithic sequential processors and provide unlimited scalability as Moore's law provides additional transistors per chip. Like all multicore processors, tiled multicores (such as Raw, TRIPS [37], Tilera's TILE64 [50, 7], and Intel's Tera-Scale Research Processor [6]) contain several small computational cores rather than a single large one. Since simpler cores are more area- and energy-efficient than larger ones, more functional units can be supported within a single chip's area and power budget [1]. Specifically, in the absence of other bottlenecks, multicores increase throughput in proportion to the number of cores for parallel workloads without the need to increase clock frequency. The multicore approach is power efficient because increasing the clock frequency requires operating at proportionally higher voltages in optimized processor designs, which can increase power by the cube of the increase in frequency. However, the key concept in tiled multicores is the way in which the processing cores are interconnected. In a tiled multicore, each core is combined with a communication network router, as shown in Fig. 1.1, to form an independent modular "tile." By replicating

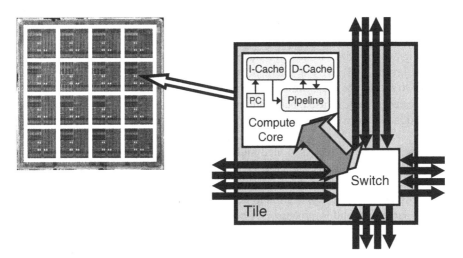

Fig. 1.1 A tiled multicore processor is composed of an array of tiles. Each tile contains an independent compute core and a communication switch to connect it to its neighbors. Because cores are *only* connected through the routers, this design can easily be scaled by adding additional tiles

tiles across the area of a chip and connecting neighboring routers together, a complete on-chip communication network is created.

The use of a general network router in each tile distinguishes tiled multicores from other mainstream multicores such as Intel's Core processors, Sun's Niagara [21], and the Cell Broadband Engine [18]. Most of these multicores have distributed processing elements but still connect cores together using non-scalable centralized structures such as bus interconnects, crossbars, and shared caches. The Cell processor uses ring networks that are physically scalable but can suffer from significant performance degradation due to congestion as the number of cores increases. Although these designs are adequate for small numbers of cores, they will not scale to the thousand-core chips we will see within the next decade.

Tiled multicores distribute both computation *and* communication structures providing advantages in efficiency, scalability, design costs, and versatility. As mentioned previously, smaller simpler cores are faster and more efficient due to the scaling properties of certain internal processor structures. In addition, they provide fast, cheap access to local resources (such as caches) and incur extra cost only when additional distant resources are required. Centralized designs, on the other hand, force every access to incur the costs of using a single large, distant resource. This is true to a lesser extent even for other multicore designs with centralized interconnects. Every access that leaves a core must use the single large interconnect. In a tiled multicore, an external access is routed through the on-chip network and uses only the network segments between the source and destination.

Tiled multicore architectures are specifically designed to scale easily as improvements in process technology provide more transistors on each chip. Because tiled multicores use distributed communication structures as well as distributed computation, processors of *any* size can be built by simply laying down additional tiles. Moving to a new process generation does not require any redesign or re-verification of the tile design. Besides future scalability, this property has enormous advantages for design costs today. To design a huge billion-transistor chip, one only needs to design, layout, and verify a small, relatively simple tile and then replicate it as needed to fill the die area. Multicores with centralized interconnect allow much of the core design to be re-used, but still require some customized layout for each core. In addition, the interconnect may need to be completely redesigned to add additional cores.

As we will see in Sect. 1.5, tiled multicores are also much more versatile than traditional general-purpose processors. This versatility stems from the fact that, much like FPGAs, tiled multicores provide large quantities of general processing resources and allow the application to decide how best to use them. This is in contrast to large monolithic processors where the majority of die area is consumed by special-purpose structures that may not be needed by all applications. If an application does need a complex function, it can dedicate some of the resources to emulating it in software. Thus, tiled multicores are, in a sense, more general than general-purpose processors. They can provide competitive performance on single-threaded ILP (instruction-level parallelism) applications as well as applications that are traditionally the domain of DSPs, FPGAs, and ASICs. As demonstrated in Sect. 1.4, they do so by supporting multiple models of computation such as ILP, streaming, TLP (task or thread-level parallelism), and DLP (data-level parallelism).

1.1.3 Raw: A Prototype Tiled Multicore Processor

The Raw processor is a prototype tiled multicore. Developed in the Computer Architecture Group at MIT from 1997 to 2002, it is one of the first multicore processors.[1] The design was initially motivated by the increasing importance of managing wire delay and the desire to expand the domain of "general-purpose" processors into the realm of applications traditionally implemented in ASICs. To obtain some intuition on how to approach this challenge, we conducted an early study [5, 48] on the factors responsible for the significantly better performance of application-specific VLSI chips. We identified four main factors: specialization; exploitation of parallel resources (gates, wires, and pins); management of wires and wire delay; and management of pins.

1. **Specialization:** ASICs specialize each "operation" at the gate level. In both the VLSI circuit and microprocessor context, an operation roughly corresponds to the unit of work that can be done in one cycle. A VLSI circuit forms operations by combinational logic paths, or "operators," between flip-flops. A microprocessor, on the other hand, has an instruction set that defines the operations that can be performed. Specialized operators, for example, for implementing an incompatible floating-point operation, or implementing a linear feedback shift register, can yield an order of magnitude performance improvement over an extant general-purpose processor that may require many instructions to perform the same one-cycle operation as the VLSI hardware.
2. **Exploitation of Parallel Resources:** ASICs further exploit plentiful silicon area to implement enough operators and communications channels to sustain a tremendous number of parallel operations in each clock cycle. Applications that merit direct digital VLSI circuit implementations typically exhibit massive, operation-level parallelism. While an aggressive VLIW implementation like Intel's Itanium II [32] executes six instructions per cycle, graphics accelerators may perform hundreds or thousands of word-level operations per cycle. Because they operate on very small word operands, logic emulation circuits such as Xilinx II Pro FPGAs can perform hundreds of thousands of operations each cycle. Clearly the presence of many physical execution units is a minimum prerequisite to the exploitation of the same massive parallelism that ASICs are able to exploit.

---

[1] However, the term "multicore" was not coined until more recently when commercial processors with multiple processing cores began to appear in the marketplace.

3. **Management of Wires and Wire Delay:** ASIC designers can place and wire communicating operations in ways that minimize wire delay, minimize latency, and maximize bandwidth. ASIC designers manage wire delay inherent in large distributed arrays of function units in multiple steps. First, they place close together operations that need to communicate frequently. Second, when high bandwidth is needed, they create multiple customized communication channels. Finally, they introduce pipeline registers between distant operators, thereby converting propagation delay into pipeline latency. By doing so, the designer acknowledges the inherent trade-off between parallelism and latency: leveraging more resources requires signals to travel greater distances.
4. **Management of Pins:** ASICs customize the usage of their pins. Rather than being bottlenecked by a cache-oriented multi-level hierarchical memory system (and subsequently by a generic PCI-style I/O system), ASICs utilize their pins in ways that fit the applications at hand, maximizing realizable I/O bandwidth or minimizing latency. This efficiency applies not just when an ASIC accesses external DRAMs, but also in the way that it connects to high-bandwidth input devices like wide-word analog-to-digital converters, CCDs, and sensor arrays. There are currently few easy ways to arrange for these devices to stream data into a general-purpose microprocessor in a high-bandwidth way, especially since DRAM must almost always be used as an intermediate buffer.

The goal of the Raw project is to build a general-purpose microprocessor that can leverage the above four factors while still running existing ILP-based sequential applications with reasonable performance. In addition, the design must be scalable in the face of ever-increasing wire delays. It needs to implement the gamut of general-purpose features that we expect in a microprocessor such as functional unit virtualization, unpredictable interrupts, instruction virtualization, data caching, and context switching. To achieve good performance it also needs to exploit ILP in sequential programs and allow multiple threads of control to communicate and coordinate efficiently. Raw takes the following approach to leveraging the four factors behind the success of ASICs:

1. Raw implements the most common operations needed by ILP, stream, or TLP applications in specialized hardware mechanisms. Most of the primitive mechanisms are exposed to software through a new ISA. These mechanisms include the usual integer and floating-point operations; specialized bit manipulation ops; scalar operand routing between adjacent tiles; operand bypass between functional units, registers and I/O queues; and data cache access (e.g., load with tag check).
2. Raw implements a large number of these operators which exploit the copious VLSI resources—including gates, wires, and pins—and exposes them through a new ISA, such that the software can take advantage of them for both ILP and highly parallel applications.
3. Raw manages the effect of wire delays by exposing the wiring channel operators to the software, so that the software can account for latencies by orchestrating

both scalar and stream data transport. By orchestrating operand flow on the interconnect, Raw can also create customized communications patterns. Taken together, the wiring channel operators provide the abstraction of a scalar operand network [44] that offers very low latency for scalar data transport and enables the exploitation of ILP.
4. Raw software manages the pins for cache data fetches and for specialized stream interfaces to DRAM or I/O devices.

1.1.4 Chapter Overview

The rest of this chapter describes the Raw processor in more detail and evaluates the extent to which it succeeds in serving as a more versatile general-purpose processor. Section 1.2 provides an overview of the Raw architecture and its mechanisms for specialization, exploitation of parallel resources, orchestration of wires, and management of pins. Section 1.3 describes the specific implementation of the prototype Raw chip. Section 1.4 evaluates the performance of Raw on applications drawn from several classes of computation including ILP, streams and vectors, server, and bit-level embedded computation. Section 1.5 introduces a new metric called "versatility" that folds into a single scalar number the performance of an architecture over many application classes. Section 1.6 follows with a detailed discussion of other related processor architectures. Finally, Section 1.7 concludes with a summary of our findings.

1.2 Raw Architecture Overview

The Raw architecture supports an ISA that provides a parallel interface to the gate, pin, and wiring resources of the chip through suitable high-level abstractions. As illustrated in Fig. 1.2, the Raw processor exposes the copious gate resources of the

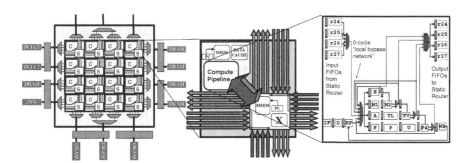

Fig. 1.2 The Raw microprocessor comprises 16 tiles. Each tile has a compute processor, routers, network wires, and instruction and data memories. The compute processor uses a conventional eight-stage MIPS pipeline with interfaces to the network routers tied directly into the bypass paths

chip by dividing the usable silicon area into an array of 16 identical, programmable tiles. Each tile contains a processing core and communication routers to connect it to neighboring tiles. The processing core contains an eight-stage in-order single-issue MIPS-style processing pipeline (right side of Fig. 1.2), a four-stage single-precision pipelined FPU, a 32 KB data cache, and a 32 KB software-managed instruction cache [30] for the processing pipeline. There are two types of communication routers (static and dynamic), with a 64 KB software-managed instruction cache for the static router. Each tile is sized so that the amount of time for a signal to travel through a small amount of logic and across the tile is one clock cycle. Future Raw-like processors will likely have hundreds or even thousands of tiles.

1.2.1 On-Chip Networks

The tiles are interconnected by four 32-bit full-duplex on-chip networks, consisting of over 12,500 wires (see Fig. 1.2). Two of the networks are static (routes are specified at compile time) and two are dynamic (routes are specified at run time). Each tile is connected only to its four neighbors. Every wire is registered at the input to its destination tile. This means that *the longest wire in the system is no longer than the length or width of a tile*. This property ensures high clock speeds, and the continued scalability of the architecture.

The design of Raw's on-chip interconnect and its interface with the processing pipeline are its key innovative features. These on-chip networks are exposed to the software through the Raw ISA, thereby giving the programmer or compiler the ability to directly program the wiring resources of the processor, and to carefully orchestrate the transfer of data values between the computational portions of the tiles—much like the routing in an ASIC. Effectively, the wire delay is exposed to the user as network hops. A route between opposite corners of the processor takes six hops, which corresponds to approximately six cycles of wire delay. To minimize the latency of inter-tile scalar data transport (which is critical for ILP) the on-chip networks are not only register-mapped but also integrated directly into the bypass paths of the processor pipeline. The register-mapped ports allow an instruction to place a value on the network with no overhead. Similarly, instructions using values from the network simply read from the appropriate register. The programmable switches bear the responsibility of routing operands through the network.

Raw's on-chip interconnects belong to the class of scalar operand networks [44], which provide an interesting way of looking at modern day processors. The register file used to be the central communication mechanism between functional units in a processor. Starting with the first pipelined processors, the bypass network has become largely responsible for the communication of active values, and the register file is more of a checkpointing facility for inactive values. The Raw networks (in particular the static networks) serve as 2-D bypass networks, bridging between the bypass networks of separate cores.

The static router in each tile contains a 64 KB software-managed instruction cache and a pair of routing crossbars. Compiler-generated routing instructions are 64 bits and encode a small command (e.g., conditional branch with/without decrement) and several routes, one for each crossbar output. Each Raw static router, also known as a switch processor, contains a four-way crossbar, with each way corresponding to one of the cardinal directions (north, east, south, and west). The single-cycle routing instructions are one example of Raw's use of specialization. Because the router program memory is cached, there is no practical architectural limit on the number of simultaneous communication patterns that can be supported in a computation. This feature, coupled with the extremely low latency and low occupancy of the in-order inter-tile ALU-to-ALU operand delivery (three cycles nearest neighbor) distinguishes Raw from prior systolic or message passing systems [4, 14, 24].

Table 1.1 shows the breakdown of costs associated with sending a single-word message on Raw's static network. Because the network interfaces are register-mapped, the send occupancy is zero—the instruction that produces the result to be sent can directly place it on the network without requiring any additional instructions. Similarly, the receiving instruction can pull its input value directly off of the network. If the switch processor is already waiting for the next value to come from the ALU, that value will be placed on the network immediately. It then takes one cycle for each tile it passes through on the way to its destination. At the receiving end, it takes two cycles to pull the value off the network and queue it up for the ALU.

Table 1.1 Breakdown of the end-to-end costs for a one-word message on Raw's static network

	Cycles
Sending processor occupancy	0
Latency from ALU output to network	0
Latency per hop	1
Latency from network to ALU input	2
Receiving processor occupancy	0

Raw's two dynamic networks support cache misses, interrupts, dynamic messages, and other asynchronous events. The two networks use dimension-ordered routing and are structurally identical. One network, the memory network, follows a deadlock-avoidance strategy to avoid end-point deadlock. It is used in a restricted manner by trusted clients such as data caches, DMA, and I/O. The second network, the general network, is used by untrusted clients, and relies on a deadlock recovery strategy [24].

Raw fully supports context switches. On a context switch, the contents of the processor registers and the general and static networks on a subset of the Raw chip occupied by the process (possibly including multiple tiles) are saved off. The process and its network data can then be restored to the same position or a new offset on the Raw grid.

1.2.2 Direct I/O Interfaces

On the edges of the network, the network channels are multiplexed down onto the pins of the chip to form flexible I/O ports that can be used for DRAM accesses or external device I/O. To toggle a pin, the user programs one of the on-chip networks to route a value off the side of the array. The 1657-pin CCGA (ceramic column-grid array) package used on Raw provides 14 full-duplex, 32-bit I/O ports. Raw implementations with fewer pins are made possible via logical channels (as is already the case for two out of the 16 logical ports), or simply by bonding out only a subset of the ports.

The static and dynamics networks, the data cache of the compute processors, and the external DRAMs connected to the I/O ports comprise Raw's memory system. The memory network is used for cache-based memory traffic while the static and general dynamic networks are used for stream-based memory traffic. Systems designed for memory-intensive applications can have up to 14 full-duplex full-bandwidth DRAM banks by placing one on each of the chip's 14 physical I/O ports. Minimal embedded Raw systems may eliminate DRAM altogether: booting from a single ROM and executing programs entirely out of the on-chip memories. In addition to transferring data directly to the tiles, off-chip devices connected to the I/O ports can route data through the on-chip networks to other devices in order to perform glueless DMA and peer-to-peer communication.

1.2.3 ISA Analogs to Physical Resources

By creating first class architectural analogs to the physical chip resources, Raw attempts to minimize the ISA gap—that is, the gap between the resources that a VLSI chip has available and the amount of resources that are usable by software. Unlike conventional ISAs, Raw exposes the quantity of all three underlying physical resources (gates, wires, and pins) in the ISA. Furthermore, it does this in a manner that is backwards-compatible—the instruction set does not change with varying degrees of resources.

Table 1.2 contrasts the ways that the Raw ISA and conventional ISAs expose physical resources to the programmer. Because the Raw ISA has more direct interfaces, Raw processors can have more functional units and more flexible and efficient pin utilization. High-end Raw processors will typically have more pins, because the architecture is better at turning pin count into performance and functionality.

Table 1.2 How Raw converts increasing quantities of physical entities into ISA entities

Physical entity	Raw ISA analog	Conventional ISA analog
Gates	Tiles, new instructions	New instructions
Wires, Wire delay	Routes, Network hops	None
Pins	I/O ports	None

Finally, Raw processors are more predictable and have higher frequencies because of the explicit exposure of wire delay.

This approach makes Raw scalable. Creating subsequent, more powerful generations of the processor is straightforward and is as simple as stamping out as many tiles and I/O ports as the silicon die and package allow. The design has no centralized resources, no global buses, and no structures that get larger as the tile or pin count increases. Finally, the longest wire, the design complexity, and the verification complexity are all independent of transistor count.

1.3 Raw Chip Implementation

The Raw chip is a 16-tile prototype implemented in IBM's 180 nm, 1.8 V, six-layer, CMOS 7SF SA-27E copper ASIC process. A die photograph of the Raw chip is shown in Fig. 1.3. One hundred and twenty chips were received from IBM in October 2002. Although the Raw array is only 16×16 mm, an 18.2×18.2 mm die is used to allow for the high pin-count package. The 1657-pin ceramic column grid array (CCGA) package provides 1080 high-speed transceiver logic (HSTL) I/O pins. Measurements indicate that the chip core averages 18.2 watts at 425 MHz (with unused functional units and memories quiesced and unused data I/O pins tri-stated). The target clock frequency was 225 MHz under worst-case conditions, which is competitive with other 180 nm lithography ASIC processors, such as VIRAM [22], Imagine [19], and Tensilica's Xtensa series. The nominal running frequency is typically higher—the Raw chip core, running at room temperature, reaches 425 MHz at 1.8 V, and 500 MHz at 2.2 V. This compares favorably to IBM-implemented microprocessors in the same process: the PowerPC 405 GP runs at 266–400 MHz, while the follow-on PowerPC 440 GP reaches 400–500 MHz.

Fig. 1.3 Die photo of the Raw chip *(left)* and photo of the Raw prototype motherboard *(right)*

The processor is aggressively pipelined, with conservative treatment of the control paths in order to ensure that only reasonable efforts would be required to close timing in the backend. Despite these efforts, wire delay inside a tile was still large enough to warrant the creation of a special infrastructure to place the cells in the timing and congestion-critical datapaths. More details on the Raw implementation are available in [43].

As one can infer from Sect. 1.5, moving from a single-issue compute processor to a dual-issue compute processor would have likely improved performance on low-ILP applications. Estimates indicate that such a compute processor would have easily fit in the remaining empty space within a tile. The frequency impact of transitioning from single-issue to dual-issue is generally held to be small.

A prototype motherboard (shown in Fig. 1.3) using the Raw chip was designed in collaboration with the Information Sciences Institute (ISI) East. It includes a single Raw chip, SDRAM chips, I/O interfaces, and interface FPGAs. A larger system consisting of separate processor and I/O boards has also been developed with ISI. Each processor board contains four Raw chips connected in a 2×2 array. Multiple processor boards can be connected to form systems consisting of up to 64 Raw chips, thereby forming virtual Raw processors with up to 1,024 tiles.

1.4 Results

This section presents measurement and experimental results of the Raw microprocessor. We begin by explaining our experimental methodology. Then we present some basic hardware statistics. The remainder of the section focuses on evaluating how well Raw supports a range of programming models and application types. The domains we examine include ILP computation, stream and embedded computation, TLP server workloads, and bit-level computation. The performance of Raw in these individual areas are presented as comparison to a reference 600 MHz Pentium III.

As you will see, in some cases Raw achieves greater than $16\times$ speedup (either vs. the Pentium III or vs. a single tile) even though it contains (at most) 16 times as many functional units. This super-linear speedup can be attributed to compounding or additive effects from several different factors. These factors are listed in Table 1.3 along with the maximum possible speedup that each can enable and are briefly discussed here.

Table 1.3 Sources of speedup for Raw over Pentium III

Factor responsible	Max. Speedup
Tile parallelism (exploitation of gates)	$16\times$
Load/store elimination (management of wires)	$4\times$
Streaming mode vs. cache thrashing (management of wires)	$15\times$
Streaming I/O bandwidth (management of pins)	$60\times$
Increased cache/register size (exploitation of gates)	$\sim 2\times$
Bit manipulation instructions (specialization)	$3\times$

1. When all 16 tiles can be used, the speedup can be 16-fold.
2. If a, b, and c are variables in memory, then an operation of the form $c = a+b$ in a traditional load–store RISC architecture will require a minimum of four instructions—two loads, one add, and one store. Stream architectures such as Raw can accomplish the operation with a single instruction (for a speedup of $4\times$) by eliminating the loads and stores. For long streams of data this is done by issuing bulk data stream requests to the memory controller and then processing data directly from the network without loading and storing it in the cache. This effect also works on a smaller scale if an intermediate value is passed directly over the network from a neighboring tile instead of being passed through memory.
3. When vector lengths exceed the cache size, streaming data from off-chip DRAM directly into the ALU achieves $7.5\times$ the throughput of cache accesses (each cache miss transports eight words in 60 cycles, while streaming can achieve one word per cycle). The streaming effect is even more powerful with strided requests that use only part of a full cache line. In this case, streaming throughput is 15 times greater than going through the cache.
4. Raw has $60\times$ the I/O bandwidth of the Pentium III. Furthermore, Raw's direct programmatic interface to the pins enables more efficient utilization.
5. When multiple tiles are used in a computation, the effective number of registers and cache lines is increased, allowing a greater working set to be accessed without penalty. We approximate the potential speedup from this effect as two-fold.
6. Finally, specialized bit manipulation instructions can optimize table lookups, shifts, and logical operations. We estimate the potential speedup from these instructions as three-fold.

1.4.1 Experimental Methodology

The evaluation presented here makes use of a validated cycle-accurate simulator of the Raw chip. Using the validated simulator as opposed to actual hardware facilitates the normalization of differences with a reference system, e.g., DRAM memory latency, and instruction cache configuration. It also allows for exploration of alternative motherboard configurations. The simulator was meticulously verified against the gate-level RTL netlist to have *exactly* the same timing and data values for all 200,000 lines of the hand-written assembly test suite, as well as for a number of C applications and randomly generated tests. Every stall signal, register file write, SRAM write, on-chip network wire, cache state machine transition, interrupt signal, and chip signal pin matches in value on every cycle between the two. This gate-level RTL netlist was then shipped to IBM for manufacturing. Upon receipt of the chip, a subset of the tests was compared to the actual hardware to verify that the chip was manufactured according to specifications.

Reference Processor: To have a meaningful evaluation of a new architecture, it is important to compare the new empirical data to an existing commercial processor.

For this evaluation, a 600 MHz Pentium III (P3) was selected as the reference processor. The 600 MHz P3, is implemented in the same process generation as the Raw chip (180 nm) and represents the middle of the initial production frequency range, before extensive process or speedpath tuning. This levels the playing field somewhat when comparing the P3 to the "first-silicon" prototype Raw chip. In addition, the P3's functional unit latencies and level of pipelining are nearly identical to Raw, making direct cycle-count comparisons meaningful. Table 1.4 lists a few key characteristics of Raw and the P3 for comparison. For a more detailed comparison and discussion of why the 600 MHz P3 is an appropriate reference processor see [46].

Table 1.4 Comparison of processor characteristics for Raw and P3-Coppermine

Parameter	Raw (IBM ASIC)	Pentium III (Intel)
Lithography generation	180 nm	180 nm
Metal layers	6 Cu	6 Al
Frequency	425 MHz	600 MHz
Die area[a]	331 mm^2	106 mm^2
Signal pins	~ 1100	~ 190
Sustained issue width	1 in-order (per tile)	3 out-of-order
ALU latency/occupancy	1 / 1	1 / 1
Load latency/occupancy (hit)	3 / 1	3 / 1
Store latency/occupancy (hit)	– / 1	– / 1
Mispredict penalty	3	10–15
L1 data cache size	32 K	16 K
L1 instruction cache size	32 K	16 K
L1 cache associativities	2-way	4-way
L2 cache size	none	256 K
DRAM access latency	54 cycles	86 cycles

[a]Note that despite the area penalty for an ASIC implementation, Raw is almost certainly a larger design than the P3. This evaluation does not aim to make a cost-normalized comparison, but rather seeks to demonstrate the scalability of the tiled multicore approach for future microprocessors.

System Models: After the selection of a reference processor comes the selection of an enclosing system. A pair of 600 MHz Dell Precision 410 machines were used to run the reference benchmarks. These machines were outfitted with identical 100 MHz 2-2-2 PC100 256 MB DRAMs, and several microbenchmarks were used to verify that the memory system timings matched.

To provide a fair comparison between the Raw and Dell systems, the Raw simulator extension language was used to implement a cycle-matched PC100 DRAM model and a chipset. This model has the same wall-clock latency and bandwidth as the Dell 410. However, since Raw runs at a slower frequency than the P3 (425 MHz vs. 600 MHz), the latency, measured in cycles, is less. The term **RawPC** is used to describe a simulation which uses eight PC100 DRAMs, occupying four ports on the left hand side of the chip, and four on the right hand side.

Because Raw is designed for streaming applications, it is necessary to measure applications that use the full pin bandwidth of the chip. In this case, a simulation of CL2 PC 3500 DDR DRAM, which provides enough bandwidth to saturate both directions of a Raw port, was used. This is achieved by attaching 16 PC 3500 DRAMs to all 16 logical ports on the chip, in conjunction with a memory controller, implemented in the chipset, that supports a number of stream requests. A Raw tile can send a message over the general dynamic network to the chipset to initiate large bulk transfers from the DRAMs into and out of the static network. Simple interleaving and striding is supported, subject to the underlying access and timing constraints of the DRAM. This configuration is called **RawStreams**.

The placement of a DRAM on a Raw port does not preclude the use of other devices on that port—the chipsets have a simple demultiplexing mechanism that allows multiple devices to connect to a single port.

Other Normalizations: Except where otherwise noted, gcc 3.3 –O3 was used to compile C and Fortran code for both Raw[2] and the P3[3]. For programs that do C or Fortran stdio calls, newlib 1.9.0 was used for both Raw and the P3. Finally, to eliminate the impact of disparate file and operating systems, the results of I/O system calls for the Spec benchmarks were captured and embedded into the binaries as static data using [41].

One final normalization was performed to enable comparisons with the P3. The cycle-accurate simulator was augmented to employ conventional two-way set-associative hardware instruction caching. These instruction caches are modeled cycle-by-cycle in the same manner as the rest of the hardware. Like the data caches, they service misses over the memory dynamic network. Resource contention between the caches is modeled accordingly.

1.4.2 ILP Computation

This section examines how well Raw is able to support conventional sequential applications. Typically, the only form of parallelism available in these applications is instruction-level parallelism (ILP). For this evaluation, a range of benchmarks that encompasses a wide spectrum of program types and degrees of ILP is used.

Much like a VLIW architecture, Raw is designed to rely on the compiler to find and exploit ILP. However, unlike a VLIW, it does so by distributing instructions across multiple processor cores with independent program counters. This process of distributing ILP across multiple cores is called distributed ILP or DILP. We have developed Rawcc [8, 25, 26] to explore DILP compilation issues. Rawcc takes sequential C or Fortran programs and orchestrates them across the Raw tiles

---

[2]The Raw gcc backend targets a single tile's compute and network resources.

[3]For the P3, the –march=pentium3 and –mfpmath=sse flags were added.

in two steps. First, Rawcc distributes the data and code across the tiles in a way that attempts to balance the trade-off between locality and parallelism. Then, it schedules the computation and communication to maximize parallelism and minimize communication stalls.

The speedups attained in Table 1.5 show the potential of automatic parallelization and ILP exploitation on Raw. Rawcc is a prototype research compiler and is, therefore, not robust enough to compile every application in standard benchmark suites. Of the benchmarks Rawcc can compile, Raw is able to outperform the P3 for all the scientific benchmarks and several irregular applications.

Table 1.5 Performance of sequential programs on Raw and on a P3

				Speedup vs. P3	
Benchmark	Source	# Raw tiles	Cycles on Raw	Cycles	Time
Dense-matrix scientific applications					
Swim	Spec95	16	14.5 M	4.0	2.9
Tomcatv	Nasa7:Spec92	16	2.05 M	1.9	1.3
Btrix	Nasa7:Spec92	16	516 K	6.1	4.3
Cholesky	Nasa7:Spec92	16	3.09 M	2.4	1.7
Mxm	Nasa7:Spec92	16	247 K	2.0	1.4
Vpenta	Nasa7:Spec92	16	272 K	9.1	6.4
Jacobi	Raw benchmark suite	16	40.6 K	6.9	4.9
Life	Raw benchmark suite	16	332 K	4.1	2.9
Sparse-matrix/integer/irregular applications					
SHA	Perl Oasis	16	768 K	1.8	1.3
AES Decode	FIPS-197	16	292 K	1.3	0.96
Fpppp-kernel	Nasa7:Spec92	16	169 K	4.8	3.4
Unstructured	CHAOS	16	5.81 M	1.4	1.0

Table 1.6 shows the speedups achieved by Rawcc as the number of tiles varies from two to 16. The speedups are relative to the performance of a single Raw tile. Overall, the improvements are primarily due to increased parallelism, but several of the dense-matrix benchmarks benefit from increased cache capacity as well (which explains the super-linear speedups). In addition, Fpppp-kernel benefits from increased register capacity, which leads to fewer spills.

For completeness, we also compiled a selection of the Spec2000 benchmarks with gcc for a single tile, and ran them using MinneSPEC's [20] *LgRed* data sets. The results, shown in Table 1.7, represent a lower bound for the performance of those codes on Raw, as they only use 1/16 of the resources on the Raw chip. The numbers are quite surprising; on average, the simple in-order Raw tile with no L2 cache is only $1.4\times$ slower by cycles and $2\times$ slower by time than the full P3. This suggests that in the event that the parallelism in these applications is too small to be exploited across Raw tiles, a simple two-way Raw compute processor might be sufficient to make the performance difference negligible.

Table 1.6 Speedup of the ILP benchmarks relative to single-tile Raw processor

	Number of tiles				
Benchmark	1	2	4	8	16
Dense-matrix scientific applications					
Swim	1.0	1.1	2.4	4.7	9.0
Tomcatv	1.0	1.3	3.0	5.3	8.2
Btrix	1.0	1.7	5.5	15.1	33.4
Cholesky	1.0	1.8	4.8	9.0	10.3
Mxm	1.0	1.4	4.6	6.6	8.3
Vpenta	1.0	2.1	7.6	20.8	41.8
Jacobi	1.0	2.6	6.1	13.2	22.6
Life	1.0	1.0	2.4	5.9	12.6
Sparse-matrix/integer/irregular applications					
SHA	1.0	1.5	1.2	1.6	2.1
AES decode	1.0	1.5	2.5	3.2	3.4
Fpppp-kernel	1.0	0.9	1.8	3.7	6.9
Unstructured	1.0	1.8	3.2	3.5	3.1

Table 1.7 Performance of SPEC2000 programs on one Raw tile

				Speedup vs. P3	
Benchmark	Source	No. of Raw tiles	Cycles on Raw	Cycles	Time
172.mgrid	SPECfp	1	240 M	0.97	0.69
173.applu	SPECfp	1	324 M	0.92	0.65
177.mesa	SPECfp	1	2.40B	0.74	0.53
183.equake	SPECfp	1	866 M	0.97	0.69
188.ammp	SPECfp	1	7.16B	0.65	0.46
301.apsi	SPECfp	1	1.05B	0.55	0.39
175.vpr	SPECint	1	2.52B	0.69	0.49
181.mcf	SPECint	1	4.31B	0.46	0.33
197.parser	SPECint	1	6.23B	0.68	0.48
256.bzip2	SPECint	1	3.10B	0.66	0.47
300.twolf	SPECint	1	1.96B	0.57	0.41

1.4.3 Stream Computation

Stream computations arise naturally out of real-time I/O applications as well as from embedded applications. The data sets for these applications are often large and may even be a continuous stream in real-time, which makes them unsuitable for traditional cache-based memory systems. Raw provides more natural support for stream-based computation by allowing data to be fetched efficiently through a register-mapped, software-orchestrated network.

Stream computation can be mapped to Raw in several different ways including pipelining, task-level parallelism, and data parallelism [12]. A detailed discussion of these techniques is beyond the scope of this chapter; however, in general, stream computation is performed by defining kernels of code that run on different tiles. These kernels pull their input data directly off the on-chip networks (using the register-mapped network interfaces), perform some computation using the data, and then send their output data directly to a kernel on another tile. The streams of data between tiles (or between tiles and I/O devices) are never stored in memory except as may be required for a kernel's internal calculations. This technique allows many tiles to work together on a problem and greatly reduces the overhead of communicating data between them compared to cache-based sharing.

We present two sets of results for stream computation on Raw. First, we show the performance of programs written in a high-level stream language called StreamIt. The applications are automatically compiled to Raw. Then, we show the performance of some hand-written streaming applications.

1.4.3.1 StreamIt

StreamIt is a high-level, architecture-independent language for high-performance streaming applications. StreamIt contains language constructs that improve programmer productivity for streaming, including hierarchical structured streams, graph parameterization, and circular buffer management. These constructs also expose information to the compiler and enable novel optimizations [47]. We have developed a Raw backend for the StreamIt compiler, which includes fully automatic load balancing, graph layout, communication scheduling, and routing [13, 12].

We evaluate the performance of RawPC on several StreamIt benchmarks, which represent large and pervasive DSP applications. Table 1.8 summarizes the performance of 16 Raw tiles vs. a P3. For both architectures, we use StreamIt versions of the benchmarks; we do not compare to hand-coded C on the P3 because StreamIt performs at least 1–2× better for four of the six applications (this is due to aggressive unrolling and constant propagation in the StreamIt compiler). The comparison reflects two distinct influences: (1) the scaling of Raw performance as the number of tiles increases, and (2) the performance of a Raw tile vs. a P3 for the same StreamIt code. To distinguish between these influences, Table 1.9 shows detailed speedups relative to StreamIt code running on a 1-tile Raw configuration.

Table 1.8 StreamIt performance results

Benchmark	Cycles per output on raw	Speedup vs. P3	
		Cycles	Time
Beamformer	2074.5	7.3	5.2
Bitonic Sort	11.6	4.9	3.5
FFT	16.4	6.7	4.8
Filterbank	305.6	15.4	10.9
FIR	51.0	11.6	8.2
FMRadio	2614.0	9.0	6.4

Table 1.9 Speedup (in cycles) of StreamIt benchmarks relative to a 1-tile Raw configuration. From left, the columns show the results on Raw configurations with 1–16 tiles and on a P3

Benchmark	StreamIt on n raw tiles					StreamIt on P3
	1	2	4	8	16	
Beamformer	1.0	4.1	4.5	5.2	21.8	3.0
Bitonic sort	1.0	1.9	3.4	4.7	6.3	1.3
FFT	1.0	1.6	3.5	4.8	7.3	1.1
Filterbank	1.0	3.3	3.3	11.0	23.4	1.5
FIR	1.0	2.3	5.5	12.9	30.1	2.6
FMRadio	1.0	1.0	1.2	4.0	10.9	1.2

The primary result illustrated by Table 1.9 is that StreamIt applications scale effectively for increasing sizes of the Raw configuration. For FIR, FFT, and Bitonic, the scaling is approximately linear across all tile sizes (FIR is actually super-linear due to decreasing register pressure in larger configurations). For Beamformer, Filterbank, and FMRadio, the scaling is slightly inhibited for small configurations. This is because (1) these applications are larger, and IMEM constraints prevent an unrolling optimization for small tile sizes and (2) they have more data parallelism, yielding speedups for large configurations but inhibiting small configurations due to a constant control overhead.

The second influence is the performance of a P3 vs. a single Raw tile on the same StreamIt code, as illustrated by the last column in Table 1.9. In most cases, performance is similar. The P3 performs significantly better in two cases because it can exploit ILP: Beamformer has independent real/imaginary updates in the inner loop, and FIR is a fully unrolled multiply-accumulate operation. In other cases, ILP is obscured by circular buffer accesses and control dependences.

In all, StreamIt applications benefit from Raw's exploitation of parallel resources and management of wires. The abundant parallelism and regular communication patterns in stream programs are an ideal match for the parallelism and tightly orchestrated communication on Raw. As stream programs often require high bandwidth, register-mapped communication serves to avoid costly memory accesses. Also, autonomous streaming components can manage their local state in Raw's distributed data caches and register banks, thereby improving locality. These aspects are key to the scalability demonstrated in the StreamIt benchmarks.

1.4.3.2 Hand-Optimized Stream Applications

Various students and researchers in the MIT Computer Architecture Group, the MIT Oxygen Project, and at ISI East have coded and manually tuned a wide range of streaming applications to take advantage of Raw as an embedded processor. These include a set of linear algebra routines implemented as Stream Algorithms, the STREAM benchmark, and several other embedded applications including a real-time 1020-node acoustic beamformer. The benchmarks are typically written in C and compiled with gcc, with inline assembly for a subset of inner loops. Some of the simpler benchmarks (such as the STREAM and FIR benchmarks) were small enough that coding entirely in assembly was the most expedient approach. This section presents the results.

Streaming Algorithms: Table 1.10 presents the performance of a set of linear algebra algorithms on RawPC vs. the P3.

Table 1.10 Performance of linear algebra routines

Benchmark	Problem size	MFlops on Raw	Speedup vs. P3	
			Cycles	Time
Matrix multiplication	256 × 256	6310	8.6	6.3
LU factorization	256 × 256	4300	12.9	9.2
Triangular solver	256 × 256	4910	12.2	8.6
QR factorization	256 × 256	5170	18.0	12.8
Convolution	256 × 16	4610	9.1	6.5

The Raw implementations are coded as Stream Algorithms [17], which emphasize computational efficiency in space and time and are designed specifically to take advantage of tiled microarchitectures like Raw. They have three key features. First, stream algorithms operate directly on data from the interconnect and achieve an asymptotically optimal 100% compute efficiency for large numbers of tiles. Second, stream algorithms use no more than a small, bounded amount of storage on each processing element. Third, data are streamed through the compute fabric from and to peripheral memories.

With the exception of convolution, we compare against the P3 running single precision Lapack (linear algebra package). We use clapack version 3.0 [3] and a tuned BLAS implementation, ATLAS [51], version 3.4.2. We disassembled the ATLAS library to verify that it uses P3 SSE extensions appropriately to achieve high performance. Since Lapack does not provide a convolution, we compare against the Intel integrated performance primitives (IPP).

As can be seen in Table 1.10, Raw performs significantly better than the P3 on these applications even with optimized P3 SSE code. Raw's better performance is due to load/store elimination (see Table 1.3), and the use of parallel resources. Stream Algorithms operate directly on values from the network and avoid loads and stores, thereby achieving higher utilization of parallel resources than the blocked code on the P3.

STREAM Benchmark: The STREAM benchmark was created by John McCalpin to measure sustainable memory bandwidth and the corresponding computation rate for vector kernels [29]. Its performance has been documented on thousands of machines, ranging from PCs and desktops to MPPs and other supercomputers.

We hand-coded an implementation of STREAM on RawStreams. We also tweaked the P3 version to use single-precision SSE floating point, improving its performance. The Raw implementation employs 14 tiles and streams data between 14 processors and 14 memory ports through the static network. Table 1.11 displays

Table 1.11 Performance on the STREAM benchmark

Problem size	Bandwidth (GB/s)			Speedup Raw vs. P3
	Pentium III	Raw	NEC SX-7	
Copy	0.57	47.6	35.1	84×
Scale	0.51	47.3	34.8	92×
Add	0.65	35.6	35.3	55×
Scale & add	0.62	35.5	35.3	59×

the results. As shown in the right-most column, Raw is 55×–92× better than the P3. The table also includes the performance of STREAM on NEC SX-7 Super-computer, which has the highest reported STREAM performance of any single-chip processor. Note that Raw surpasses that performance. This extreme single-chip per-formance is achieved by taking advantage of three Raw architectural features: its ample pin bandwidth, the ability to precisely route data values in and out of DRAMs with minimal overhead, and a careful match between floating point and DRAM bandwidth.

Other stream-based applications: Table 1.12 presents the performance of some hand-optimized stream applications on Raw. A 1020-microphone real-time Acous-tic Beamformer has been developed which uses the Raw system for processing. On this application, Raw runs 16 instantiations of the same set of instructions (code) and the data from the microphones is striped across the array in a data-parallel manner. Raw's software-exposed I/O allows for much more efficient transfer of stream data than the DRAM-based I/O on the P3. The assumption that the stream data for the P3 is coming from DRAM represents a best-case situation. The P3 results would be worse in a typical system where the input data is brought in over a general-purpose I/O bus. For the FIR, we compared to the Intel IPP. Results for Corner Turn, Beam Steering, and CSLC are discussed in the previously published [40].

Table 1.12 Performance of hand-optimized stream applications

Benchmark	Machine configuration	Cycles on Raw	Speedup vs. P3	
			Cycles	Time
Acoustic beamforming	RawStreams	7.83 M	9.7	6.9
512-pt Radix-2 FFT	RawPC	331 K	4.6	3.3
16-tap FIR	RawStreams	548 K	10.9	7.7
CSLC	RawPC	4.11 M	17.0	12.0
Beam steering	RawStreams	943 K	65	46
Corner Turn	RawStreams	147 K	245	174

1.4.4 Server

As large numbers of processor cores are integrated onto a single chip, one potential application of them is as a "server-farm-on-a-chip." In this application area, independent applications are run on each of the processor cores.To measure the performance of Raw on such server-like workloads, we conduct the following experiment on RawPC to obtain SpecRate-like metrics. For each of a subset of Spec2000 applications, an independent copy of it is executed on each of the 16 tiles, and the overall throughput of that workload is measured relative to a single run on the P3.

Table 1.13 presents the results. Note that the speedup of Raw vs. P3 is equivalent to the throughput of Raw relative to P3's throughput. As anticipated, RawPC outperforms the P3 by a large margin, with an average throughput advantage of $10.8\times$ (by cycles) and $7.6\times$ (by time). The key Raw feature that enables this performance is the high pin bandwidth available to off-chip memory. RawPC contains eight separate memory ports to DRAM. This means that even when all 16 tiles are running applications, each memory port and DRAM is only shared between two applications.

Table 1.13 shows the efficiency of RawPC's memory system for each server workload. Efficiency is the ratio between the actual throughput and the ideal $16\times$ speedup attainable on 16 tiles. Less than the ideal throughput is achieved because of interference among memory requests originating from tiles that share the same DRAM banks and ports. We see that the efficiency is high across all the workloads, with an average of 93%.

Table 1.13 Performance of Raw on server workloads relative to the P3

Benchmark	Cycles on Raw	Speedup vs. P3		Efficiency (%)
		Cycles	Time	
172.mgrid	0.24B	15.0	10.6	96
173.applu	0.32B	14.0	9.9	96
177.mesa	2.40B	11.8	8.4	99
183.equake	0.87B	15.1	10.7	97
188.ammp	7.16B	9.1	6.5	87
301.apsi	1.05B	8.5	6.0	96
175.vpr	2.52B	10.9	7.7	98
181.mcf	4.31B	5.5	3.9	74
197.parser	6.23B	10.1	7.2	92
256.bzip2	3.10B	10.0	7.1	94
300.twolf	1.96B	8.6	6.1	94

1.4.5 Bit-Level Computation

Many applications have computational requirements that make them ill-suited to conventional general-purpose processors. In particular, applications which perform many operations on very small operands do not make efficient use of the small number of wide datapaths found in most processors. Currently, these applications are implemented using specialized processors, ASICs, or FPGAs. Tiled multicore processors (with their large numbers of functional units and efficient on-chip networks) have the potential to bridge the gap and enable a broader range of applications on general-purpose processors. To investigate this potential, we study the performance of Raw on applications that manipulate individual bits and are usually implemented in ASICs.

We measure the performance of RawStreams on two bit-level computations: an 802.11a convolutional encoder and an 8b/10b encoder [49]. Table 1.14 presents the results for the P3, Raw, FPGA, and ASIC implementations. The FPGA implementations use a Xilinx Virtex-II 3000-5 FPGA, which is built using the same process generation as the Raw chip. The ASIC implementations were synthesized to the IBM SA-27E process that the Raw chip is implemented in. For each benchmark, we present three problem sizes: 1024, 16,384, and 65,536 samples. These problem sizes are selected to fit in the L1, L2, and miss in the cache on the P3, respectively. We use a randomized input sequence in all cases.

Table 1.14 Performance of two bit-level applications: 802.11a convolutional encoder and 8b/10b encoder. The hand-coded Raw implementations are compared to reference sequential implementations on the P3

| | | | Speedup vs. P3 | | | |
| | | | Raw | | FPGA | ASIC |
	Problem size	Cycles on Raw	Cycles	Time	Time	Time
802.11a	1,024 bits	1,048	11.0	7.8	6.8	24
ConvEnc	16,408 bits	16,408	18.0	12.7	11.0	38
	65,536 bits	65,560	32.8	23.2	20.0	68
8b/10b	1,024 bytes	1,054	8.2	5.8	3.9	12
Encoder	16,408 bytes	16,444	11.8	8.3	5.4	17
	65,536 bytes	65,695	19.9	14.1	9.1	29

On these two applications, Raw is able to excel by exploiting fine-grain pipeline parallelism. To do this, the computations were spatially mapped across multiple tiles. Both applications benefited by more than 2× from Raw's specialized bit-level manipulation instructions, which reduce the latency of critical feedback loops. Another factor in Raw's high performance on these applications is Raw's exposed streaming I/O. This I/O model is in sharp contrast to having to move data through the cache hierarchy on a P3.

Whereas the implementations from Table 1.14 are optimized to use the entire chip to provide the best performance for a single stream of data, Table 1.15 presents results for an implementation designed to maximize throughput on 16 parallel streams. This is to simulate a potential workload for a base-station communications chip that needs to encode multiple simultaneous connections. For this throughput test, a more area-efficient implementation of each encoder was used on Raw. This implementation has lower performance on a single stream but utilizes fewer tiles, achieving a higher per-area throughput. Instantiating 16 copies of this implementation results in the maximum total throughput.

Table 1.15 Performance of two bit-level applications for 16 streams: 802.11a convolutional encoder and 8b/10b encoder. This test simulates a possible workload for a base-station that processes multiple communication streams

Benchmark	Problem size	Cycles on Raw	Speedup vs. P3	
			Cycles	Time
802.11a	16×64 bits	259	45	32
ConvEnc	16×1024 bits	4,138	71	51
	16×4096 bits	16,549	130	92
8b/10b	16×64 bytes	257	34	24
Encoder	16×1024 bytes	4,097	47	33
	16×4096 bytes	16,385	80	56

These results also highlight one of the inherent strengths of tiled multicores such as Raw: they can be easily reconfigured to exploit different types of parallelism. In the first case, the high-performance on-chip network was used to exploit pipeline parallelism to accelerate computation for a single stream. In the second case, each input stream was assigned to a single tile and task-level parallelism was used to handle 16 streams. Since parallelizing these algorithms introduces some overhead, the second case provides higher total throughput at the expense of longer latency for a particular stream. The flexible nature of tiled multicores allows them to support a wide variety of types and levels of parallelism, and switch between them as needed.

1.5 Analysis

Sections 1.4.2 through 1.4.5 presented performance results of Raw for several application classes, and showed that Raw's performance was within a factor of 2× of the P3 for low-ILP applications, 2×–9× better than the P3 for high-ILP applications, and 10×–100× better for stream or embedded computations. Table 1.16 summarizes the primary features that are responsible for performance improvements on Raw.

In this section, we compare the performance of Raw to other machines that have been designed specifically with streams or embedded computation in mind. We also attempt to explore quantitatively the degree to which Raw succeeds in being a more versatile general-purpose processor. To do so, we selected a representative subset of applications from each of our computational classes, and obtained performance results for Raw, P3, and machines especially suited for each of those applications. We note that these results are exploratory in nature and not meant to be taken as any sort of proof of Raw's versatility, rather as an early indication of the possibilities.

Figure 1.4 summarizes these results. We can make several observations from the figure. First, it is easy to see that the P3 does well relative to Raw for applications with low degrees of ILP, while the opposite is true for applications with

Table 1.16 Raw feature utilization table. **S** = Specialization, **R** = Exploiting Parallel Resources, **W** = Management of Wire Delays, **P** = Management of Pins

Category	Benchmarks	S	R	W	P
ILP	Swim, Tomcatv, Btrix, Cholesky, Vpenta, Mxm, Life, Jacobi, Fpppp-kernel, SHA, AES Encode, Unstructured, 172.mgrid, 173.applu, 177.mesa, 183.equake, 188.ammp, 301.apsi, 175.vpr, 181.mcf, 197.parser, 256.bzip2, 300.twolf	X	X	X	
Stream:StreamIt	Beamformer, Bitonic Sort, FFT, Filterbank, FIR, FMRadio	X	X	X	
Stream:Stream Algorithms	Mxm, LU fact., Triang. solver, QR fact., Conv.			X	X
Stream:STREAM	Copy, Scale, Add, Scale & Add			X	X
Stream:Other	Acoustic Beamforming, FIR, FFT, Beam Steering	X	X	X	
	Corner Turn			X	X
	CSLC	X	X		
Server	172.mgrid, 173.applu, 177.mesa, 183.equake, 188.ammp, 301.apsi, 175.vpr, 181.mcf, 197.parser, 256.bzip2, 300.twolf		X		X
Bit-level	802.11a ConvEnc, 8b/10b Encoder	X	X	X	

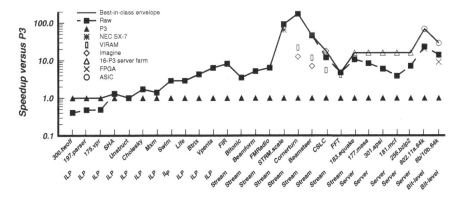

Fig. 1.4 Performance of various architectures over several applications classes. Each point in the graph represents the speedup of an architecture over the P3 for a given application. The best-in-class envelope connects the best speedups for each application. The *dashed line* connects the speedups of Raw for all the applications. A versatile architecture has speedups close to the best-in-class envelope for all application classes. Imagine and VIRAM results are obtained from [40] and [34]. Bit-level results for FPGA and ASIC implementations are obtained from [49]

higher degrees of ILP, such as Vpenta. For streams and vectors, the performance of Raw is comparable to that of stream and vector architectures like VIRAM and Imagine. All three outperform the P3 by factors of $10\times - 100\times$. Raw, using the RawStreams configuration, beats the highest reported single-chip STREAM memory bandwidth champion, the NEC SX-7 Supercomputer, and is $55\times$–$92\times$ better than the P3. Essential to Raw's performance on this benchmark is the ample pin bandwidth, the ability to precisely route data values in and out of DRAM with minimal overhead, and a careful match between floating point and DRAM bandwidth.

We chose a server-farm with 16 P3s as our best-in-class server system. Notice that a single-chip Raw system comes within a factor of three of this server-farm for most applications. (Note that this is only a pure performance comparison, and we have not attempted to normalize for cost.) We chose FPGAs and ASICs as the best in class for our embedded bit-level applications. Raw's performance is comparable to that of an FPGA for these applications, and is a factor of $2\times - 3\times$ off from an ASIC. (Again, note that we are only comparing performance—ASICs use significantly lower area and power than Raw [49].) Raw performs well on these applications for the same reasons that FPGAs and ASICs do—namely, a careful orchestration of the wiring or communication patterns.

Thus far, our analysis of Raw's flexibility has been qualitative. Given the recent interest in flexible, versatile, or polymorphic architectures such as Scale [23], Grid [33], and SmartMemories [27], which attempt to perform well over a wider range of applications than extant general-purpose processors, it is intriguing to search for a metric that can capture the notion of versatility. We would like to offer up a candidate and use it to evaluate the versatility of Raw quantitatively. In a manner similar to the computation of SpecRates, *we define the versatility of a machine M as the geometric mean over all applications of the ratio of machine M's speedup for a given application relative to the speedup of the best machine for that application.*[4]

For the application set graphed in Fig. 1.4, Raw's versatility is 0.72, while that of the P3 is 0.14. The P3's relatively poor performance on stream benchmarks hurts its versatility. Although Raw's 0.72 number is relatively good, even our small sample of applications highlights two clear areas which merit additional work in the design of polymorphic processors. One is for embedded bit-level designs, where ASICs perform $2\times$–$3\times$ better than Raw for our small application set. Certainly there are countless other applications for which ASICs outstrip Raw by much higher factors. Perhaps the addition of small amounts of bit-level programmable logic à la PipeRench [11] or Garp [15] can bridge the gap.

Computation with low levels of ILP is another area for further research. We will refer to Fig. 1.5 to discuss this in more detail. The figure plots the speedups *(in cycles)* of Raw and a P3 with respect to execution on a single Raw tile. The applications are listed on the *x*-axis and sorted roughly in the order of increasing ILP. The figure indicates that Raw is able to convert ILP into performance when ILP exists in reasonable quantities. This indicates the scalability of Raw's scalar

---

[4]Since we are taking ratios, the individual machine speedups can be computed relative to any one machine, since the effect of that machine cancels out. Accordingly, the speedups in Fig. 1.4 are expressed relative to the P3 without loss of generality. Further, like SpecRates' rather arbitrary choice of the Sun Ultra5 as a normalizing machine, the notion of versatility can be generalized to future machines by choosing equally arbitrarily the best-in-class machines graphed in Fig. 1.4 as our reference set for all time. Thus, since the best-in-class machines are fixed, the versatilities of future machines can become greater than 1.0.

operand network. The performance of Raw is lower than the P3 by about 33 percent for applications with low degrees of ILP for several reasons. First, the three left-most applications in Fig. 1.5 were run on a single tile. We hope to continue tuning our compiler infrastructure and do better on some of these applications. Second, a near-term commercial implementation of a Raw-like processor might likely use a two-way superscalar in place of our single-issue processor which would be able to match the P3 for integer applications with low ILP. See [31] for details on grain-size trade-offs in Raw processors.

Fig. 1.5 Speedup (in cycles) achieved by Raw and the P3 over executing on a single Raw tile

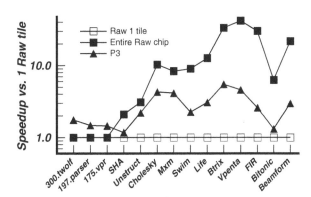

1.6 Related Architectures

Raw distinguishes itself from other multicore and parallel processors by being a mode-less architecture and supporting all forms of parallelism, including ILP, DLP, TLP, and streams. Several other projects have attempted to exploit specific forms of parallelism. These include systolic (iWarp [14]), vector (VIRAM [22]), and stream (Imagine [19]). These machines, however, were not designed for ILP. In contrast, Raw was designed to exploit ILP effectively in addition to these other forms of parallelism. ILP presents a difficult challenge because it requires that the architecture be able to transport scalar operands between logic units with very low latency, even when there are a large number of highly irregular communication patterns.

A recent paper [44] employs a 5-tuple to characterize the cost of sending operands between functional units in a number of architectures (see Table 1.1 for a list of the components in this 5-tuple). Qualitatively, larger 5-tuple values represent proportionally more expensive operand transport costs. The large values in the network 5-tuples for iWarp <1,6,5,0,1> and shared memory <1,18,2,14,1>, compared to the low numbers in the 5-tuples of machines that can exploit ILP (e.g., superscalar <0,0,0,0,0>, Raw <0,0,1,2,0>, and TRIPS <0,0,1,0,0>) quantitatively demonstrate the difference. The low 5-tuple of Raw's scalar operand network compared to that of iWarp enables Raw to exploit diverse forms of parallelism, and is a direct consequence of the integration of the interconnect into Raw's pipeline and Raw's early pipeline commit point. We will further discuss the comparison with iWarp here, but see [44] and [42] for more details on comparing networks for ILP.

Raw supports statically orchestrated communication like iWarp and NuMesh [38]. iWarp and NuMesh support a small number of fixed communication patterns, and can switch between these patterns quickly. However, establishing a new pattern is more expensive. Raw supports statically orchestrated communication by using a programmable switch that issues an instruction during each cycle. The instruction specifies the routes through the switch during that cycle. Because the switch program memory in Raw is large, and virtualized through caching, there is no practical architectural limit on the number of simultaneous communication patterns that can be supported in a computation. This virtualization becomes particularly important for supporting ILP, because switch programs become as large or even larger than the compute programs.

The TRIPS processor [37, 33] is targeted specifically for ILP and proposes using low-latency scalar operand networks. Raw shares in its ILP philosophy but implements a static-transport, point-to-point scalar operand network, whereas TRIPS uses a dynamic-transport, point-to-point network. Both Raw and TRIPS perform compile time instruction assignment to compute nodes. Raw uses compile-time operand matching, while TRIPS uses dynamic associative operand matching queues. Accordingly, using the *AsTrO* categorization (Assignment, Transport, Ordering) from [45], Raw and TRIPS can be classified as SSS and SDD architectures, respectively, where S stands for static and D for dynamic. TRIPS and Raw represent two distinct points in the scalar operand network design space, with Raw taking a more compile-time oriented approach and TRIPS taking a more dynamic approach.

Raw took inspiration from the Multiscalar processor [39], which uses a separate one-dimensional network to forward register values between ALUs. Raw generalizes the basic idea, and supports a two-dimensional programmable mesh network both to forward operands and for other forms of communication.

Both Raw and SmartMemories [27] share the philosophy of an exposed communication architecture, and represent two design points in the space of tiled architectures that can support multiple forms of parallelism. Raw uses homogeneous, programmable static and dynamic mesh networks, while SmartMemories uses programmable static communication within a local collection of nodes, and a dynamic network between these collections of nodes. The node granularities are also different in the two machines. Perhaps the most significant architectural difference, however, is that Raw (like Scale [23]) is *mode-less*, while SmartMemories and TRIPS have modes for different application domains. Raw's research focus is on discovering and implementing a minimal set of primitive mechanisms (e.g., scalar operand network) useful for all forms of parallelism, while the modal approach implements special mechanisms for each form of parallelism. We believe the mode-less approach is more area efficient and significantly less complex. We believe the issue of modal vs. mode-less architectures for versatile processors is likely to be a controversial topic of debate in the forthcoming years.

Like VIRAM and Imagine, Raw supports vector and stream computations, but does so very differently. Both VIRAM and Imagine sport large memories or stream register files on one side of the chip connected via a crossbar interconnect to multiple, deep compute pipelines on the other. The computational model is one that extracts data streams from memory, pipes them through the compute pipelines, and then deposits them back in memory. In contrast, Raw implements many co-located distributed memories and compute elements, interconnected by a mesh network. The Raw computational model is more ASIC-like in that it streams data through the pins and on-chip network to the ALUs, continues through the network to more ALUs, and finally through the network to the pins. Raw's ALUs also can store data temporarily in the local memories if necessary. We believe the lower latencies of the memories in Raw, together with the tight integration of the on-chip network with the compute pipelines, make Raw more suitable for distributed ILP.

Finally, the Tile architecture [50, 7] from Tilera Corporation is a commercial outgrowth of the Raw project. Tile's first implementation, the TILE64, integrates 64 power-efficient tiles as a single SoC (system on chip) in 90 nm technology, clocks at speeds of up to 1 GHz, and consumes on the order of 12 watts in its 64-core fabric running real applications. Each of the tiles contains a three-way VLIW processor, caches, and a switch router. The tiles are coupled together using a point-to-point scalable mesh interconnect. TILE64 uses homogeneous, full-featured processor cores, implements virtual memory, and executes SMP Linux across all cores. It runs programs written in ANSI C and C++ and supports standard parallel programming APIs such as pthreads, MPI, and iLib, a novel light-weight sockets-like streaming API.

1.7 Conclusion

This chapter describes and evaluates the Raw microprocessor, a multicore processor based on a tiled architecture. Raw's exposed ISA allows parallel applications to exploit all of the chip resources, including gates, wires, and pins. Raw supports multiple models of computation including ILP, streams, TLP, and DLP. Raw implements a distributed ILP technique by scheduling operands over a scalar operand network that offers very low latency for scalar data transport. Raw supports the streaming model by using its general-purpose processing cores which can run large, independent programs, and which can stream data over the on-chip interconnect with low processor occupancy. Raw's compiler manages the effect of wire delays by orchestrating both scalar and stream data transport. The Raw processor demonstrates that existing architectural abstractions like interrupts, caches, and context-switching can continue to be supported in this environment, even as applications take advantage of the low-latency scalar operand network and the large number of ALUs.

Our results demonstrate that the Raw processor performs at or close to the level of the best specialized machine for each application class. When compared to a Pentium III, Raw displays one to two orders of magnitude better performance for stream applications, while performing within a factor of two for low-ILP applications.

We believe that tiled architectures like Raw provide the solutions needed for future generations of massively parallel processors which have been variously called manycores and massive multicores. The use of decentralized, distributed communication structures allows these architectures to scale simply and efficiently as additional on-chip resources become available. In addition, the software-exposed interface to these communication resources allows for the highly efficient core-to-core communication that will be required to harness the true power of multicore designs. It is our hope that the Raw research will provide insight for architects who are looking for ways to build versatile processors that leverage vast silicon resources.

Acknowledgments We thank our StreamIt collaborators, specifically M. Gordon, J. Lin, and B. Thies for the StreamIt backend and the corresponding section of this chapter. We are grateful to our collaborators from ISI East including C. Chen, S. Crago, M. French, L. Wang and J. Suh for developing the Raw motherboard, firmware components, and several applications. T. Konstantakopoulos, L. Jakab, F. Ghodrat, M. Seneski, A. Saraswat, R. Barua, A. Ma, J. Babb, M. Stephenson, S. Larsen, V. Sarkar, and several others too numerous to list also contributed to the success of Raw. The Raw chip was fabricated in cooperation with IBM. Raw is funded by DARPA, NSF, ITRI, and the Oxygen Alliance.

References

1. A. Agarwal and M. Levy. Going multicore presents challenges and opportunities. *Embedded Systems Design*, 20(4), April 2007.
2. V. Agarwal, M. S. Hrishikesh, S. W. Keckler, and D. Burger. Clock Rate versus IPC: The End of the Road for Conventional Microarchitectures. In *ISCA '00: Proceedings of the 27th Annual International Symposium on Computer Architecture*, pages 248–259, 2000.
3. E. Anderson, Z. Bai, J. Dongarra, A. Greenbaum, A. McKenney, J. Du Croz, S. Hammerling, J. Demmel, C. Bischof, and D. Sorensen. LAPACK: A Portable Linear Algebra Library for High-Performance Computers. In *Supercomputing '90: Proceedings of the 1990 ACM/IEEE Conference on Supercomputing*, pages 2–11, 1990.
4. M. Annaratone, E. Arnould, T. Gross, H. T. Kung, M. Lam, O. Menzilicioglu, and J. A. Webb. The Warp Computer: Architecture, Implementation and Performance. *IEEE Transactions on Computers*, 36(12):1523–1538, December 1987.
5. J. Babb, M. Frank, V. Lee, E. Waingold, R. Barua, M. Taylor, J. Kim, S. Devabhaktuni, and A. Agarwal. The RAW Benchmark Suite: Computation Structures for General Purpose Computing. In *Proceedings of the IEEE Workshop on FPGAs for Custom Computing Machines (FCCM)*, pages 134–143, 1997.
6. M. Baron. Low-key Intel 80-core Intro: The tip of the iceberg. *Microprocessor Report*, April 2007.
7. M. Baron. Tilera's cores communicate better. *Microprocessor Report*, November 2007.
8. R. Barua, W. Lee, S. Amarasinghe, and A. Agarwal. Maps: A Compiler-Managed Memory System for Raw Machines. In *ISCA '99: Proceedings of the 26th Annual International Symposium on Computer Architecture*, pages 4–15, 1999.
9. M. Bohr. Interconnect Scaling – The Real Limiter to High Performance ULSI. In *1995 IEDM*, pages 241–244, 1995.

10. P. Bose, D. H. Albonesi, and D. Marculescu. Power and complexity aware design. *IEEE Micro: Guest Editor's Introduction for Special Issue on Power and Complexity Aware Design*, 23(5):8–11, Sept/Oct 2003.

11. S. Goldstein, H. Schmit, M. Moe, M. Budiu, S. Cadambi, R. R. Taylor, and R. Laufer. PipeRench: A Coprocessor for Streaming Multimedia Acceleration. In *ISCA '99: Proceedings of the 26th Annual International Symposium on Computer Architecture*, pages 28–39, 1999.

12. M. Gordon, W. Thies, and S. Amarasinghe. Exploiting coarse-grained task, data, and pipeline parallelism in stream programs. In *ASPLOS-XII: Proceedings of the 12th International Conference on Architectural Support for Programming Languages and Operating Systems*, pages 75–86, October 2006.

13. M. I. Gordon, W. Thies, M. Karczmarek, J. Lin, A. S. Meli, A. A. Lamb, C. Leger, J. Wong, H. Hoffmann, D. Maze, and S. Amarasinghe. A Stream Compiler for Communication-Exposed Architectures. In *ASPLOS-X: Proceedings of the Tenth International Conference on Architectural Support for Programming Languages and Operating Systems*, pages 291–303, 2002.

14. T. Gross and D. R. O'Halloron. *iWarp, Anatomy of a Parallel Computing System*. The MIT Press, Cambridge, MA, 1998.

15. J. R. Hauser and J. Wawrzynek. Garp: A MIPS Processor with Reconfigurable Coprocessor. In *Proceedings of the IEEE Workshop on FPGAs for Custom Computing Machines (FCCM)*, pages 12–21, 1997.

16. R. Ho, K. W. Mai, and M. A. Horowitz. The Future of Wires. *Proceedings of the IEEE*, 89(4):490–504, April 2001.

17. H. Hoffmann, V. Strumpen, A. Agarwal, and H. Hoffmann. Stream Algorithms and Architecture. Technical Memo MIT-LCS-TM-636, MIT Laboratory for Computer Science, 2003.

18. H. P. Hofstee. Power efficient processor architecture and the Cell processor. In *HPCA '05: Proceedings of the 11th International Symposium on High Performance Computer Architecture*, pages 258–262, 2005.

19. U. Kapasi, W. J. Dally, S. Rixner, J. D. Owens, and B. Khailany. The Imagine Stream Processor. In *ICCD '02: Proceedings of the 2002 IEEE International Conference on Computer Design*, pages 282–288, 2002.

20. A. KleinOsowski and D. Lilja. MinneSPEC: A New SPEC Benchmark Workload for Simulation-Based Computer Architecture Research. *Computer Architecture Letters*, 1, June 2002.

21. P. Kongetira, K. Aingaran, and K. Olukotun. Niagara: A 32-Way Multithreaded Sparc Processor. *IEEE Micro*, 25(2):21–29, 2005.

22. C. Kozyrakis and D. Patterson. A New Direction for Computer Architecture Research. *IEEE Computer*, 30(9):24–32, September 1997.

23. R. Krashinsky, C. Batten, M. Hampton, S. Gerding, B. Pharris, J. Casper, and K. Asanovic. The Vector-Thread Architecture. In *ISCA '04: Proceedings of the 31st Annual International Symposium on Computer Architecture*, June 2004.

24. J. Kubiatowicz. *Integrated Shared-Memory and Message-Passing Communication in the Alewife Multiprocessor*. PhD thesis, Massachusetts Institute of Technology, 1998.

25. W. Lee, R. Barua, M. Frank, D. Srikrishna, J. Babb, V. Sarkar, and S. Amarasinghe. Space-Time Scheduling of Instruction-Level Parallelism on a Raw Machine. In *ASPLOS-VIII: Proceedings of the Eighth International Conference on Architectural Support for Programming Languages and Operating Systems*, pages 46–54, 1998.

26. W. Lee, D. Puppin, S. Swenson, and S. Amarasinghe. Convergent Scheduling. In *MICRO-35: Proceedings of the 35th Annual International Symposium on Microarchitecture*, pages 111–122, 2002.

27. K. Mai, T. Paaske, N. Jayasena, R. Ho, W. J. Dally, and M. Horowitz. Smart Memories: A Modular Reconfigurable Architecture. In *ISCA '00: Proceedings of the 27th Annual International Symposium on Computer Architecture*, pages 161–171, 2000.

28. D. Matzke. Will Physical Scalability Sabotage Performance Gains? *IEEE Computer*, 30(9):37–39, September 1997.
29. J. McCalpin. STREAM: Sustainable Memory Bandwidth in High Performance. Computers. http://www.cs.virginia.edu/stream.
30. J. E. Miller. *Software Instruction Caching*. PhD thesis, Massachusetts Institute of Technology, Cambridge, MA, June 2007. http://hdl.handle.net/1721.1/40317.
31. C. A. Moritz, D. Yeung, and A. Agarwal. SimpleFit: A Framework for Analyzing Design Tradeoffs in Raw Architectures. *IEEE Transactions on Parallel and Distributed Systems*, pages 730–742, July 2001.
32. S. Naffziger, G. Hammond, S. Naffziger, and G. Hammond. The Implementation of the Next-Generation 64b Itanium Microprocessor. In *Proceedings of the IEEE International Solid-State Circuits Conference*, pages 344–345, 472, 2002.
33. R. Nagarajan, K. Sankaralingam, D. Burger, and S. W. Keckler. A Design Space Evaluation of Grid Processor Architectures. In *MICRO-34: Proceedings of the 34th Annual International Symposium on Microarchitecture*, pages 40–51, 2001.
34. M. Narayanan and K. A. Yelick. Generating Permutation Instructions from a High-Level Description. TR UCB-CS-03-1287, UC Berkeley, 2003.
35. S. Palacharla. *Complexity-Effective Superscalar Processors*. PhD thesis, University of Wisconsin–Madison, 1998.
36. J. Sanchez and A. Gonzalez. Modulo Scheduling for a Fully-Distributed Clustered VLIW Architecture. In *MICRO-33: Proceedings of the 33rd Annual International Symposium on Microarchitecture*, pages 124–133, December 2000.
37. K. Sankaralingam, R. Nagarajan, R. McDonald, R. Desikan, S. Drolia, M. S. Govindan, P. Gratz, D. Gulati, H. Hanson, C. Kim, H. Liu, N. Ranganathan, S. Sethumadhavan, S. Sharif, P. Shivakumar, S. W. Keckler, and D. Burger. Distributed microarchitectural protocols in the TRIPS prototype processor. In *MICRO-39: Proceedings of the 39th Annual International Symposium on Microarchitecture*, pages 480–491, Dec 2006.
38. D. Shoemaker, F. Honore, C. Metcalf, and S. Ward. NuMesh: An Architecture Optimized for Scheduled Communication. *Journal of Supercomputing*, 10(3):285–302, 1996.
39. G. Sohi, S. Breach, and T. Vijaykumar. Multiscalar Processors. In *ISCA '95: Proceedings of the 22nd Annual International Symposium on Computer Architecture*, pages 414–425, 1995.
40. J. Suh, E.-G. Kim, S. P. Crago, L. Srinivasan, and M. C. French. A Performance Analysis of PIM, Stream Processing, and Tiled Processing on Memory-Intensive Signal Processing Kernels. In *ISCA '03: Proceedings of the 30th Annual International Symposium on Computer Architecture*, pages 410–419, June 2003.
41. M. B. Taylor. Deionizer: A Tool for Capturing and Embedding I/O Calls. Technical Report MIT-CSAIL-TR-2004-037, MIT CSAIL/Laboratory for Computer Science, 2004. http://cag.csail.mit.edu/~mtaylor/deionizer.html.
42. M. B. Taylor. *Tiled Processors*. PhD thesis, Massachusetts Institute of Technology, Cambridge, MA, Feb 2007.
43. M. B. Taylor, J. Kim, J. Miller, D. Wentzlaff, F. Ghodrat, B. Greenwald, H. Hoffman, J.-W. Lee, P. Johnson, W. Lee, A. Ma, A. Saraf, M. Seneski, N. Shnidman, V. Strumpen, M. Frank, S. Amarasinghe, and A. Agarwal. The Raw Microprocessor: A Computational Fabric for Software Circuits and General-Purpose Programs. *IEEE Micro*, pages 25–35, Mar 2002.
44. M. B. Taylor, W. Lee, S. Amarasinghe, and A. Agarwal. Scalar Operand Networks: On-Chip Interconnect for ILP in Partitioned Architectures. In *HPCA '03: Proceedings of the 9th International Symposium on High Performance Computer Architecture*, pages 341–353, 2003.
45. M. B. Taylor, W. Lee, S. Amarasinghe, and A. Agarwal. Scalar Operand Networks. *IEEE Transactions on Parallel and Distributed Systems (Special Issue on On-chip Networks)*, Feb 2005.
46. M. B. Taylor, W. Lee, J. E. Miller, D. Wentzlaff, I. Bratt, B. Greenwald, H. Hoffmann, P. Johnson, J. Kim, J. Psota, A. Saraf, N. Shnidman, V. Strumpen, M. Frank, S. Amarasinghe, and A. Agarwal. Evaluation of the Raw microprocessor: An exposed-wire-delay architecture

for ILP and streams. In *ISCA '04: Proceedings of the 31st Annual International Symposium on Computer Architecture*, pages 2–13, June 2004.

47. W. Thies, M. Karczmarek, and S. Amarasinghe. StreamIt: A Language for Streaming Applications. In *2002 Compiler Construction*, pages 179–196, 2002.
48. E. Waingold, M. Taylor, D. Srikrishna, V. Sarkar, W. Lee, V. Lee, J. Kim, M. Frank, P. Finch, R. Barua, J. Babb, S. Amarasinghe, and A. Agarwal. Baring it All to Software: Raw Machines. *IEEE Computer*, 30(9):86–93, Sep 1997.
49. D. Wentzlaff. Architectural Implications of Bit-level Computation in Communication Applications. Master's thesis, Massachusetts Institute of Technology, 2002.
50. D. Wentzlaff, P. Griffin, H. Hoffmann, L. Bao, B. Edwards, C. Ramey, M. Mattina, C.-C. Miao, J. F. Brown, and A. Agarwal. On-Chip Interconnection Architecture of the Tile Processor. *IEEE Micro*, 27(5):15–31, Sept–Oct 2007.
51. R. Whaley, A. Petitet, J. J. Dongarra, and Whaley. Automated Empirical Optimizations of Software and the ATLAS Project. *Parallel Computing*, 27(1–2):3–35, 2001.

Chapter 2
On-Chip Networks for Multicore Systems

Li-Shiuan Peh, Stephen W. Keckler, and Sriram Vangal

Abstract With Moore's law supplying billions of transistors, and uniprocessor architectures delivering diminishing performance, multicore chips are emerging as the prevailing architecture in both general-purpose and application-specific markets. As the core count increases, the need for a scalable on-chip communication fabric that can deliver high bandwidth is gaining in importance, leading to recent multicore chips interconnected with sophisticated on-chip networks. In this chapter, we first present a tutorial on on-chip network architecture fundamentals including on-chip network interfaces, topologies, routing, flow control, and router microarchitectures. Next, we detail case studies on two recent prototypes of on-chip networks: the UT-Austin TRIPS operand network and the Intel TeraFLOPS on-chip network. This chapter organization seeks to provide the foundations of on-chip networks so that readers can appreciate the different design choices faced in the two case studies. Finally, this chapter concludes with an outline of the challenges facing research into on-chip network architectures.

2.1 Introduction

The combined pressures from ever-increasing power consumption and the diminishing returns in performance of uniprocessor architectures have led to the advent of multicore chips. With growing number of transistors at each technology generation, and multicore chips' modular design enabling a reduction in design complexity, this multicore wave looks set to stay, in both general-purpose computing chips as well as application-specific systems-on-a-chip. Recent years saw every industry chip vendor releasing multicore products, with higher and higher core counts.

L.-S. Peh (✉)
Princeton University, Princeton, New Jersey 08544, USA
e-mail: peh@princeton.edu

S.W. Keckler et al. (eds.), *Multicore Processors and Systems*, Integrated Circuits and
Systems, DOI 10.1007/978-1-4419-0263-4_2, © Springer Science+Business Media, LLC 2009

As the number of on-chip cores increases, a scalable and high-bandwidth communication fabric to connect them becomes critically important [34]. As a result, packet-switched on-chip networks are fast replacing buses and crossbars to emerge as the pervasive communication fabric in many-core chips. Such on-chip networks have routers at every node, connected to neighbors via short local interconnects, while multiplexing multiple packet flows over these interconnects to provide high, scalable bandwidth. This evolution of interconnect, as core count increases, is clearly illustrated in the choice of a crossbar interconnect in the 8-core Sun Niagara [29], two packet-switched rings in the 9-core IBM Cell [23], and five packet-switched meshes in the 64-core Tilera TILE64 [3].

While on-chip networks can leverage ideas from prior multi chassis interconnection networks in supercomputers [1], clusters of workstations [21], and Internet routers [14], the design requirements facing on-chip networks differ so starkly in magnitude from those of prior multi chassis networks that novel designs are critically needed. On the plus side, by moving on-chip, the I/O bottlenecks that faced prior multi chassis interconnection networks are alleviated substantially: on-chip interconnects supply bandwidth that is orders of magnitude higher than off-chip I/Os, yet obviating the inherent overheads of off-chip I/O transmission. Unfortunately, while this allows on-chip networks to deliver scalable, high-bandwidth communications, challenges are abound due to highly stringent constraints. Specifically, on-chip networks must supply high bandwidth at ultra-low latencies, with a tight power envelope and area budget.

2.2 Interconnect Fundamentals

An on-chip network architecture can be defined by four parameters: its topology, routing algorithm, flow control protocol, and router micro architecture. The topology of a network dictates how the nodes and links are connected. While the topology determines all possible paths a message can take through the network, the routing algorithm selects the specific path a message will take from source to destination. The flow control protocol determines how a message actually traverses the assigned route, including when a message leaves a router through the desired outgoing link and when it must be buffered. Finally, the micro architecture of a router realizes the routing and flow control protocols and critically shapes its circuits implementation.

Throughout this section, we will discuss how different choices of the above four parameters affect the overall cost–performance of an on-chip network. Clearly, the cost–performance of an on-chip network depends on the requirements faced by its designers. Latency is a key requirement in many on-chip network designs, where network latency refers to the delay experienced by messages as they traverse from source to destination. Most on-chip networks must also ensure high throughput, where network throughput is the peak bandwidth the network can handle. Another metric that is particularly critical in on-chip network design is network power, which approximately correlates with the activity in the network as well as its complexity. Network complexity naturally affects its area footprint, which is a concern for on-chip networks where area devoted to communication detracts from the area available for processing and memory. Tolerating process variability and faults while delivering deterministic or bounded performance guarantees (including network reliability and quality-of-service) is a design constraint of increasing importance.

2.2.1 Interfacing an On-Chip Network to a Multicore System

Many different ways of interfacing the on-chip network with the multicore system have been developed. Explicit interfaces involve modifying the instruction set architecture (ISA) so that software (either the programmer or the compiler) can explicitly indicate when and where messages are to be delivered, and conversely when and from whom messages are to be received. For instance, adding *send* and *receive* instructions to the ISA allows a programmer/compiler to indicate message deliveries and receipts to the network. Implicit interfaces, on the other hand, infer network sends/receives without the knowledge of software. An example of an implicit interface is a shared-memory multicore system, in which memory loads and stores at one core implicitly trigger network messages to other cores for maintaining coherence among the caches.

Another dimension where network interfaces differ lies in where the network physically connects to a multicore system, which can be any point along the datapath of a multicore system. For example, a network can connect into the bypass paths between functional units, to registers, to level-1 cache, to level-2 and higher level caches, to the memory bus, or to the I/O bus. The further the network interface is from the functional units, the longer the delay involved in initiating a message. A network that interfaces through registers allows messages to be injected into the network once the register is written in the processor pipeline, typically with a latency on the order of a couple of cycles. A network that interfaces through the I/O bus requires heavier weight operations including interrupt servicing and pipeline flushing, requiring latencies on the order of microseconds.

Finally, while the logical unit of communication for a multicore system is a message, most on-chip networks handle communications in units of packets. So, when a message is injected into the network, it is first segmented into packets, which are then divided into fixed-length flits, short for flow control digits. For instance, a 128-byte cache line sent from a sharer to a requester will be injected as a message, and divided into multiple packets depending on the maximum packet size. Assuming a maximum packet size of 32 bytes, the entire cache line will be divided into four packets. Each packet consists of a head flit that contains the destination address, multiple body flits, and a tail flit that indicate the end of a packet; the head, body, and tail flits each contain a portion of the cache line. For instance, for a flit size of 64 bits, the packet will consist of $(32 \times 8)/64 = 4$ flits: 1 head, 2 body and 1 tail, ignoring the extra bits needed to encode the destination, flow control information for the network, and the bits needed to indicate the action to be performed upon receiving the packet. Since only the head flit contains the destination, all flits of a packet follow the same route.

2.2.2 Topology

The effect of a topology on overall network cost–performance is profound. A topology determines the number of hops (or routers) a message must traverse as well as the interconnect lengths between hops, thus influencing network latency significantly. As traversing routers and links incurs energy, a topology's effect on hop count also directly affects network energy consumption. As for its effect on throughput, since a topology dictates the total number of alternate paths between nodes, it affects how well the network can spread out traffic and thus the effective bandwidth a network can support. Network reliability is also greatly influenced by the topology as it dictates the number of alternative paths for routing around faults. The implementation complexity cost of a topology depends on two factors: the number of links at each node (node degree) and the ease of laying out a topology on a chip (wire lengths and the number of metal layers required).

Figure 2.1 shows three commonly used on-chip topologies. For the same number of nodes, and assuming uniform random traffic where every node has an equal probability of sending to every other node, a ring (Fig. 2.1a) will lead to higher hop count than a mesh (Fig. 2.1b) or a torus [11] (Fig. 2.1c). For instance, in the figure shown, assuming bidirectional links and shortest-path routing, the maximum hop count of the ring is 4, that of a mesh is also 4, while a torus improves it to 2. A ring topology also offers fewer alternate paths between nodes than a mesh or torus, and thus saturates at a lower network throughput for most traffic patterns. For instance, a message between nodes A and B in the ring topology can only traverse one of two paths in a ring, but in a 3×3 mesh topology, there are six possible paths. As for network reliability, among these three networks, a torus offers the most tolerance to faults because it has the highest number of alternative paths between nodes.

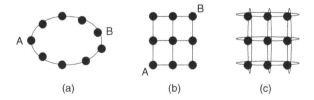

(a)   (b)   (c)

Fig. 2.1 Common on-chip network topologies: (**a**) ring, (**b**) mesh, and (**c**) torus

While rings have poorer performance (latency, throughput, energy, and reliability) when compared to higher-dimensional networks, they have lower implementation overhead. A ring has a node degree of 2 while a mesh or torus has a node degree of 4, where node degree refers to the number of links in and out of a node. A higher node degree requires more links and higher port counts at routers. All three topologies featured are two-dimensional planar topologies that map readily to a single metal layer, with a layout similar to that shown in the figure, except for tori which should be arranged physically in a folded manner to equalize wire lengths (see Fig. 2.2), instead of employing long wrap-around links between edge nodes. A torus illustrates the importance of considering implementation details in comparing alternative topologies. While a torus has lower hop count (which leads to lower delay and energy) compared to a mesh, wire lengths in a folded torus are twice that in a mesh of the same size, so per-hop latency and energy are actually higher. Furthermore, a torus requires twice the number of links which must be factored into the wiring budget. If the available wiring bisection bandwidth is fixed, a torus will be restricted to narrower links than a mesh, thus lowering per-link bandwidth, and increasing transmission delay. Determining the best topology for an on-chip network subject to the physical and technology constraints is an area of active research.

Fig. 2.2 Layout of a 8×8 folded torus

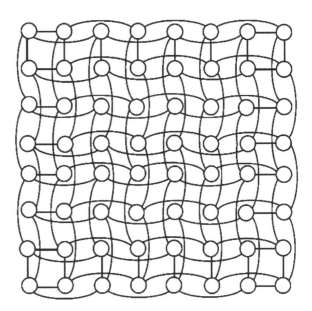

2.2.3 Routing

The goal of the routing algorithm is to distribute traffic evenly among the paths supplied by the network topology, so as to avoid hotspots and minimize contention, thus improving network latency and throughput. In addition, the routing algorithm is the critical component in fault tolerance: once faults are identified, the routing algorithm must be able to skirt the faulty nodes and links without substantially affecting network performance. All of these performance goals must be achieved while adhering to tight constraints on implementation complexity: routing circuitry can stretch critical path delay and add to a router's area footprint. While energy overhead of routing circuitry is typically low, the specific route chosen affects hop count directly, and thus substantially affects energy consumption.

While numerous routing algorithms have been proposed, the most commonly used routing algorithm in on-chip networks is dimension-ordered routing (DOR) due to its simplicity. With DOR, a message traverses the network dimension-by-dimension, reaching the coordinate matching its destination before switching to the next dimension. In a two-dimensional topology such as the mesh in Fig. 2.3, dimension-ordered routing, say X–Y routing, sends packets along the X-dimension first, followed by the Y-dimension. A packet traveling from (0,0) to (2,3) will first traverse two hops along the X-dimension, arriving at (2,0), before traversing three hops along the Y-dimension to its destination. Dimension-ordered routing is an example of a deterministic routing algorithm, in which all messages from node A to B will always traverse the same path. Another class of routing algorithms are oblivious ones, where messages traverse different paths from A to B, but the path is selected without regards to the actual network situation at transmission time. For instance, a router could randomly choose among alternative paths prior to sending a message. Figure 2.3 shows an example where messages from (0,0) to (2,3) can be randomly sent along either the Y–X route or the X–Y route. A more sophisticated routing algorithm can be adaptive, in which the path a message takes from A to B depends on network traffic situation. For instance, a message can be going along the X–Y route, sees congestion at (1,0)'s east outgoing link and instead choose to take the north outgoing link towards the destination (see Fig. 2.3).

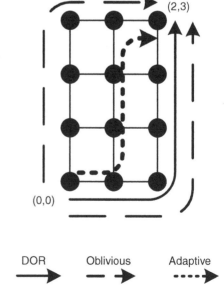

Fig. 2.3 *DOR* illustrates an X–Y route from (0,0) to (2,3) in a mesh, while *Oblivious* shows two alternative routes (X–Y and Y–X) between the same source–destination pair that can be chosen obliviously prior to message transmission. *Adaptive* shows a possible adaptive route that branches away from the X–Y route if congestion is encountered at (1,0)

In selecting or designing a routing algorithm, not only must its effect on delay, energy, throughput, and reliability be taken into account, most applications also require the network to guarantee deadlock freedom. A deadlock occurs when a cycle exists among the paths of multiple messages. Figure 2.4 shows four gridlocked (and deadlocked) messages waiting for links that are currently held by other messages and prevented from making forward progress: The packet entering router A from the South input port is waiting to leave through the East output port, but another packet is holding onto that exact link while waiting at router B to leave via the South output port, which is again held by another packet that is waiting at router C to leave via the West output port and so on.

Fig. 2.4 A classic network deadlock where four packets cannot make forward progress as they are waiting for links that other packets are holding on to

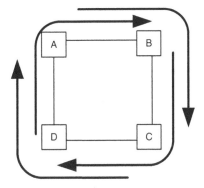

Deadlock freedom can be ensured in the routing algorithm by preventing cycles among the routes generated by the algorithm, or in the flow control protocol by preventing router buffers from being acquired and held in a cyclic manner [13]. Using the routing algorithms we discussed above as examples, dimension-ordered routing is deadlock-free since in X–Y routing, there cannot be a turn from a Y link to an X link, so cycles cannot occur. Two of the four turns in Fig. 2.4 will not be permitted, so a cycle is not possible. The oblivious algorithm that randomly chooses between X–Y or Y–X routes is not deadlock-free because all four turns from Fig. 2.4 are possible leading to potential cycles in the link acquisition graph. Likewise, the adaptive route shown in Fig. 2.3 is a superset of oblivious routing and is subject to potential deadlock. A network that uses a deadlock-prone routing algorithm requires a flow control protocol that ensures deadlock freedom, as discussed in Sect. 2.2.4.

Routing algorithms can be implemented in several ways. First, the route can be embedded in the packet header at the source, known as source routing. For instance, the X–Y route in Fig. 2.3 can be encoded as <E, E, N, N, N, Eject>, while the Y–X route can be encoded as <N, N, N, E, E, Eject>. At each hop, the router will read the left-most direction off the route header, send the packet towards the specified outgoing link, and strip off the portion of the header corresponding to the current hop. Alternatively, the message can encode the coordinates of the destination, and comparators at each router determine whether to accept or forward the message. Simple routing algorithms are typically implemented as combinational circuits within the router due to the low overhead, while more sophisticated algorithms are realized using routing tables at each hop which store the outgoing link a message should take to reach a particular destination. Adaptive routing algorithms need mechanisms to track network congestion levels, and update the route. Route adjustments can be implemented by modifying the header, employing combinational circuitry that accepts as input these congestion signals, or updating entries in a routing table. Many congestion-sensitive mechanisms have been proposed, with the simplest being tapping into information that is already captured and used by the flow control protocol, such as buffer occupancy or credits.

2.2.4 Flow Control

Flow control governs the allocation of network buffers and links. It determines when buffers and links are assigned to which messages, the granularity at which they are allocated, and how these resources are shared among the many messages using the network. A good flow control protocol lowers the latency experienced by messages at low loads by not imposing high overhead in resource allocation, and pushes network throughput through enabling effective sharing of buffers and links across messages. In determining the rate at which packets access buffers (or skip buffer access altogether) and traverse links, flow control is instrumental in determining network energy and power. Flow control also critically affects network quality-of-service since it determines the arrival and departure time of packets at each hop. The implementation complexity of a flow control protocol includes the complexity of the router microarchitecture as well as the wiring overhead imposed in communicating resource information between routers.

In *store-and-forward* flow control [12], each node waits until an entire packet has been received before forwarding any part of the packet to the next node. As a result, long delays are incurred at each hop, which makes them unsuitable for on-chip networks that are usually delay-critical. To reduce the delay packets experience at each hop, *virtual cut-through* flow control [24] allows transmission of a packet to begin before the entire packet is received. Latency experienced by a packet is thus drastically reduced, as shown in Fig. 2.5. However, bandwidth and storage are still allocated in packet-sized units. Packets still only move forward if there is enough storage to hold the entire packet. On-chip networks with tight area and power constraints find it difficult to accommodate the large buffers needed to support virtual cut-through (assuming large packets).

Like virtual cut-through flow control, *wormhole* flow control [11] cuts through flits, allowing flits to move on to the next router as soon as there is sufficient buffering for this flit. However, unlike store-and-forward and virtual cut-through flow control, wormhole flow control allocates storage and bandwidth to flits that are smaller than a packet. This allows relatively small flit-buffers to be used in each router, even for large packet sizes. While wormhole flow control uses buffers effectively, it makes inefficient use of link bandwidth. Though it allocates storage and bandwidth

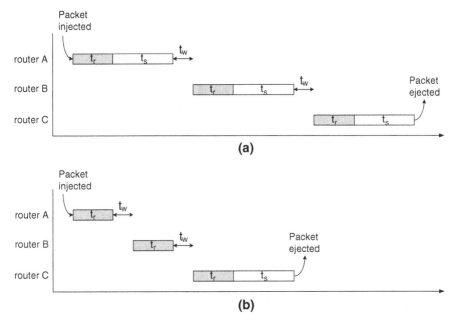

Fig. 2.5 Timing for (**a**) store-and-forward and (**b**) cut-through flow control at low loads, where t_r refers to the delay routing the head flit through each router, t_s refers to the serialization delay transmitting the remaining flits of the packet through each router, and t_w refers to the time involved in propagating bits across the wires between adjacent routers. Wormhole and virtual-channel flow control have the same timing as cut-through flow control at low loads

in flit-sized units, a link is held for the duration of a packet's lifetime in the router. As a result, when a packet is blocked, all of the physical links held by that packet are left idle. Throughput suffers because other packets queued behind the blocked packet are unable to use the idle physical links.

Virtual-channel flow control [9] improves upon the link utilization of wormhole flow control, allowing blocked packets to be passed by other packets. A virtual channel (VC) consists merely of a separate flit queue in the router; multiple VCs share the physical wires (physical link) between two routers. Virtual channels arbitrate for physical link bandwidth on a flit-by-flit basis. When a packet holding a virtual channel becomes blocked, other packets can still traverse the physical link through other virtual channels. Thus, VCs increase the utilization of the critical physical links and extend overall network throughput. Current on-chip network designs overwhelmingly adopt wormhole flow control for its small area and power footprint, and use virtual channels to extend the bandwidth where needed. Virtual channels are also widely used to break deadlocks, both within the network, and for handling system-level deadlocks.

Figure 2.6 illustrates how two virtual channels can be used to break a cyclic deadlock in the network when the routing protocol permits a cycle. Here, since each VC is associated with a separate buffer queue, and since every VC is time-multiplexed

Fig. 2.6 Two virtual channels (denoted by *solid* and *dashed lines*) and their associated separate buffer queues (denoted as two *circles* at each router) used to break the cyclic route deadlock in Fig. 2.4

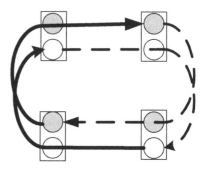

onto the physical link cycle-by-cycle, holding onto a VC implies holding onto its associated buffer queue rather than locking down a physical link. By enforcing an order on VCs, so that lower-priority VCs cannot request and wait for higher-priority VCs, there cannot be a cycle in resource usage. At the system level, messages that can potentially block on each other can be assigned to different message classes that are mapped onto different virtual channels within the network, such as request and acknowledgment messages of coherence protocols. Implementation complexity of virtual channel routers will be discussed in detail next in Sect. 2.2.5 on router microarchitecture.

Buffer backpressure: Unlike broadband networks, on-chip network are typically not designed to tolerate dropping of packets in response to congestion. Instead, buffer backpressure mechanisms stall flits from being transmitted when buffer space is not available at the next hop. Two commonly used buffer backpressure mechanisms are credits and on/off signaling. Credits keep track of the number of buffers available at the next hop, by sending a credit to the previous hop when a flit leaves and vacates a buffer, and incrementing the credit count at the previous hop upon receiving the credit. On/off signaling involves a signal between adjacent routers that is turned off to stop the router at the previous hop from transmitting flits when the number of buffers drop below a threshold, with the threshold set to ensure that all in-flight flits will have buffers upon arrival.

2.2.5 Router Microarchitecture

How a router is built determines its critical path delay, per-hop delay, and overall network latency. Router microarchitecture also affects network energy as it determines the circuit components in a router and their activity. The implementation of the routing and flow control and the actual router pipeline will affect the efficiency at which buffers and links are used and thus overall network throughput. In terms of reliability, faults in the router datapath will lead to errors in the transmitted

Portions reprinted, with permission, from "Express Virtual Channels: Towards the Ideal Interconnection Fabric", A. Kumar, L-S. Peh, P. Kundu and N. K. Jha, In Proceedings of International Symposium on Computer Architecture (ISCA), June 2007, ACM.

message, while errors in the control circuitry can lead to lost and mis-routed messages. The area footprint of the router is clearly highly determined by the chosen router microarchitecture and underlying circuits.

Figure 2.7 shows the microarchitecture of a state-of-the-art credit-based virtual channel (VC) router to explain how typical routers work. The example assumes a two-dimensional mesh, so that the router has five input and output ports corresponding to the four neighboring directions and the local processing element (PE) port. The major components which constitute the router are the input buffers, route computation logic, virtual channel allocator, switch allocator, and the crossbar switch. Most on-chip network routers are input-buffered, in which packets are stored in buffers only at the input ports.

Figure 2.8a shows the corresponding pipeline for this basic router (BASE). A head flit, on arriving at an input port, is first decoded and buffered according to its input VC in the buffer write (BW) pipeline stage. In the next stage, the routing logic performs route computation (RC) to determine the output port for the packet. The header then arbitrates for a VC corresponding to its output port in the

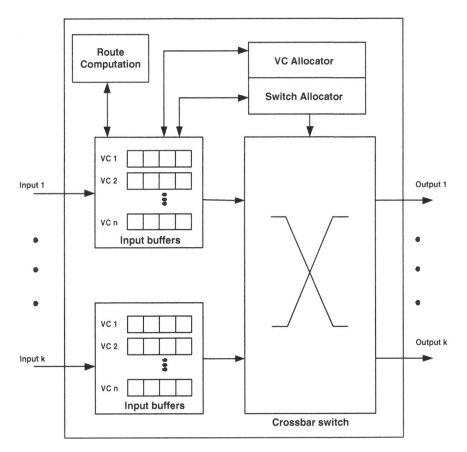

Fig. 2.7 A credit-based virtual channel router microarchitecture

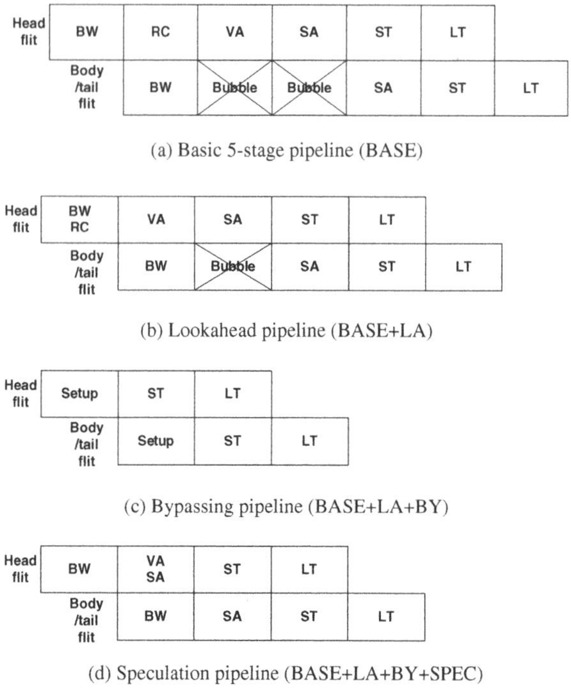

Fig. 2.8 Router pipeline
[BW: Buffer Write, RC:
Route Computation, VA:
Virtual Channel Allocation,
SA: Switch Allocation, ST:
Switch Traversal, LT: Link
Traversal]

Head flit	BW	RC	VA	SA	ST	LT	
Body /tail flit	BW	Bubble	Bubble	SA	ST	LT

(a) Basic 5-stage pipeline (BASE)

Head flit	BW RC	VA	SA	ST	LT	
Body /tail flit	BW	Bubble	SA	ST	LT

(b) Lookahead pipeline (BASE+LA)

Head flit	Setup	ST	LT
Body /tail flit	Setup	ST	LT

(c) Bypassing pipeline (BASE+LA+BY)

Head flit	BW	VA SA	ST	LT
Body /tail flit	BW	SA	ST	LT

(d) Speculation pipeline (BASE+LA+BY+SPEC)

VC allocation (VA) stage. Upon successful allocation of a VC, the header flit proceeds to the switch allocation (SA) stage where it arbitrates for the switch input and output ports. On winning the output port, the flit then proceeds to the switch traversal (ST) stage, where it traverses the crossbar. Finally, the flit is passed to the next node in the link traversal (LT) stage. Body and tail flits follow a similar pipeline except that they do not go through RC and VA stages, instead inheriting the VC allocated by the head flit. The tail flit, on leaving the router, deallocates the VC reserved by the header.

To remove the serialization delay due to routing, prior work has proposed *lookahead routing* [16] where the route of a packet is determined one hop in advance. Precomputing the route enables flits to compete for VCs immediately after the BW stage, with the RC stage essentially responsible for just the decoding of the route header encoded within the head flit. Figure 2.8b shows the router pipeline with lookahead routing (BASE+LA). *Pipeline bypassing* is another technique that is commonly used to further shorten the router critical path by allowing a flit to speculatively enter the ST stage if there are no flits ahead of it in the input buffer queue. Figure 2.8c shows the pipeline where a flit goes through a single stage of switch setup, during which the crossbar is set up for flit traversal in the next cycle while simultaneously allocating a free VC corresponding to the desired output port, followed by ST and LT (BASE+LA+BY). The allocation is aborted upon a port conflict. When the router ports are busy, thereby disallowing pipeline bypassing, aggressive *speculation* can be used to cut down the critical path [31, 32, 36].

Figure 2.8d shows the router pipeline, where VA and SA take place in parallel in the same cycle (BASE+LA+BY+SPEC). A flit enters the SA stage speculatively after BW and arbitrates for the switch port while at the same time trying to acquire a free VC. If the speculation succeeds, the flit directly enters the ST pipeline stage. However, when speculation fails, the flit must go through some of these pipeline stages again, depending on where the speculation failed.

The basic router pipeline is implemented using a set of microarchitectural building blocks whose designs are described below.

Router buffers: Router buffers tend to be built using SRAM or register file cells, depending on the buffer size and access timing requirements. The storage cells are then typically organized into queues for each VC, either as a physical circular buffer or a logical linked list, with head and tail pointers maintained dynamically.

Switch design: The hardware cost of crossbars in terms of area and power scale is $O((pw)^2)$, where p is the number of crossbar ports and w is the crossbar port width in bits. Different switch designs can be chosen to drive down p and/or w. Dimension slicing a crossbar [10] in a two-dimensional mesh uses two 3×3 crossbars instead of one 5×5 crossbar, with the first crossbar for traffic remaining in the X-dimension, and the second crossbar for traffic remaining in the Y-dimension. A port on the first crossbar connects with a port on the second crossbar, so traffic that turns from the X to Y dimension traverses both crossbars while those remaining within a dimension only traverses one crossbar. This is particularly suitable for the dimension-ordered routing protocol where traffic mostly stays within a dimension. Bit interleaving the crossbar (or double-pumping as explained in Sect. 2.5) targets w instead. It sends alternate bits of a link on the two phases of a clock on the same line, thus halving w.

Allocators and arbiters: The virtual-channel allocator (VA) resolves contention for output virtual channels and grants them to input virtual channels, while the switch allocator (SA) matches and grants crossbar switch ports to input virtual channels. An allocator essentially matches N requests to M resources, where the resources are VCs for the VA, and crossbar switch ports for the SA. An allocator that delivers high matching probability translates to more packets succeeding in obtaining virtual channels and passage through the crossbar switch, thereby leading to higher network throughput. Allocators must also be fast and pipelinable so that they can accommodate high clock frequencies.

Due to complexity concerns, simple separable allocators are typically used, with an allocator being composed of arbiters, where an arbiter is one that chooses one out of N requests to a single resource. Figure 2.9 shows how each stage of a 3:4 separable allocator (an allocator matching three requests to four resources) is composed of arbiters. For instance, a separable VA will have the first stage of the allocator (comprising four 3:1 arbiters) selecting one of the eligible output VCs, with the winning requests from the first stage of allocation then arbitrating for an output VC in the second stage (comprising three 4:1 arbiters). Different arbiters have been used in practice, with round-robin arbiters being the most popular due to their simplicity.

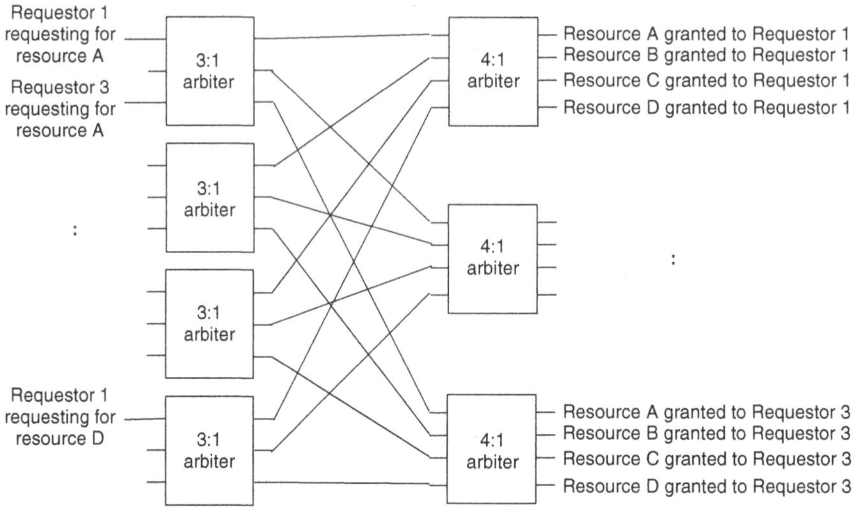

Fig. 2.9 A separable 3:4 allocator (three requestors, four resources) which consists of four 3:1 arbiters in the first stage and three 4:1 arbiters in the second. The 3:1 arbiters in the first stage decides which of the three requestors win a specific resource, while the 4:1 arbiters in the second stage ensure a requestor is granted just one of the four resources

2.2.6 Summary

In this section, we sought to present the reader with the fundamental principles of the field of on-chip networks, with a highly selective subset of basic architectural concepts chosen largely to allow the reader to appreciate the two case studies that follow. Substantial ongoing research is investigating all aspects of network architectures, including topology, routing, flow control, and router microarchitecture optimizations. For example, recent novel topologies proposed include the flattened butterfly [26], STmicroelectronics' spidergon topology [8], and Ambric's hierarchical mesh [6]. Several papers explored novel adaptive routing algorithms for on-chip networks [40, 17] and using routing to avoid thermal hotspots [39]. For flow control protocols, recent research investigated ways to bypass intermediate routers even at high loads [30,15], improving energy-delay-throughput simultaneously. Substantial work has gone into designing new router microarchitectures: decoupling arbitration thus enabling smaller crossbars within a router [27], dynamically varying the number of virtual channels depending on traffic load [33], and microarchitectural techniques that specifically target a reduction in router power [46].

2.3 Case Studies

The next two sections detail two chip prototypes of on-chip networks that are driven by vastly different design requirements. The TRIPS operand network is the principal interconnect between 25 arithmetic units, memory banks, and register banks in a distributed microprocessor. In particular, this network replaces both the operand bypass bus and the level-1 cache bus of a conventional processor. Because instruction interaction and level-1 cache delays are critical, this network is optimized for low-latency transmission with one hop per cycle. Furthermore, both sides of the network interface are integrated tightly into the pipelines of the execution units to enable fast pipeline forwarding.

In contrast, the Intel TeraFLOP chip's network connects the register files of 80 simple cores at a very high clock rate. The router in each tile uses five ports to connect to its four neighbors and the local PE, using point-to-point links that can deliver data at 20 GB/s. These links support mesochronous or phase-tolerant communication across tiles, paying a synchronization latency penalty for the benefit of a lightweight global clock distribution. An additional benefit of the low-power mesochronous clock distribution is its scalability, which enabled a high level of integration and single-chip realization of the TeraFLOPS processor. At the system level, software directly controls the routing of the messages and application designers orchestrate fine-grained message-based communication in conjunction with the computation. Table 2.1 summarizes the specific design choices of these networks for each aspect of on-chip network design, including the messaging layer, topology, routing, flow control, and router microarchitecture.

Both of these on-chip networks are quite different from interconnection networks found in large-scale multi-chip parallel systems. For example, the IBM BlueGene networks [1] housed within the multi-chassis supercomputer are faced with design requirements that differ by orders of magnitude in delay, area, power, and through-

Table 2.1 Summary of the characteristics of UT-Austin TRIPS operand network and the Intel TeraFLOPS network

	UT-Austin TRIPS OPN	Intel TeraFLOPS
Process technology	130 nm ASIC	65 nm custom
Clock frequency	400 MHz	5 GHz
Channel width	140 bits	38 bits
Interface	Explicit in instruction targets	Explicit send/receive instructions
	Interfaces with bypass paths	Interfaces with registers
	Single-flit messages	Unlimited message size
Topology	5×5 mesh	8×10 mesh
Routing	Y–X dynamic	Static source routing via software
Flow control	Control-data reservation	Two virtual channels
	On/off backpressure	On/off backpressure
Microarchitecture	Single-stage lookahead pipeline	Six-stage pipeline (including link traversal)
	Two 4-flit input buffer queues	Two 16-flit input buffer queues
	Round-robin arbiters	Separable Two-stage round-robin allocator
	Two 4×4 crossbars	5×5 double-pumped crossbar

put. The design choices in the BlueGene networks reflect this difference. First, BlueGene's network interfaces through the message passing interface (MPI) standard, whose high software overheads may not be compatible with an on-chip network with much tighter delay requirements. BlueGene's main network adopts a three-dimensional torus, which maps readily to its three-dimensional chassis form factor, but would create complexities in layout for two-dimensional on-chip networks. BlueGene's choice to support adaptive routing with a bubble escape channel and hints for source routing can lead to a complex pipeline that is hard for on-chip networks to accommodate. Its flow control protocol is virtual cut-through, with four virtual channels, each with 1 KB buffering, and its router pipeline is eight stages deep. Such a complex router whose area footprint is larger than a PowerPC 440 core and its associated double FPU unit is reasonable for a supercomputer, but not viable for on-chip networks.

2.4 UT-Austin TRIPS Operand Network

The TRIPS processor chip is an example of an emerging scalable architecture in which the design is partitioned into distributed tiles connected via control and data networks. TRIPS aims to provide a design scalable to many execution units with networks tightly integrated into the processing cores to enable instruction-level and thread-level parallelism. This section describes in detail the TRIPS operand network (OPN), which transports operands among processor tiles. Each processor's OPN is a 5×5 dynamically routed 2D mesh network with 140-bit links. The OPN connects a total of 25 distributed execution, register file, and data cache tiles. Each tile is replicated and interacts only with neighboring tiles via the OPN and other control networks. The OPN subsumes the role of several traditional microprocessor interconnect buses, including the operand bypass network, register file read and write interface, and the level-1 memory system bus. In order to serve in this function, we designed the OPN to have single-cycle per-hop latency and split control/data delivery to speed instruction wakeup at the destination. The OPN router also is small, with relatively little buffering and no virtual channels so that the effect on overall area is minimized.

2.4.1 TRIPS Overview

TRIPS is a distributed processor consisting of multiple tiles connected via several control and data networks. Figure 2.10 shows a photo micrograph of the TRIPS prototype processor chip (fabricated in a 130 nm ASIC process) and a block diagram

---

Portions reprinted, with permission, from "Implementation and Evaluation of a Dynamically Routed Processor Operand Network," P. Gratz, K. Sankaralingam, H. Hanson, P. Shivakumar, R. McDonald, S.W. Keckler, and D. Burger, International Symposium on Networks-on-Chips©2007, IEEE.

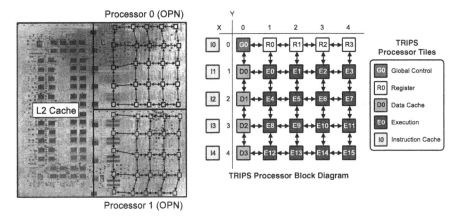

Fig. 2.10 TRIPS die photo and processor block diagram

of a TRIPS processor core. The on-chip memory system (configurable as a level-2 cache) appears on the left side of the chip micrograph and includes an on-chip network (OCN) tailored to deliver cache-lined sized items among the processors and data banks [18]. On the right side of the figure are the two processors, each with its own separate operand network, as indicated by the superimposed diagram. The right side of Fig. 2.10 shows a tile-level diagram of the processor with its OPN links. The processor contains five types of tiles: execution tiles (ET) which contain integer and floating-point ALUs and 10 KB of reservation stations; register tiles (RT) each of which contains a fraction of the processor register file; data tiles (DT) each of which contains a fraction of the level-1 data cache; instruction tiles (IT) each of which contains a fraction of the level-1 instruction cache; and a global control tile (GT) which orchestrates instruction fetch, execution, and commit. In addition, the processor contains several control networks for implementing protocols such as instruction fetch, completion, and commit in a distributed fashion. Tiles communicate directly only with their nearest neighbors to keep wires short and mitigate the effect of wire delay.

During block execution, the TRIPS operand network (OPN) has responsibility for delivering operands among the tiles. The TRIPS instruction formats contain target fields indicating to which consumer instructions a producer sends its values. At runtime, the hardware resolves those targets into coordinates to be used for network routing. An operand passed from producer to consumer on the same ET can be bypassed directly without delay, but operands passed between instructions on different tiles must traverse a portion of the OPN. The TRIPS execution model is inherently dynamic and data driven, meaning that operand arrival drives instruction execution, even if operands are delayed by unexpected or unknown memory or network latencies. Because of the data-driven nature of execution and because multiple blocks execute simultaneously, the OPN must dynamically route the operand across the processor tiles. Additional details on the TRIPS instruction set architecture and microarchitecture can be found in [5,37].

2.4.2 Operand Network Architecture

The operand network (OPN) is designed to deliver operands among the TRIPS processor tiles with minimum latency, and serves a similar purpose as the RAW scalar operand network [41]. While tight integration of the network into the processor core reduces the network interface latency, two primary aspects of the TRIPS processor architecture simplify the router design and reduce routing latency. First, because of the block execution model, reservation stations for all operand network packets are pre-allocated, guaranteeing that all OPN messages can be consumed at the targets. Second, all OPN messages are of fixed length, and consist of a single flit broken into header (control) and payload (data) phits.

The OPN is a 5×5 2D routed mesh network as shown in Fig. 2.10. Flow control is on/off based, meaning that the receiver tells the transmitter when there is enough buffer space available to send another flit. Packets are routed through the network in Y–X dimension-order with one cycle taken per hop. A packet arriving at a router is buffered in an input FIFO prior to being launched onward towards its destination. Due to dimension-order routing and the guarantee of consumption of messages, the OPN is deadlock free without requiring virtual channels. The absence of virtual channels reduces arbitration delay and makes a single-cycle router and link transmission possible.

Each operand network message consists of a control phit and a data phit. The control phit is 30 bits and encodes OPN source and destination node coordinates, along with identifiers to indicate which instruction to select and wakeup in the target ET. The data phit is 110 bits, with room for a 64-bit data operand, a 40-bit address for store operations, and 6 bits for status flags. We initially considered narrow channels and longer messages to accommodate address and data delivered by a store instruction to the data tiles. However, the single-flit message design was much simpler to implement and not prohibitively expensive given the abundance of wires.

The data phit always trails the control phit by one cycle in the network. The OPN supports different physical wires for the control and data phit; so one can think of each OPN message consisting of one flit split into a 30-bit control phit and a 110-bit data phit. Because of the distinct control and data wires, two OPN messages with the same source and destination can proceed through the network separated by a single cycle. The data phit of the first message and the control phit of the second are on the wires between the same two routers at the same time. Upon arrival at the destination tile, the data phit may bypass the input FIFO and be used directly, depending on operation readiness. This arrangement is similar to flit-reservation flow control, although here the control phit contains some payload information and does not race ahead of the data phit [35]. In all, the OPN has a peak injection bandwidth of 175 GB/s when all nodes are injecting packets every cycle at its designed frequency of 400 MHz. The network's bisection bandwidth is 70 GB/s measured horizontally or vertically across the middle of the OPN.

Figure 2.11 shows a high-level block diagram of the OPN router. The OPN router has five inputs and five outputs, one for each ordinal direction (N, S, E, and W) and one for the local tile's input and output. The ordinal directions' inputs each have two

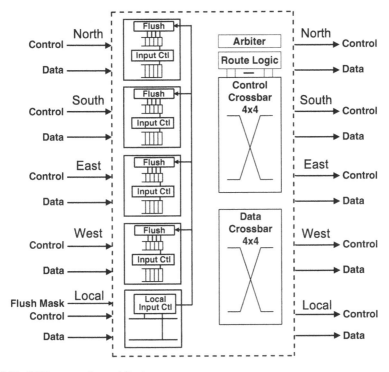

Fig. 2.11   OPN router microarchitecture

four-entry deep FIFOs, one 30 bits wide for control phits and one 110 bits wide for data phits. The local input has no FIFO buffer. The control and data phits of the OPN packet have separate 4×4 crossbars. Although the router has five input and output ports, the crossbar only needs four inputs and outputs because several input/output connections are not allowed, such as entering and exiting the router on the north port. All arbitration and routing is done on the control phit, in round-robin fashion among all incoming directions. The data phit follows one cycle behind the control phit in lock step, using the arbitration decision from its control phit.

2.4.3 OPN/Processor Integration

The key features of the TRIPS OPN that distinguish it from general purpose data transport networks are its integration into the pipelines of the ALUs and its support for speculative instruction execution.

ALU pipeline integration: Figure 2.12 shows the operand network datapath between the ALUs in two adjacent ETs. The instruction selection logic and the output latch of the ALU are both connected directly to the OPN's local input port, while the instruction wakeup logic and bypass network are both connected to the

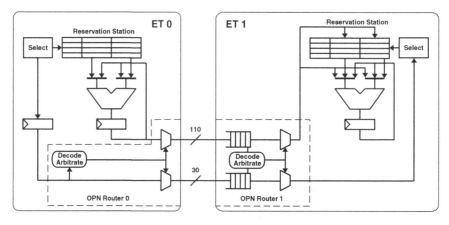

Fig. 2.12 Operand datapath between two neighboring ETs

OPN's local output. The steps below describe the use of the OPN to bypass data between two adjacent ALUs and feature the tight integration of the OPN and instruction execution. Operands transmitted greater distances travel through the network without disturbing the intervening execution unit. The timeline for message injection/reception between two adjacent ETs proceeds as follows:

- Cycle 0: Instruction wakeup/select on ET 0

  - ET0 selects a ready instruction and sends it to the ALU.
  - ET0 recognizes that the instruction target is on ET1 and creates the control phit.

- Cycle 1: Instruction execution on ET0

  - ET0 executes the instruction on the ALU.
  - ET0 delivers the control phit to router FIFO of ET1.

- Cycle 2: Instruction wakeup/select on ET1

  - ET0 delivers the data phit to ET1, bypassing the FIFO and depositing the data in a pipeline latch.
  - ET1 wakes up and selects the instruction depending on the data from ET0.

- Cycle 3: Instruction execution ET1

  - ET1 selects the data bypassed from the network and executes the instruction.

The early wakeup, implemented by delivering the control phit in advance of the data phit, overlaps instruction pipeline control with operand data delivery. This optimization reduces the remote bypass time by a cycle (to one cycle) and improves processor performance by approximately 11% relative to a design where the wakeup occurs when the data arrives. In addition, the separation of the control and data phits onto separate networks with shared arbitration and routing eliminates arbitration for the data phit and reduces network contention relative to a network that sends the header and payload on the same wires in successive cycles. This optimization is inexpensive in an on-chip network due to the high wire density.

The OPN employs round-robin arbitration among all of the inputs, including the local input. If the network is under load and chooses not to accept the control phit, the launching node captures the control phit and later the data phit in a local output buffer. The ET will stall if the instruction selected for execution needs the OPN and the ET output buffer is already full. However, an instruction that needs only to deliver its result to another instruction on the same ET does not stall due to OPN input contention. While OPN contention can delay instruction execution on the critical path of program execution, the compiler's instruction placement phase is effective at placing instructions to mitigate the distance that operands must travel and the contention they encounter.

Selective OPN message invalidation: Because the TRIPS execution model uses both instruction predication and branch prediction, some of the operand messages are actually speculative. On a branch misprediction or a block commit, the processor must flush all in-flight state for the block, including state in any of the OPN's routers. The protocol must selectively flush only those messages in the routers that belong to the flushed block. The GT starts the flush process by multicasting a flush message to all of the processor tiles using the global control network (GCN). This message starts at the GT and propagates across the GCN within 10 cycles. The GCN message contains a block mask indicating which blocks are to be flushed. Tiles that receive the GCN flush packet instruct their routers to invalidate from their FIFOs any OPN messages with block-identifiers matching the flushed block mask. As the invalidated packets reach the head of the associated FIFOs they are removed. While we chose to implement the FIFOs using shift registers to simplify invalidation, the protocol could also be implemented for circular buffer FIFOs. A collapsing FIFO that immediately eliminates flushed messages could further improve network utilization, but we found that the performance improvement did not outweigh the increased design complexity. In practice, very few messages are actually flushed.

2.4.4 Area and Timing

The TRIPS processor is manufactured using a 130 nm IBM ASIC technology and returned from the foundry in September 2006. Each OPN router occupies approximately 0.25 mm^2, which is similar in size to a 64-bit integer multiplier. Table 2.2 shows a breakdown of the area consumed by the components of an OPN router. The router FIFOs dominate the area in part because of the width and depth of the FIFOs. Each router includes a total of 2.2 kilobits of storage, implemented using standard cell flip-flops rather than generated memory or register arrays. Using shift FIFOs added some area overhead due to extra multiplexors. We considered using ASIC library-generated SRAMs instead of flip-flops, but the area overhead turned out to

Table 2.2 Area occupied by the components of an OPN router

Component	% Router Area	% E-Tile Area
Router input FIFOs	74.6(%)	7.9(%)
Router crossbar	20.3(%)	2.1(%)
Router arbiter logic	5.1(%)	0.5(%)
Total for single router	–	10.6(%)

Table 2.3 Critical path timing for OPN control and data phit

Component	Latency	(%) Path
Control phit path		
Read from instruction buffer	290 ps	13
Control phit generation	620 ps	27
ET0 router arbitration	420 ps	19
ET0 OPN output mux	90 ps	4
ET1 OPN FIFO muxing and setup time	710 ps	31
Latch setup + clock skew	200 ps	9
Total	2.26 ns	–
Data phit path		
Read from output latch	110 ps	7
Data phit generation	520 ps	32
ET0 OPN output mux	130 ps	8
ET1 router muxing/bypass	300 ps	19
ET1 operand buffer muxing/setup	360 ps	22
Latch setup + clock skew	200 ps	12
Total	1.62 ns	–

be greater given the small size of each FIFO. Custom FIFO design would be more efficient than either of these approaches.

A single OPN router takes up approximately 10% of the ET's area and all the routers together form 14% of a processor core. While this area is significant, the alternative of a broadcast bypass network across all 25 tiles would consume considerable area and not be feasible. We could have reduced router area by approximately 1/3 by sharing the FIFO entries and wires for the control and data phits. However, the improved OPN bandwidth and overall processor performance justifies the additional area.

We performed static timing analysis on the TRIPS design using Synopsys Primetime to identify and evaluate critical paths. Table 2.3 shows the delay for the different elements of the OPN control and data critical paths, matching the datapath of Fig. 2.12. We report delays using a nominal process corner, which matches the clock period of 2.5 ns obtained in the prototype chip. A significant fraction of the clock cycle time is devoted to overheads such as flip-flop read and setup times as well as clock uncertainty (skew and jitter). A custom design would likely be able to drive these overheads down. On the logic path, the control phit is much more constrained than the data phit due to router arbitration delay. We were a little surprised by the delay associated with creating the control phit, which involves decoding and encoding. This path could be improved by performing the decoding and encoding in a previous cycle and storing the control phit with the instruction before execution. We found that wire delay was small in our 130 nm process given the relatively short transmission distances. Balancing router delay and wire delay may be more challenging in future process technologies.

2.4.5 Design Optimizations

We considered a number of OPN enhancements but chose not to implement them in the prototype in order to simplify the design. One instance where performance can be improved is when an instruction must deliver its result to multiple consumers. The TRIPS ISA allows an instruction to specify up to four consumers, and in the current implementation, the same value is injected in the network once for each consumer. Multicast in the network would automatically replicate a single message in the routers at optimal bifurcation points. This capability would reduce overall network contention and latency while increasing ET execution bandwidth, as ETs would spend less time blocking for message injection. Another optimization would give network priority to those OPN messages identified to be on the program's critical path. We have also considered improving network bandwidth by having multiple, identical operand networks by replicating the routers and wires. Finally, the area and delay of our design was affected by the characteristics of the underlying ASIC library. While the trade-offs may be somewhat different with a full-custom design, our results are relevant because not all on-chip networked systems will be implemented using full-custom silicon. ASIC design for networks would also benefit from new ASIC cells, such as small but dense memory arrays and FIFOs.

2.4.6 Summary

The TRIPS Operand Network (OPN) is an effective replacement for both the global ALU bypass bus and the L1-cache bus in a tiled multicore processor. The OPN is designed for single-cycle per-hop latency and is tightly integrated into the tile pipelines to minimize network interface latency. Because the interconnection topology is exposed to software for instruction placement optimizations, the communication distances are low, averaging about two hops per message but with many nearest neighbor messages. Experiments on the prototype confirm the expectation that distributed processor performance is quite sensitive to network latency, as just one additional cycle per hop results in a 20% drop in performance. Increasing the link width in bits does not help this network since the messages already consist of only one flit. Tailoring the on-chip network to its particular use provided an efficient and high-performance design, a lesson that will carry over to multicore networks in general.

2.5 Intel TeraFLOPS Chip

The Intel TeraFLOPS processor was designed to demonstrate a monolithic on-chip network architecture capable of a sustained performance of 10^{12} floating-point operations per second (1.0 TFLOPS) while dissipating less than 100 W. The TeraFLOPS processor extends work on key network building blocks and integrates them into a 80-core on-chip network design using an effective tiled-design methodology. The $275\,mm^2$ prototype was designed to provide specific insights in new silicon design methodologies for large-scale on-chip networks (100+ cores), high-bandwidth interconnects, scalable clocking solutions, and effective energy management techniques. The core was designed to be small and efficient and to deliver a high level of floating-point performance. The number of cores reflects a trade-off between die size and reaching teraflop performance in a power-efficient manner.

The TeraFLOPS processor architecture contains 80 tiles (Fig. 2.13) arranged as a 8×10 2D array and connected by a mesh network that is designed to operate at 5 GHz [44]. A tile consists of a processing engine (PE) connected to a five-port router which forwards packets between tiles. The PE in each tile (Fig. 2.13) contains two independent, fully pipelined, single-precision floating-point multiply-accumulator (FPMAC) units, 3 KB of single-cycle instruction memory (IMEM), and 2 KB of data memory (DMEM). A 96-bit very long instruction word (VLIW) encodes up to eight operations per cycle. With a 10-port (6-read, 4-write) register file, the architecture allows scheduling to both FPMACs, simultaneous DMEM loads and stores, packet send/receive from mesh network, program control, synchronization primitives for data transfer between PEs, and dynamic sleep instructions.

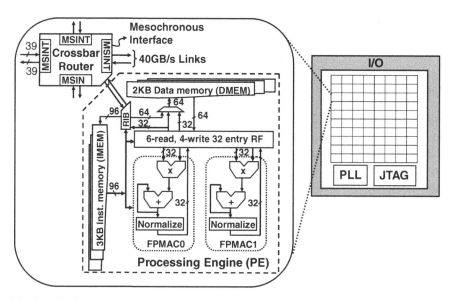

Fig. 2.13 Eighty tiles network-on-chip tile diagram and chip architecture

A router interface block (RIB) handles packet encapsulation between the PE and router. The fully symmetric architecture lets any PE send or receive instruction and data packets to or from any other PE.

The nine-stage pipelined FPMAC architecture uses a single-cycle accumulate algorithm [43], with base-32 and internal carry-save arithmetic with delayed addition. The algorithm reduces latency of dependent FPMAC instructions and enables a sustained multiply–add result (2 FLOPS) every cycle. Several optimizations allow accumulation to be implemented in just 15 fan-out-of-4 (FO4) logic stages. The dual FPMACs in each PE provide 20 GigaFLOPS of aggregate performance and are critical in achieving the goal of TeraFLOP performance.

2.5.1 On-Die Interconnect Architecture

The TeraFLOPS processor uses a 2D mesh because of the large number of cores, and operates at full core speed to provide high bisection bandwidth of 320 GB/s and low-average latency between nodes. The mesh network utilizes a five-port router based on wormhole switching and supports two identical logical lanes (better known as virtual channels), typically separated as instruction and data lanes to prevent short instruction packets from being held up by long data packets. The router in each tile uses five ports to connect to its four neighbors and the local PE, using point-to-point links that can forward data at 20 GB/s in each direction. These links support mesochronous or phase-tolerant communication across tiles, but at the cost of synchronization latency. This approach enables low-power global clock distribution which is critical in ensuring a scalable on-die communication fabric.

2.5.2 Packet Format, Routing, and Flow Control

The on-chip network architecture employs packets that are subdivided into "flits" or "Flow control unITs." Each flit contains six control bits and 32 data bits. Figure 2.14 describes the packet structure. The control bits for each flit include bits for specifying which lane to use for valid flits, and bits for specifying the start and end of a packet. The packet header (FLIT_0) enables a flexible source-directed routing scheme, where a user-specified sequence of 3-bit destination IDs (DIDs) determines the router exit port for each hop on the route. This field is updated at each hop. Each header flit supports a maximum of 10 hops. A chained header (CH) bit in the packet enables a larger number of hops. FLIT_1, that follows the header flit, specifies processing engine control information including sleep and wakeup control bits. The minimum packet size required by the protocol is two flits. The router architecture places no restriction on the maximum packet size.

The implementation choice of source-directed routing (or source routing as discussed in Sect. 2.2.3 enables the use of any of the multiple static routing schemes such as dimension-ordered routing. Throughout its static route, a packet is also

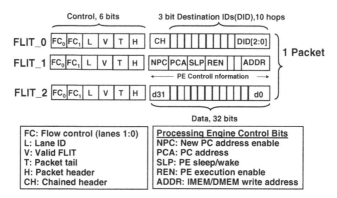

Fig. 2.14 TeraFLOPS processor on-chip network packet format and FLIT description

statically assigned to a particular lane, so virtual channels or lanes are not leveraged for flow control here, but used to avoid system-level deadlocks. Such static assignments result in simple routing and flow control logic, enabling a high frequency implementation.

Flow control and buffer management between routers utilizes an on/off scheme using almost-full signals, which the downstream router signals via two flow control bits (FC0–FC1), one for each lane, when its buffer occupancy reaches a specified threshold. The buffer almost-full threshold can be programmed via software.

2.5.3 Router Microarchitecture

A five-port two-lane pipelined packet-switched router core (Fig. 2.15) with phase-tolerant mesochronous links forms the key communication block for the 80-tile on-chip network architecture. Each port has two 39-bit unidirectional point-to-point links implementing a mesochronous interface (MSINT) with first-in, first-out (FIFO) based synchronization at the receiver. The 32 data bits enable transfer of a single-precision word in a single flit. The router uses two logical lanes (lane 0-1) and a fully non-blocking crossbar switch with a total bandwidth of 100 GB/s (32 bits × 5 GHz × 5 ports). Each lane has a 16 flit queue, arbiter, and flow control logic. A shared datapath allows crossbar switch re-use across both lanes on a per-flit basis.

The router uses a five-stage pipeline with input buffering, route computation, switch arbitration, and switch traversal stages. Figure 2.16 shows the data and control pipelines, including key signals used in the router design. The shaded portion between stages four and five represents the physical crossbar switch. To ensure 5 GHz operation, the router pipeline employs a maximum of 15 FO4 logic levels between pipe stages. The fall-through latency through the router is thus 1 ns at 5 GHz operation. Control overhead for switch arbitration can result in two additional cycles of latency, one cycle for port arbitration and one cycle for lane arbitration. Data flow through the router is as follows: (1) The synchronized input data

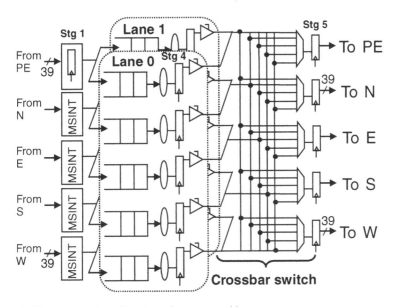

Fig. 2.15 Five-port two-lane shared crossbar router architecture

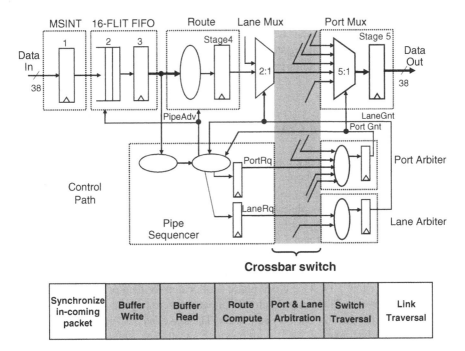

Fig. 2.16 Router data and control pipeline diagram and pipelined routing of a packet

is first buffered into the queues. (2) Data coming out of the queues is examined by routing and sequencing logic to generate output binding requests based on packet header information. (3) A separable, two-stage, round-robin arbitration scheme is used, with arbiters replicated at each port for speed. (4) With five ports and two lanes, the crossbar switch requires a 10-to-1 arbiter and is implemented as a 5-to-1 port arbiter followed by a 2-to-1 lane arbiter. The first stage of arbitration within a particular lane essentially binds one input port to an output port, for the entire duration of the packet. If packets in different lanes are contending for the same physical output resource, then a second stage of arbitration occurs at a flit level granularity.

2.5.4 Interconnect Design Details

Several circuit and design techniques were employed to achieve the objective of high performance, low power, and short design cycle. By following a tiled method-ology where each tile is completely self-contained, including C4 bumps, power, and global clock routing, all tiles could be seamlessly arrayed at the top level simply by abutment. This methodology enabled rapid turnaround of the fully custom designed processor. To enable 5 GHz operation, the entire core uses hand-optimized data path macros. For quick turnaround, CMOS static gates are used to implement most of the logic. However, critical registers in the FPMAC and at the router crossbar out-put utilize implicit-pulsed semi-dynamic flip-flops (SDFF) [44].

2.5.5 Double-Pumped Crossbar

Crossbar switch area increases as a square function $O(n^2)$ of the total number of I/O ports and number of bits per port. A large part of the router area is dominated by the crossbar, in some cases as much as 53% of the total [47]. In this design, the crossbar data buses are double-pumped [45], also known as bit-interleaved, reduc-ing the crossbar hardware cost by half. A full clock cycle is allocated for commu-nication between stages 4 and 5 of the router pipeline. The schematic in Fig. 2.17 shows the 36-bit crossbar data bus double-pumped at the fourth pipe-stage of the router by interleaving alternate data bits using dual edge-triggered flip-flops, reduc-ing crossbar area by 50%. One clock phase is allocated for data propagation across the crossbar channel. The master latch M0 for data bit zero ($i0$) is combined with the slave latch S1 for data bit one ($i1$) and so on. The slave latch S0 is placed halfway across the crossbar channel. A 2:1 multiplexer, controlled by clock, is used to select between the latched output data on each clock phase. At the receiver, the double-pumped data is first latched by slave latch (S0), which retains bit corresponding to $i0$ prior to capture by pipe-stage 5. Data bit corresponding to $i1$ is easily extracted using a rising edge flip-flop.

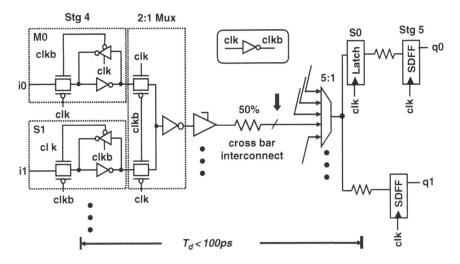

Fig. 2.17 Double-pumped crossbar switch schematic

2.5.6 Mesochronous Communication

The mesochronous interface (Fig. 2.18) allows for efficient, low-power global clock distribution as well as clock-phase-insensitive communication across tiles. Four of the five 2 mm long router links are source synchronous, each providing a strobe (Tx_clk) with 38 bits of data. To reduce active power, Tx_clk is driven at half the clock rate. A four deep circular FIFO built using transparent latches captures data on

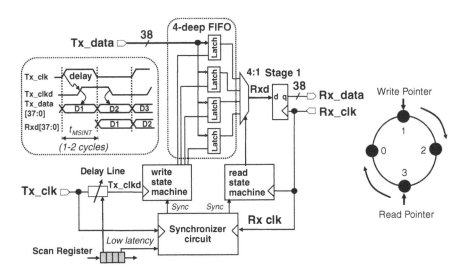

Fig. 2.18 Phase-tolerant mesochronous interface and timing diagram

both edges of the delayed link strobe at the receiver. The strobe delay and duty-cycle can be digitally programmed using the on-chip scan chain. A synchronizer circuit sets the latency between the FIFO write and read pointers to one or two cycles at each port, depending on the phase of the arriving strobe with respect to the local clock. A more aggressive low-latency setting reduces the synchronization penalty by one cycle.

2.5.7 Fine-Grain Power Management

Fine-grain clock gating and sleep transistor circuits [42] are used to reduce active and standby leakage power, which are controlled at full-chip, tile-slice, and individual tile levels based on workload. About 90% of FPMAC logic and 74% of each PE is sleep-enabled. The design uses NMOS sleep transistors to reduce frequency penalty and area overhead. The average sleep transistor area overhead (of the 21 sleep regions in each tile) is 5.4% of the fub area where it is placed. Each tile is partitioned into 21 smaller sleep regions with dynamic control of individual blocks based on instruction type. For instance, each FPMAC can be controlled through NAP/WAKE instructions. The router is partitioned into 10 smaller sleep regions with control of individual router ports (through software) (Fig. 2.19), depending on network traffic patterns. Enable signals gate the clock to each port, MSINT, and the links. In addition, the enable signals also activate the NMOS sleep transistors in the input queue arrays of both lanes. In other words, stages 1–5 of the entire port logic are clock gated, while within each lane, only the FIFOs can be independently power-gated (to reduce leakage power), based on port inactivity. The 360 μm NMOS sleep device in the queue is sized via circuit simulations to provide single-cycle wakeup and a 4.3× reduction in array leakage power with a 4% frequency

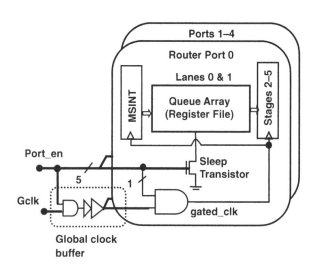

Fig. 2.19 Router and on-die network power management

impact. The global clock buffer feeding the router is finally gated at the tile level
based on port activity.

2.5.8 Experimental Results

The TeraFLOPS processor is fabricated in a 65 nm process technology with eight
layers of copper interconnect [2]. The 275 mm² fully custom design contains 100
million transistors. Each 3 mm² tile contains 1.2 million transistors with the processing engine accounting for 1 million (83%) and the router 17% of the total tile
device count. The functional blocks of the tile and router are identified in the die
photographs in Fig. 2.20. The tiled nature of the physical implementation is easily
visible. The five router ports within each lane are stacked vertically, and the two
lanes sandwich the crossbar switch. Note the locations of the inbound FIFOs and
their relative area compared to the routing channel in the crossbar.

Silicon chip measurements at 80°C demonstrate that the chip achieves
1.0 TFLOPS of average performance at 4.27 GHz and 1.07 V supply with a total
power dissipation of just 97 W. Silicon data confirms that the 2D mesh operates at
a maximum frequency of 5.1 GHz at 1.2 V. Corresponding measured on-chip network power per-tile with all router ports active is 924 mW. This number includes
power dissipated in the router, MSINT, and links. Power reduction due to selective
router port activation with combined clock gating and sleep transistor usage is given
in Fig. 2.21a. Measured at 1.2 V, 80°C, and 5.1 GHz operation, the total network
power per-tile can be lowered from a maximum of 924 mW with all router ports
active to 126 mW, resulting in a 7.3× or 86.6% reduction. The 126 mW number
represents the network leakage power per-tile with all ports and global router clock
buffers disabled and accounts for 13.4% of total network power.

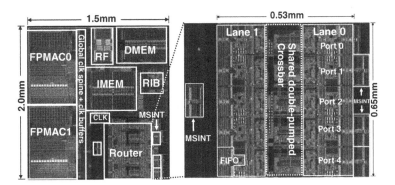

Fig. 2.20 Individual tile and five-port two-lane shared crossbar router layout in 65 nm CMOS
technology. The mesochronous interfaces (MSINT) on four of the five ports are also shown

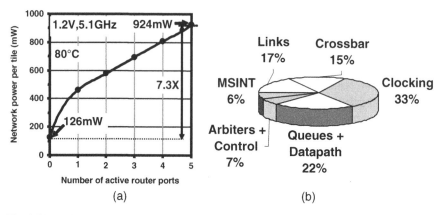

Fig. 2.21 (**a**) Measured on-die network power reduction benefit. (**b**) Estimated on-die communication power breakdown at 4 GHz, 1.2 V, and 110°C

The pie-chart in Fig. 2.21b shows the estimated on-die network power breakdown obtained at 4 GHz, 1.2 V supply, and at 110°C. At the tile level, the communication power, which includes the router, MSINT, and links is significant and dissipates 28% of the total tile power. Clocking power is the largest component due to the deep pipelines used and accounts for 33% of the total network power. The input queues on both lanes and datapath circuits is the second major component dissipating 22% of the communication power. The MSINT blocks result in 6% power overhead.

2.5.9 Summary

This section presented an 80-tile high-performance multicore processor implemented in a 65 nm process technology. The prototype integrates 160 low-latency FPMAC cores and features a single-cycle accumulator architecture for high throughput. Each tile contains a fast and compact router operating at core speed where the 80 tiles are interconnected using a 2D mesh topology providing a high bisection bandwidth of over 2.5 Tera-bits/s. When operating at 5 GHz, the router combines 100 GB/s of raw bandwidth with a per-hop fall-through latency of 1 ns. The design uses co-design of micro architecture, logic, circuits, and a 65 nm process to reach target performance. The proposed mesochronous clocking scheme provides a scalable and low-power alternative to conventional synchronous clocking. The fine-grained on-chip network power management features, controllable on a router per-port basis, allow on-demand performance while reducing active and stand-by leakage power. The processor delivers TeraFLOPS performance while consuming less than 100 W.

2.6 Conclusions

Research into on-chip networks emerged only in the past decade and substantial challenges clearly remain. To conclude this chapter, we discuss fundamental challenges for the design of on-chip networks and how developments in the underlying technology drivers and overarching multicore chip architectures can significantly affect the requirements for future on-chip networks.

2.6.1 Fundamental Challenges

Throughout this chapter, we have highlighted how on-chip network design faces immense pressure from tight design requirements and constraints: the need to deliver high bandwidth while satisfying stringent constraints in delay, area, power, and reliability. Techniques that extend bandwidth usually add complexity and thus area and power overhead, making on-chip network design difficult in the face of such tight constraints. Recent work has characterized the substantial energy-delay-throughput gap that exists between today's state-of-the-art on-chip network design and the ideal interconnect fabric of dedicated wires directly connecting every pair of cores [30]. Substantial innovations are needed in all aspects of on-chip network architecture (topology, routing, flow control, microarchitecture) to narrow this gap, as on-chip networks stand as a major stumbling block to realizing many-core chips.

2.6.2 Underlying Technology Drivers

While research in addressing the fundamental requirements facing on-chip network architectures is critically needed, one should keep in mind that innovations in underlying technology can further alter the landscape of on-chip network architectures. While current on-chip networks tend to utilize full-swing repeated interconnects, recent developments in interconnect circuits such as those tackling low-swing global signaling [19, 22, 25] can significantly alter current trade-offs and prompt new network architectures. Other more disruptive interconnect technologies for on-chip communications include optics [28, 38] and RF interconnects [7], which will significantly change the demands on the on-chip network architecture. Further, developments in off-chip I/O technologies, such as 3-D chip integration [4] and capacitive I/Os [20], can again have a large effect on on-chip network design, as the demand for on-chip network bandwidth can grow by orders of magnitude.

2.6.3 Overarching Multi-core Chip Architectures

While we have covered the fundamentals of on-chip networks in this chapter, the range of multicore architectures demanding such networks is immense. The debate over the granularity of the cores (large versus small) in multicore systems is far from settled. While many multicore systems are largely homogeneous, opportunities abound for chip-based systems composed of different types of cores with different communication demands. While some of today's systems employ shared memory with cache-to-cache transactions carried on the network, others employ a message-passing approach. Different types of systems will likely support different communication paradigms, resulting in different communication substrates. Different points on this large multicore chip design space require different on-chip networks, further challenging the on-chip network architect.

2.6.4 Summary

In this chapter, we introduced readers to the fundamentals of on-chip networks for multicore systems and presented case studies of two recent on-chip network implementations. Because of their different purposes, these two networks differed across many design aspects, indicating a strong need to customize on-chip network design. The field of on-chip networks is in its infancy providing ripe opportunities for research innovations. As we are only able to briefly motivate and discuss the challenges here, we refer interested readers to a recent article that summarizes the key findings of a workshop initiated by NSF in December 2006 to chart the research challenges in on-chip networks [34].

Acknowledgements Dr. Peh wishes to thank her entire Princeton research group, as well as students who have taken the ELE580 graduate course on interconnection networks as those research and teaching experiences helped significantly in the writing of this chapter. Her research has been kindly supported by the National Science Foundation, Intel Corporation, and the MARCO Gigascale Systems Research Center. Dr. Keckler thanks the entire TRIPS team, in particular Doug Burger, Paul Gratz, Heather Hanson, Robert McDonald, and Karthikeyan Sankarlingam, for their contributions to the design and implementation of the TRIPS operand network. The TRIPS project was supported by the Defense Advanced Research Projects Agency under contract F33615-01-C-4106 and by NSF CISE Research Infrastructure grant EIA-0303609. Dr. Vangal thanks the entire TeraFLOPS processor design team at Circuit Research Laboratories, Intel Corporation, for flawless execution of the design.

References

1. N. R. Adiga, M. A. Blumrich, D. Chen, P. Coteus, A. Gara, M. E. Giampapa, P. Heidelberger, S. Singh, B. D. Steinmacher-Burow, T. Takken, M. Tsao, and P. Vranas. Blue Gene/L torus interconnection network. *IBM Journal of Research and Development*, 49(2/3):265–276, 2005.
2. P. Bai, C. Auth, S. Balakrishnan, M. Bost, R. Brain, V. Chikarmane, R. Heussner, M. Hussein, J. Hwang, D. Ingerly, R. James, J. Jeong, C. Kenyon, E. Lee, S.-H. Lee, N. Lindert, M. Liu, Z. Ma, T. Marieb, A. Murthy, R. Nagisetty, S. Natarajan, J. Neirynck, A. Ott, C. Parker, J. Sebastian, R. Shaheed, S. Sivakumar, J. Steigerwald, S. Tyagi, C. Weber, B. Woolery, A.Yeoh, K. Zhang, and M. Bohr. A 65 nm Logic Technology Featuring 35 nm Gate Lengths, Enhanced Channel Strain, 8 Cu Interconnect Layers, Low-k ILD and 0.57 um^2 SRAM Cell. In *International Electron Devices Meeting (IEDM)*, pages 657–660, Dec 2004.

3. S. Bell, B. Edwards, J. Amann, R. Conlin, K. Joyce, V. Leung, J. MacKay, and M. Reif. TILE64 processor: A 64-core SoC with mesh interconnect. In *International Solid State Circuits Conference*, Feb 2008.

4. S. Borkar. Thousand core chips: a technology perspective. In *Design Automation Conference*, pages 746–749, June 2007.

5. D. Burger, S. Keckler, K. McKinley, M. Dahlin, L. John, C. Lin, C. Moore, J. Burrill, R. McDonald, and W. Yoder. Scaling to the End of Silicon with EDGE Architectures. *IEEE Computer*, 37(7):44–55, July 2004.

6. M. Butts. Synchronization through communication in a massively parallel processor array. *IEEE Micro*, 27(5):32–40, Sep/Oct 2007.

7. M. F. Chang, J. Cong, A. Kaplan, M. Naik, G. Reinman, E. Socher, and S. Tam. CMP network-on-chip overlaid with multi-band RF-interconnect. In *International Conference on High-Performance Computer Architecture*, Feb 2008.

8. M. Coppola, R. Locatelli, G. Maruccia, L. Pieralisi, and A. Scandurra. Spidergon: a novel on-chip communication network. In *International Symposium on System-on-Chip*, page 15, Nov 2004.

9. W. J. Dally. Virtual-channel flow control. In *International Symposium of Computer Architecture*, pages 60–68, May 1990.

10. W. J. Dally, A. Chien, S. Fiske, W. Horwat, R. Lethin, M. Noakes, P. Nuth, E. Spertus, D. Wallach, D. S. Wills, A. Chang, and J. Keen. Retrospective: the J-machine. In *25 years of the International Symposium on Computer Architecture (selected papers)*, pages 54–58, 1998.

11. W. J. Dally and C. L. Seitz. The torus routing chip. *Journal of Distributed Computing*, 1:187–196, 1986.

12. W. J. Dally and B. Towles. *Principles and Practices of Interconnection Networks*. Morgan Kaufmann Publishers, San Francisco, CA, 2004.

13. J. Duato, S. Yalamanchili, and L. Ni. *Interconnection Networks*. Morgan Kaufmann Publishers, San Francisco, CA, 2003.

14. W. Eatherton. The push of network processing to the top of the pyramid. Keynote speech, International Symposium on Architectures for Networking and Communications Systems.

15. N. Enright-Jerger, L.-S. Peh, and M. Lipasti. Circuit-switched coherence. In *International Symposium on Networks-on-Chip*, April 2008.

16. M. Galles. Scalable pipelined interconnect for distributed endpoint routing: The SGI SPIDER chip. In *Hot Interconnects 4*, Aug 1996.

17. P. Gratz, B. Grot, and S. Keckler. Regional congestion awareness for load balance in networks-on-chip. In *International Conference on High-Performance Computer Architecture*, pages 203–214, Feb 2008.

18. P. Gratz, C. Kim, R. McDonald, S. W. Keckler, and D. Burger. Implementation and Evaluation of On-Chip Network Architectures. In *International Conference on Computer Design*, pages 477–484, Sep 2006.

19. R. Ho, K. Mai, and M. Horowitz. The future of wires. *Proceedings of the IEEE*, 89(4), Apr 2001.

20. D. Hopkins, A. Chow, R. Bosnyak, B. Coates, J. Ebergen, S. Fairbanks, J. Gainsley, R. Ho, J. Lexau, F. Liu, T. Ono, J. Schauer, I. Sutherland, and R. Drost. Circuit techniques to enable 430 GB/s/mm^2 proximity communication. *International Solid-State Circuits Conference*, pages 368–609, Feb 2007.

21. Infiniband trade organization. http://www.infinibandta.org/

22. A. P. Jose, G. Patounakis, and K. L. Shepard. Pulsed current-mode signaling for nearly speed-of-light intrachip communication. *Proceedings of the IEEE*, 41(4):772–780, April 2006.

23. J. A. Kahle, M. N. Day, H. P. Hofstee, C. R. Johns, T. R. Maeurer, and D. Shippy. Introduction to the cell multiprocessor. *IBM Journal of Research and Development*, 49(4/5):589–604, 2005.

24. P. Kermani and L. Kleinrock. Virtual cut-through: A new computer communication switching technique. *Computer Networks*, 3:267–286, 1979.

25. B. Kim and V. Stojanovic. Equalized interconnects for on-chip networks: Modeling and optimization framework. In *International Conference on Computer-Aided Design*, pages 552–559, November 2007.

26. J. Kim, J. Balfour, and W. J. Dally. Flattened butterfly topology for on-chip networks. In *International Symposium on Microarchitecture*, pages 172–182, December 2007.

27. J. Kim, C. A. Nicopoulos, D. Park, N. Vijaykrishnan, M. S. Yousif, and C. R. Das. A gracefully degrading and energy-efficient modular router architecture for on-chip networks. In *International Symposium on Computer Architecture*, pages 4–15, June 2006.

28. N. Kirman, M. Kirman, R. K. Dokania, J. F. Martinez, A. B. Apsel, M. A. Watkins, and D. H. Albonesi. Leveraging optical technology in future bus-based chip multiprocessors. In *International Symposium on Microarchitecture*, pages 492–503, December 2006.

29. P. Kongetira, K. Aingaran, and K. Olukotun. Niagara: A 32-way multithreaded sparc processor. *IEEE Micro*, 25(2):21–29, March/April 2005.

30. A. Kumar, L.-S. Peh, P. Kundu, and N. K. Jha. Express virtual channels: Towards the ideal interconnection fabric. In *International Symposium on Computer Architecture*, pages 150–161, June 2007.

31. S. S. Mukherjee, P. Bannon, S. Lang, A. Spink, and D. Webb. The Alpha 21364 network architecture. *IEEE Micro*, 22(1):26–35, Jan/Feb 2002.

32. R. Mullins, A. West, and S. Moore. Low-latency virtual-channel routers for on-chip networks. In *International Symposium on Computer Architecture*, pages 188–197, June 2004.

33. C. A. Nicopoulos, D. Park, J. Kim, N. Vijaykrishnan, M. S. Yousif, and C. R. Das. ViChaR: A dynamic virtual channel regulator for network-on-chip routers. In *International Symposium on Microarchitecture*, pages 333–346, December 2006.

34. J. D. Owens, W. J. Dally, R. Ho, D. N. J. Jayasimha, S. W. Keckler, and L.-S. Peh. Research challenges for on-chip interconnection networks. *IEEE Micro*, 27(5):96–108, Sep/Oct 2007.

35. L.-S. Peh and W. J. Dally. Flit-reservation flow control. In *International Symposium on High-Performance Computer Architecture*, pages 73–84, Jan 2000.

36. L.-S. Peh and W. J. Dally. A delay model and speculative architecture for pipelined routers. In *International Conference on High-Performance Computer Architecture*, pages 255–266, January 2001.

37. K. Sankaralingam, R. Nagarajan, P. Gratz, R. Desikan, D. Gulati, H. Hanson, C. Kim, H. Liu, N. Ranganathan, S. Sethumadhavan, S. Sharif, P. Shivakumar, W. Yoder, R. McDonald, S. Keckler, and D. Burger. The Distributed Microarchitecture of the TRIPS Prototype Processor. In *International Symposium on Microarchitecture*, pages 480–491, December 2006.

38. A. Shacham, K. Bergman, and L. P. Carloni. The case for low-power photonic networks on chip. In *Design Automation Conference*, pages 132–135, June 2007.

39. L. Shang, L.-S. Peh, A. Kumar, and N. K. Jha. Thermal modeling, characterization and management of on-chip networks. In *International Symposium on Microarchitecture*, pages 67–78, Decemeber 2004.

40. A. Singh, W. J. Dally, A. K. Gupta, and B. Towles. Goal: a load-balanced adaptive routing algorithm for torus networks. In *International Symposium on Computer Architecture*, pages 194–205, June 2003.

41. M. B. Taylor, W. Lee, S. P. Amarasinghe, and A. Agarwal. Scalar Operand Networks: On-Chip Interconnect for ILP in Partitioned Architecture. In *International Symposium on High-Performance Computer Architecture*, pages 341–353, Feb 2003.

42. J. Tschanz, S. Narendra, Y. Ye, B. Bloechel, S. Borkar, and V. De. Dynamic Sleep Transistor and Body Bias for Active Leakage Power Control of Microprocessors. *IEEE Journal of Solid-State Circuits*, 38(11):1838–1845, Nov 2003.

43. S. Vangal, Y. Hoskote, N. Borkar, and A. Alvandpour. A 6.2-GFLOPS Floating-Point Multiply-Accumulator with Conditional Normalization. *IEEE Journal of Solid-State Circuits*, 41(10):2314–2323, Oct 2006.

44. S. Vangal, J. Howard, G. Ruhl, S. Dighe, H. Wilson, J. Tschanz, D. Finan, A. Singh, T. Jacob, S. Jain, C. Roberts, Y. Hoskote, N. Borkar, and S. Borkar. An 80-Tile Sub-100 W TeraFLOPS Processor in 65-nm CMOS. *IEEE Journal of Solid-State Circuits*, 43(1):29–41, Jan 2008.

45. S. Vangal, A. Singh, J. Howard, S. Dighe, N. Borkar, and A. Alvandpour. A 5.1 GHz 0.34 mm^2
   Router for Network-on-Chip Applications. In *International Symposium on VLSI Circuits*,
   pages 42–43, June 2007.
46. H.-S. Wang, L.-S. Peh, and S. Malik. Power-driven design of router microarchitectures in on-
   chip networks. In *International Symposium on Microarchitecture*, pages 105–116, Nov 2003.
47. H. Wilson and M. Haycock. A Six-port 30-GB/s Non-blocking Router Component Using
   Point-to-Point Simultaneous Bidirectional Signaling for High-bandwidth Interconnects. *IEEE
   Journal of Solid-State Circuits*, 36(12):1954–1963, Dec 2001.

Chapter 3
Composable Multicore Chips

Doug Burger, Stephen W. Keckler, and Simha Sethumadhavan

Abstract When designing a multicore chip, two primary concerns are how large and powerful to make each core and how many cores to include. A design with a few large cores is attractive for general purpose computing that has coarse-grained threads which depend on instructional-level parallelism for performance. At the other end of the processor granularity spectrum, a large collection of small processors is suited to applications with ample thread-level parallelism. While much of the spectrum between large cores and tiny cores is being pursued by different vendors, each of these fixed architectures are suited only to the range of workloads that maps well to its granularity of parallelism. An alternative is a class of architectures in which the processor granularity (and the number of processors) can be dynamically configured–at runtime. Called composable lightweight processors (CLPs), these architectures consist of arrays of simple processors that can be aggregated together to form larger more powerful processors, depending on the demand of the running application. Experimental results show that matching the granularity of the processors to the granularity of the tasks improves both performance and efficiency over fixed multicore designs.

Portions reprinted, with permission, from "Distributed Microarchitectural Protocols in the TRIPS Prototype Processor," K. Sankaralingam, R. Nagarajan, R. McDonald, R. Desikan, S. Drolia, M. S. S. Govindan, P. Gratz, D. Gulati, H. Hanson, C. Kim, H. Liu, N. Ranganathan, S. Sethumadhavan, S. Sharif, P. Shivakumar, S. W. Keckler, and D. Burger, *International Symposium on Microarchitecture* ©2006, IEEE. and "Composable Lightweight Processors," C. Kim, S. Sethumadhavan, M. Govindan, N. Ranganathan, D. Gulati, D. Burger, and S. W. Keckler, *International Symposium on Microarchitecture* ©2007, IEEE.

D. Burger (✉)
The University of Texas at Austin, 1 University Station C0500, Austin, TX 78712-0233, USA
e-mail: dburger@cs.utexas.edu

3.1 Introduction

Two fundamental questions when designing a multicore chip are (1) how many processor cores to place on a chip and (2) how much area to devote for the structures required to exploit parallelism in each processor core. An apt – if whimsical – metaphorical taxonomy in use for small, medium, and large processor cores is *termites*, *chainsaws*, and *bulldozers*. A design with a few large cores (or bulldozers) is attractive for general purpose computing that has few coarse-grained threads which exploit instructional-level parallelism for performance. At the other end of the spectrum, a large collection of small processors (or termites) is suited to applications with ample thread-level parallelism. Other designs represent a compromise, having a moderate number of moderately complex processors (or chainsaws).

To date, the issue of core granularity – whether to build termites, chainsaws or bulldozers – is driven by the application space and power envelope chosen by a vendor. Many points in the granularity spectrum are represented by commercial multicore chips. Table 3.1 lists a collection of recent multicore chips along with the approximate size of each core. The chips are ordered by core size, largest to smallest, with the bulldozers appearing at the top and termites appearing at the bottom. The last column shows all of the core sizes scaled to a 65 nm technology for those chips implemented with larger feature sizes. The data in that column is computed by linearly scaling the core size in each of the X and Y dimensions. The core sizes and

Table 3.1 Recent multicore chips with estimates of core sizes

Chip	Technology	No. of cores	Die size (mm^2)	Core size (mm^2)	Normalized to 65 nm
Bulldozers					
Intel Itanium [36]	65 nm	4	699	87.3	87.3
IBM Power6 [5]	65 nm	2	341	48.8	48.8
TFlex-16 [16]	—	16	—	—	37.9
AMD Barcelona [4]	65 nm	4	285	35.6	35.6
TRIPS [29]	130 nm	2	336	94.5	23.6
Chainsaws					
Sun Rock [39]	65 nm	16	396	15.1	15.1
Sun T2 [24]	65 nm	8	342	14.3	14.3
TFlex-4 [16]	—	16	—	—	9.5
IBM Cell [12]	90 nm	9	220	14.5	7.6
Termites					
Raw [38]	180 nm	16	331	18.6	3.7
Intel Polaris [40]	65 nm	80	275	2.5	2.5
TFlex-1 [16]	—	16	—	—	2.4
Tilera [2]	90 nm	64	330	3.6	1.9
NVIDIA GeForce 8800 [20]	90 nm	128	470	<3	<1.6
SPI Storm-1 [14]	130 nm	16	164	5.0	1.3
Clearspeed CS301 [9]	130 nm	64	72	1.13	0.28

chip sizes are estimates taken from published papers and examinations of published die photos.

We list cores larger than $20\,mm^2$ in $65\,nm$ technology as "bulldozer cores," and cores larger than $5\,mm^2$ as "chainsaw cores," with anything smaller being "termite cores." The chip containing the largest bulldozer core in Table 3.1 is the quad-core Itanium chip, which is designed for high single-threaded performance across a range of desktop and server applications. A typical chainsaw core is found on the Sun Ultrasparc T2 chip, which targets threaded throughput computing for server workloads. Termite cores can be exemplified by those on the NVIDIA GeForce 8800 chip, which includes 128 stream processors and targets data parallel graphics and rendering algorithms.

The problem with fixing the core granularity at design time, of course, is that a fixed number and granularity of cores will run workload mixes poorly, if the mixes differ from the design-time decisions about core granularity. While such concerns may be irrelevant in domain-specific designs, such as graphics processors, general-purpose designs are much more likely to have workload mixes that vary over time, and thus map poorly onto a fixed multicore design at least some of the time.

The only architecture listed in Table 3.1 that has not been implemented in silicon is the TFlex microarchitecture, an example of a composable multicore architecture described in detail later in this chapter. A composable processor can address the fixed-core granularity problem, by aggregating small cores to form medium- or large-grained logical processors, with no change needed in the running software. Thus, while one TFlex core (estimated at $2.4\,mm^2$ at $65\,nm$) is in the size range of termite cores, four can be aggregated into a logical $9.5\,mm^2$ core, and even 16 can be aggregated to form a logical $37.9\,mm^2$ core that increases average performance over finer granularities for many workloads. Thus, an array of TFlex cores can be dynamically configured as termites, chainsaws, or bulldozers, depending on the needs of the system and running software.

3.1.1 Heterogeneous Processors

Recently, architecture researchers have proposed heterogeneous multicore (also known as asymmetric multicore) processors as a means to improve power efficiency or performance over symmetric multicore architectures with many fixed granularity processors of one type. Effectively, heterogeneous multicore designs address the granularity issue by providing the software with granularity options. However, this approach fixes the processor granularity at design time, which will perform suboptimally with many workload mixes, while also complicating the task of the software scheduler, and increasing communication/coherence overheads.

Kumar et al. initially proposed a single-ISA heterogeneous multicore architecture, which consisted of instances of different generations of Compaq Alpha processors. They suggest that energy efficiency can be maximized by running a program on the least-powerful core required to meet its performance demand. Others, such as Ghiasi and Grunwald [7] and Grochowski et al. [8] also exploit heterogeneity in multicore architectures for power efficiency. Kumar et al. extended their work to the multiprogramming space, determining that the core asymmetry provided throughput benefits as well [18]. Annavaram et al. examine the viability of heterogeneity to accelerate multithreaded programs [1]. Heterogeneity is also finding its way into the commercial sphere. For example, IBM's Cell processor includes one mid-range PowerPC processor for executing system code and eight SPUs for executing threaded and data-parallel code [12]. We expect this trend to continue as specialized cores for graphics or cryptography are incorporated into processor dies. While efficient for their assigned tasks, such specialized processors do not provide adaptivity to different granularities of parallelism.

3.1.2 Composable Processors

While heterogeneous cores provide hardware specialization that can lead to efficiency improvements, the number and types of cores are fixed at design time. Composable flexible-core processors aim instead to provide adaptivity in both the number and granularity of processors, enabling the system to execute both large and small numbers of tasks efficiently. The basic approach is to aggregate a number of smaller, identical physical processors in a smaller aggregate area to form larger, logical processors. Composing logical processors from multiple simple processors gives a system the ability to adapt to ranges of instruction-level parallelism (ILP) and thread-level parallelism (TLP) dynamically and transparently to application software. Furthermore, a composable processor design that supports scaling to many cores can reduce the burden on the programmer for hard-to-parallelize applications by providing higher single-thread performance and reducing the number of threads a programmer must manage. A multicore chip composed of bulldozer processors will perform well on applications with few high-ILP threads while a swarm of termite processors will perform well on applications with many low-ILP threads. Composable processors can be configured in the field to meet both needs.

Supporting this degree of flexibility requires physical distribution of architectural structures including the register file, instruction window, L1 caches, and operand bypass network. In addition to the partitioning, various distributed control protocols are required to correctly implement instruction fetch, execute, commit, speculation recovery, and other processor actions. If some of these functions are centralized, overall scalability of the composed cores will suffer.

One example of a composable architecture is Core Fusion, which provides mechanisms to enable a limited number (up to four) of out-of-order cores to be fused into a single more powerful core [11]. Core Fusion's principal advantage is that it is ISA compatible with current industrial parts. Its main disadvantage is that its centralized register renaming function limits its scalability. Another recent composable proposal is *Federation*, which "federates" multiple in-order cores to create an out-of-order processor [37], providing more power-efficient building blocks but achieving lower composed performance. Another example of a composable system is virtual hierarchies, which enables a flexible aggregation and isolation of individual cores to form logical multiprocessors (as opposed to logical uniprocessors) [22]. Flexibility across these dimensions can offer capability and efficiency advantages to emerging multicore systems.

In this chapter, we explore the architecture and capabilities of a particular composable processor design called composable lightweight processors (or CLPs) [16]. A CLP consists of multiple simple, narrow-issue processor cores that can be aggregated dynamically to form more powerful single-threaded processors. Thus, the number and size of the processors can be adjusted on the fly to provide the target that best suits the software needs at any given time. The same software thread can run transparently – without modifications to the binary – on one core, two cores, or up to as many as 32 cores, the maximum we simulate. Low-level run-time software can decide how to best balance thread throughput (TLP), single-thread performance (ILP), and energy efficiency. Run-time software may also grow or shrink processors to match the available ILP in a thread to improve performance and power efficiency.

Figure 3.1 shows a high-level floorplan with three possible configurations of a CLP. The small squares on the left of each floorplan represent a single physical processing core, while the squares on the right half show a banked L2 cache. If a large number of threads are available, the system could run 32 threads, one on each core (Fig. 3.1a). If high single-thread performance is required and the thread has sufficient ILP, the CLP could be configured to use an optimal number of cores that maximizes performance (up to 32, as shown in Fig. 3.1c). To optimize for energy efficiency, for example in a data center or in battery-operated mode, the system could configure the CLP to run each thread at its most energy-efficient point. Figure 3.1b shows an energy (or power) optimized CLP configuration running eight threads across a range of processor granularities.

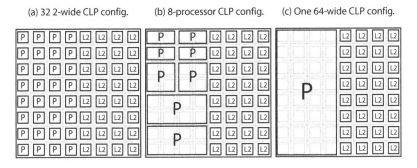

(a) 32 2-wide CLP config.   (b) 8-processor CLP config.   (c) One 64-wide CLP config.

Fig. 3.1 Three dynamically assigned CLP configurations

A fully composable processor shares no structures physically among the multiple processors. Instead, a CLP relies on distributed microarchitectural protocols to provide the necessary instruction fetch, execution, branch prediction, dependence prediction, memory access/disambiguation, and commit capabilities. Full composability is difficult in conventional ISAs, since the atomic units are individual instructions, which require that control decisions be made too frequently to coordinate across a distributed processor. Explicit Data Graph Execution (EDGE) ISAs, conversely, reduce the frequency of control decisions by employing block-based program execution [23] and explicit intra-block dataflow semantics, and have been shown to map well to distributed microarchitectures [3]. The particular CLP design that we evaluate, called TFlex, achieves the composable capability by mapping large, structured EDGE blocks across participating cores differently depending on the number of cores that are running a single thread. The TFlex CLP microarchitecture allows the dynamic aggregation of any number of cores – up to 32 for each individual thread – to find the best configuration under different operating targets: performance, area efficiency, or energy efficiency.

3.1.3 Chapter Overview

This chapter describes the instruction-set and microarchitectural support necessary to provide dynamic adaptation of multicore granularity. We first outline the EDGE instruction set architecture which is critical to scaling and composition. The chapter then describes two generations of microarchitectures that support an EDGE ISA. The first microarchitecture is TRIPS, a fixed but scalable microarchitecture that demonstrates the viability of EDGE ISAs. TRIPS has been implemented in a custom 130 nm ASIC and tested on a wide variety of applications. The program execution model and partitioned microarchitecture of TRIPS also provides a foundation for the second microarchitecture, TFlex, which is a highly scalable CLP. The TFlex description includes the features of the microarchitecture necessary for composability, and also includes details of the design of the reprogrammable mapping functions and protocols required for composition and recomposition.

We end this chapter with a set of performance estimates of the TFlex microarchitecture. To summarize the results, on a set of 26 benchmarks, including both high- and low-ILP codes, the best composed configurations range from one to 32 dual-issue cores depending on operating targets and applications. The TFlex design achieves a $1.4\times$ performance improvement, $3.4\times$ performance/area improvement, and $2.0\times$ performance^2/Watt improvement over the TRIPS processor. The TFlex CLP also shows improved parallel flexibility with the throughput of multiprogrammed workloads improving by up to 47% over the best fixed-CMP organization.

3.2 ISAs for Scalable and Multigranular Systems

Designing a composable processor that can scale to many composed cores using a conventional instruction set architecture (RISC or CISC) presents substantial challenges. First, conventional architectures have single points of instruction fetch and commit, which require centralization for those operations. Second, programs compiled for conventional architectures have frequent control-flow changes (branches and jumps occurring as frequently as every 5–6 instructions), which demand fast branch processing and wide distribution of the branch outcome. Third, adding more execution units typically requires a broadcast of ALU outputs to ALU inputs as the hardware cannot easily track all of the producers and consumers of data. Salverda and Zilles show that effective steering of instructions to multiple cores in a composable, in-order processor is difficult due to both the complexity of instruction-level data flow graphs and fetch criticality [27].

Creating larger logical microarchitectural structures from smaller ones is the principal challenge for the design of a composable architecture. Composing some structures, such as register files and level-one data caches, is straightforward as these structures in each core can be treated as address-interleaved banks of a larger aggregate structure. Changing the mapping to conform to a change in the number of composed processors merely requires adjusting the interleaving factor or function.

However, banking or distributing other structures required by a conventional instruction set architecture is less straightforward. For example, operand bypass (even when distributed) typically requires some form of broadcast, as tracking the ALUs in which producers and consumers are executing is difficult. Similarly, both instruction fetch and commit require a single point of synchronization to preserve sequential execution semantics, including features such as a centralized register rename table and load/store queues. While some of these challenges can be solved by brute force, supporting composition of a large number of processing elements – particularly in an energy- and power-efficient manner – can benefit from instruction set support.

To address the challenges of structure scalability, which inherently supports composability, we developed a new class of instruction sets, called Explicit Data Graph Execution (EDGE) architectures [3]. An EDGE architecture supports execution of serial programs by mapping compiler-generated dataflow graphs onto a distributed substrate of computation, storage, and control units. The two defining features of an EDGE ISA are block-atomic execution and direct communication of instructions within a block, which together enable efficient dataflow-like execution while supporting conventional imperative languages.

3.2.1 TRIPS Blocks

The TRIPS ISA, used both for the TRIPS prototype and for the TFlex composable lightweight processor, is an instance of EDGE ISAs that aggregates up to 128 instructions into a single block that obeys the block-atomic execution model. In this execution model, each block of instructions is logically fetched, executed, and committed as a single entity. This model amortizes the per-instruction book-keeping over a large number of instructions and reduces the number of branch predictions and register file accesses. Furthermore, this model reduces the frequency at which control decisions about what to execute must be made (such as fetch or commit), providing the additional latency tolerance to make distributed execution practical. When an interrupt arrives or an exception is raised, the processor must roll back to the beginning of the excepting block, since a block must be entirely committed or not at all. This model presents similar challenges with respect to I/O as transactional execution; blocks that produce side-effect-causing accesses to I/O (or loads with store semantics, such as a swap) must be handled carefully and guaranteed to commit once they begin execution. While these atomic blocks may resemble transactions with their notion of atomicity, they are not exposed to the programmer in an API, and are in fact an example of a large "architectural atomic unit" as described by Melvin, Shebanow, and Patt before transactions were first proposed [23]. In Melvin et al.'s taxonomy, an EDGE block corresponds to an architectural atomic unit, and the operations within each block (in the TRIPS prototype) correspond to both compiler and execution atomic units.

The TRIPS compiler constructs blocks and assigns an implicit ID (0–127) to each instruction. This ID determines an instruction's position within the block. The microarchitecture uses these instruction IDs to map each instruction to a reservation station within the execution substrate. Each block is divided into between two and five 128-byte chunks. Every block includes a header chunk which encodes up to 32 read and up to 32 write instructions that access the 128 architectural registers. The read instructions pull values out of the registers and send them to compute instructions in the block, whereas the write instructions return outputs from the block to the specified architectural registers. In the TRIPS microarchitecture, each of the 32 read and write instructions are distributed across the four register banks, as described in the next section.

The header chunk also holds three types of control state for the block: a 32-bit "store mask" that indicates which of the possible 32 memory instructions are stores, block execution flags that indicate the execution mode of the block, and the number of instruction "body" chunks in the block. The store mask is used to enable distributed detection of block completion.

A block may contain up to four body chunks – each consisting of 32 instructions – for a maximum of 128 instructions, at most 32 of which can be loads and stores. All possible executions of a given block must emit the same number of outputs (stores, register writes, and one branch) regardless of the predicated path taken through the block. This constraint is necessary to detect block completion on the distributed substrate. The compiler generates blocks that conform to these constraints [33, 21].

3.2.2 Direct Instruction Communication

With *direct instruction communication*, instructions in a block send their results directly to intra-block, dependent consumers in a dataflow fashion. This model supports distributed execution by eliminating the need for any intervening shared, centralized structures (e.g. an issue window or register file) between intra-block producers and consumers.

Figure 3.2 shows that the TRIPS ISA supports direct instruction communication by encoding the consumers of an instruction's result as targets within the producing instruction. The microarchitecture can thus determine precisely where the

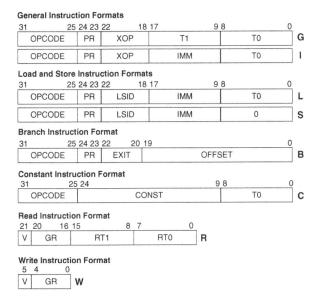

General Instruction Formats

31	25 24 23 22	18 17	9 8	0	
OPCODE	PR	XOP	T1	T0	**G**
OPCODE	PR	XOP	IMM	T0	**I**

Load and Store Instruction Formats

31	25 24 23 22	18 17	9 8	0	
OPCODE	PR	LSID	IMM	T0	**L**
OPCODE	PR	LSID	IMM	0	**S**

Branch Instruction Format

31	25 24 23 22	20 19	0	
OPCODE	PR	EXIT	OFFSET	**B**

Constant Instruction Format

31	25 24	9 8	0	
OPCODE	CONST	T0	**C**

Read Instruction Format

21 20	16 15	8 7	0	
V	GR	RT1	RT0	**R**

Write Instruction Format

5 4	0	
V	GR	**W**

INSTRUCTION FIELDS

OPCODE = Primary Opcode
XOP = Extended Opcode
PR = Predicate Field
IMM = Signed Immediate
T0 = Target 0 Specifier
T1 = Target 1 Specifier
LSID = Load/Store ID
EXIT = Exit Number
OFFSET = Branch Offset
CONST = 16-bit Constant
V = Valid Bit
GR = General Register Index
RT0 = Read Target 0 Specifier
RT1 = Read Target 1 Specifier

Fig. 3.2 TRIPS instruction formats

consumer resides and forward a producer's result directly to its target instruction(s). The nine-bit target fields (T0 and T1) each specify the target instruction with seven bits and the operand type (left, right, predicate) with the remaining two. A microarchitecture supporting this ISA maps each of a block's 128 instructions to particular coordinates, thereby determining the distributed flow of operands along the block's dataflow graph. The atomic blocks are necessary for direct instruction communication to bound the scope of the dataflow semantics, and therefore the size of each target. Outside of the blocks (i.e. between blocks), values are passed through registers and memory, similar to conventional instruction sets, and do not incorporate architecturally visible dataflow semantics. An instruction's coordinates are implicitly determined by its position in its chunk. Other non-traditional elements of this ISA include the "PR" field, which specifies an instruction's predicate and the "LSID" field, which specifies relative load/store ordering. Smith et al. [34] provide a detailed description of the prediction model in the TRIPS ISA.

3.2.3 Composable Architectures

Because they are well suited for distributed execution, EDGE ISAs are a good match for composable processors. Because of the block-atomic execution model, EDGE ISAs allow the control protocols for instruction fetch, completion, and commit to operate on large blocks – 128 instructions in the TRIPS ISA – instead of individual instructions. This amortization reduces the frequency of control decisions, reduces the overheads of bookkeeping structures, and makes the core-to-core latencies tractable [28]. For example, since only one branch prediction is made per block, a prediction in TRIPS is needed at most once every eight cycles. This slack makes it possible for one core to make a prediction and then send a message to another core to make a subsequent prediction without significantly reducing performance.

Second, the target encoding of operand destinations eliminates the need for an operand broadcast bus, since a point-to-point network can interpret the identifiers as coordinates of instruction placement within arrays of reservation stations. In a composable architecture, differently sized "logical" processors merely interpret the target instructions' coordinates differently for each composed processor size. Changing the composition of a processor employing an EDGE ISA does not require any recompilation; instead the hardware is reconfigured to interpret register, memory, and target addresses differently for each distinct processor composition. This direct instruction communication also permits out-of-order execution with energy efficiency superior to conventional superscalar architectures: both use a dataflow graph of in-flight instructions for high performance, but the majority of the dataflow arcs in an EDGE processor are specified statically in the ISA (except for a minority of arcs through registers and memory), whereas the entire graph in a superscalar processor must be built and traversed in hardware, at considerable energy expense.

3.3 TRIPS Architecture and Implementation

The first implementation of an EDGE instruction set was the TRIPS microprocessor system, an experimental computer system consisting of custom ASIC chips and custom circuit boards. The TRIPS prototype was intended to demonstrate the viability of the EDGE instruction set and compilers, as well as feasibility of scalable and distributed processor and memory systems. A TRIPS processor has many parallel execution units, no global wires, is built from a small set of reused components on routed networks, and can be extended to a wider-issue implementation without recompiling source code or changing the ISA. Although the TRIPS processors granularity cannot be adjusted after manufacturing, our experience from the design of the prototype and evaluation of applications on TRIPS inspired the development of the flexible TFlex architecture, which is described in detail in Sect. 3.4. This section on TRIPS lays the foundation for distributed and composable architectures and describes the tiles, networks, and protocols needed for a distributed processor. Further detail on the TRIPS implementation can be found in [29].

3.3.1 TRIPS Tiled Microarchitecture

Unlike homogeneous tiles chip multiprocessor architectures such as Raw [41], TRIPS consists of heterogeneous tiles that collaborate to form larger processors and a larger unified level-2 cache. Fig. 3.3 shows the tile-level block diagram of the

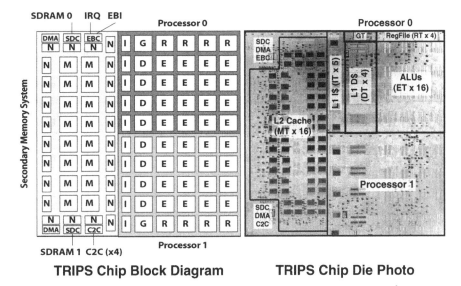

TRIPS Chip Block Diagram   **TRIPS Chip Die Photo**

Fig. 3.3 TRIPS prototype block diagram

TRIPS prototype. The three major components on the chip are two processors and the secondary memory system, each connected internally by one or more micronetworks.

Each of the processor cores is implemented using five unique tiles: one global control tile (GT), 16 execution tiles (ET), four register tiles (RT), four data tiles (DT), and five instruction tiles (IT). The main processor core micronetwork is the operand network (OPN), shown in Fig. 3.4. It connects all of the tiles except for the ITs in a two-dimensional, wormhole-routed, 5×5 mesh topology. The OPN has separate control and data channels, and can deliver one 64-bit data operand per link per cycle. A control header packet is launched one cycle in advance of the data payload packet to accelerate wakeup and select for bypassed operands that traverse the network.

Each processor core contains six other micronetworks, one for instruction dispatch – the global dispatch network (GDN) – and five for control: the global control network (GCN), for committing and flushing blocks; the global status network (GSN), for transmitting information about block completion; the global refill network (GRN), for I-cache miss refills; the data status network (DSN), for communicating store completion information; and the external store network (ESN), for determining store completion in the L2 cache or memory. Links in each of these networks connect only nearest neighbor tiles and messages traverse one tile per cycle. Figure 3.4 shows the links for four of these networks.

This type of tiled microarchitecture is *composable* at design time, permitting different numbers and topologies of tiles in new implementations with only moderate changes to the tile logic, and no changes to the software model. The particular

Global dispatch network (GDN)   **Global status network (GSN)**

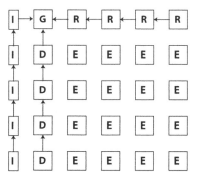

Issues block fetch command and   Signals completion of block execution, I-cache
dispatches instructions   miss refill, and block commit completion

Operand network (OPN)   **Global control network (GCN)**

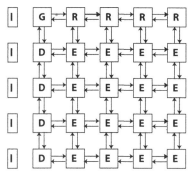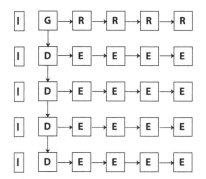

Handles transport of all data operands   Issues block commit and block flush commands

Fig. 3.4 Four TRIPS control and data micronetworks

arrangement of tiles in the prototype produces a core with 16-wide out-of-order issue, 64 KB of L1 instruction cache, and 32 KB of L1 data cache. We extended the register file and control logic to support four SMT threads as well. The microarchitecture supports up to eight TRIPS blocks in flight simultaneously, thus providing an in-flight window of up to 1024 instructions. The block control supports two modes: one non-speculative block and seven speculative blocks if a single thread is running, or two blocks per thread (one non-speculative and one speculative) if two to four threads are running.

The two processors communicate through the secondary memory system, in which the on-chip network (OCN) is embedded. The OCN is a 4×10 wormhole-routed mesh network, with 16-byte data links and four virtual channels. This network is optimized for cache-line sized transfers, although other request sizes are supported for operations such as loads and stores to uncacheable pages. The OCN is the transport fabric for all inter-processor, L2 cache, DRAM, I/O, and DMA traffic.

The physical design and implementation of the TRIPS chip were driven by the principles of partitioning and replication. The chip floorplan directly corresponds to the logical hierarchy of TRIPS tiles connected only by point-to-point, nearest-neighbor networks. The only exceptions to nearest-neighbor communication are the global reset and interrupt signals, which are latency tolerant and pipelined in multiple stages across the chip.

The TRIPS chip is implemented in the IBM CU-11 ASIC process, which has a drawn feature size of 130 nm and seven layers of metal. The chip itself includes more than 170 million transistors and 1.06 km of wire routing, in a chip area of 18.3 mm by 18.4 mm placed in a 47.5 mm square ball-grid array package. Figure 3.3 shows an annotated die photo of the TRIPS chip, showing the boundaries of the TRIPS tiles. In addition to the core tiles, TRIPS also includes six controllers that are attached to the rest of the system via the on-chip network (OCN). The two 133/266 MHz DDR SDRAM controllers (SDC) each connects to an individual 1 GB SDRAM DIMM. The chip-to-chip controller (C2C) extends the on-chip network to a four-port mesh router that gluelessly connects to other TRIPS chips. These links nominally run at one-half the core processor clock and up to 266 MHz. The two direct memory access (DMA) controllers can be programmed to transfer data to and from any two regions of the physical address space, including addresses mapped to other TRIPS processors. Finally, the external bus controller (EBC) is the interface to a board-level PowerPC control processor. To reduce design complexity, we chose to off-load much of the operating system and runtime control to this PowerPC processor.

3.3.2 TRIPS Tiles

Global Control Tile (GT): Fig. 3.5a shows the contents of the GT, which include the blocks' PCs, the instruction cache tag arrays, the I-TLB, and the next-block predictor. The GT handles TRIPS block management, including prediction, fetch, dispatch, completion detection, flush (on mispredictions and interrupts), and commit. It also holds control registers that configure the processor into different speculation, execution, and threading modes. Thus, the GT interacts with all of the control networks as well as the OPN, which provides access to the block PCs.

The GT also maintains the status of all eight in-flight blocks. When one of the block slots is free, the GT accesses the block predictor, which takes three cycles, and emits the predicted address of the next target block. Each block may emit only one "exit" branch, even though it may contain several predicated branches. The block predictor uses a branch instruction's three-bit exit field to construct exit histories of two bits per block, instead of using per-branch taken/not-taken bits. The predictor has two major parts: an exit predictor and a target predictor. The predictor uses exit histories to predict one of eight possible block exits, employing a tournament

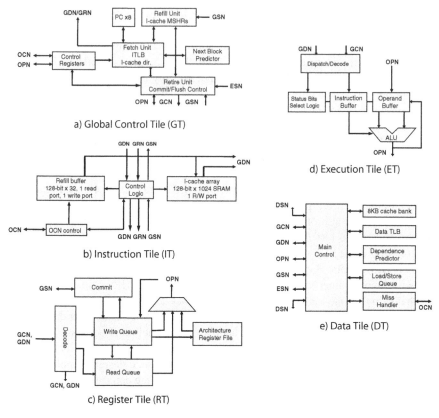

Fig. 3.5 TRIPS block-level tile microarchitectures

local/gshare predictor similar to the Alpha 21264 [13] with 9 K, 16 K, and 12 K bits in the local, global, and tournament exit predictors, respectively. The predicted exit number is combined with the current block address to access the target predictor for the next-block address. The target predictor contains four major structures: a branch target buffer (20 K bits), a call target buffer (6 K bits), a return address stack (7 K bits), and a branch-type predictor (12 K bits). The BTB predicts targets for branches, the CTB for calls, and the RAS for returns. The branch-type predictor selects among the different target predictions (call/return/branch/sequential branch). The distributed fetch protocol necessitates the type predictor; the predictor never sees the actual branch instructions, as they are sent directly from the ITs to the ETs.

Instruction Tile (IT): Figure 3.5b shows an IT, which contains a two-way, 16 KB bank of the total L1 I-cache and acts as a slave to the GT, which holds the single-tag array. Each of the five 16 KB IT banks can hold a 128-byte chunk (for a total of 640 bytes for a maximum-sized block) for each of 128 distinct blocks. Each handles refills as slaves to tag misses in the GT. A control network among the ITs informs the GT only when all of the ITs notify this network that all four refills have completed.

Register Tile (RT): To reduce power consumption and delay, the TRIPS microarchitecture partitions its many registers into banks, with one bank in each RT. The register tiles are clients on the OPN, allowing the compiler to place critical instructions that read and write from/to a given bank close to that bank. Since many def-use pairs of instructions are converted to intra-block temporaries by the compiler, they never access the register file, thus reducing total register bandwidth requirements by approximately 70%, on average, compared to a RISC or CISC processor. The four distributed banks can thus provide sufficient register bandwidth with a small number of ports; in the TRIPS prototype, each RT bank has two read ports and one write port. Each of the four RTs contains one 32-register bank for each of the four SMT threads that the core supports, for a total of 128 registers per RT and 128 registers per thread across the RTs.

In addition to the four per-thread architecture register file banks, each RT contains a read queue and a write queue, as shown in Fig. 3.5c. These queues hold up to eight read and eight write instructions from the block header for each of the eight blocks in flight, and are used to forward register writes dynamically to subsequent blocks reading from those registers. The read and write queues perform a function equivalent to register renaming for a superscalar physical register file, but are less complex to implement due to the read and write instructions in the TRIPS ISA.

Execution Tile (ET): As shown in Fig. 3.5d, each of the 16 ETs consists of a fairly standard single-issue pipeline, a bank of 64 reservation stations, an integer unit, and a floating-point unit. All units are fully pipelined except for the integer divide unit, which takes 24 cycles. The 64 reservation stations hold eight instructions for each of the eight in-flight TRIPS blocks. The reservation stations are partitioned into banks for instructions, each of two operands, and a bank for the one-bit predicate.

Data Tile (DT): Figure 3.5e shows a block diagram of a single DT. Each DT is a client on the OPN, and holds one two-way, 8 KB L1 data cache bank, for a total of 32 KB across the four DTs. Virtual addresses are interleaved across the DTs at the granularity of a 64-byte cache-line. In addition to the L1 cache bank, each DT contains a copy of the load/store queue (LSQ), a dependence predictor similar to a load-wait predictor, a one-entry backside coalescing write buffer, a data TLB, and a MSHR bank that supports up to 16 requests for up to four outstanding cache lines.[1] Because the DTs are distributed in the network, we implemented a *memory-side* dependence predictor, closely coupled with each data cache bank [30]. Although loads issue from the ETs, a dependence prediction occurs (in parallel) with the cache access only when the load arrives at the DT. An additional challenge for distributed L1-cache designs is the memory disambiguation logic (load/store queue or LSQ) to enable out-of-order execution of unrelated loads and stores. In the TRIPS prototype,

---

[1]MSHR stands for *miss status handling register* or *miss status holding register*; an MSHR maintains the state in the processor required to resume a cache access that causes a cache miss.

we sized the LSQ in each DT to 256 entries so that space is available in the unlikely event that all in-flight loads and stores target the same DT. Subsequent research has devised more area and power-efficient solutions to this difficult problem [31].

Secondary Memory System: The TRIPS prototype supports a 1 MB static NUCA [15] array, organized into 16 memory tiles (MTs), each one of which holds a four-way, 64 KB bank. Each MT also includes an OCN router and a single-entry MSHR. Each bank may be configured as an L2 cache bank or as a scratch-pad memory, by sending a configuration command across the OCN to a given MT. By aligning the OCN with the DTs, each IT/DT pair has its own private port into the secondary memory system, supporting high bandwidth into the cores for streaming applications. The network tiles (NTs) surrounding the memory system act as translation agents for determining where to route memory system requests. Each of them contains a programmable routing table that determines the destination of each memory system request. By adjusting the mapping functions within the TLBs and the network interface tiles (NTs), a programmer can configure the memory system in a variety of ways, including as a single 1 MB shared level-2 cache, as two independent 512 KB level-2 caches (one per processor), or as a 1 MB on-chip physical memory (no level-2 cache).

3.3.3 Distributed Microarchitectural Protocols

To enable concurrent, out-of-order execution on this distributed substrate, we implemented traditionally centralized microarchitectural functions, including fetch, execution, flush, and commit, with distributed protocols running across the control and data micronets.

3.3.3.1 Block Fetch Protocol

The fetch protocol retrieves a block of 128 TRIPS instructions from the ITs and distributes them into the array of ETs and RTs. In the GT, the block fetch pipeline takes a total of 13 cycles, including three cycles for prediction, one cycle for TLB and instruction cache tag access, and one cycle for hit/miss detection. On a cache hit, the GT sends eight pipelined indices out on the global dispatch network (GDN) to the ITs. Prediction and instruction cache tag lookup for the next block are overlapped with the fetch commands of the current block. Although prediction requires three cycles, as does predictor training, the eight-cycle fetch latency is sufficient to hide the six cycles per block of predictor operations. Thus, running at peak, the machine can issue fetch commands every cycle with no bubbles, beginning a new block fetch every eight cycles.

When an IT receives a block dispatch command from the GT, it accesses its I-cache bank based on the index in the GDN message. In each of the next eight cycles, the IT sends four instructions on its outgoing GDN paths to its associated row of ETs (or RTs). These instructions are written into the read and write queues of the RTs and the reservation stations in the ETs when they arrive at their respective tiles, and are available to execute as soon as they arrive. Since the fetch commands and fetched instructions are delivered in a pipelined fashion across the ITs, ETs, and RTs, the furthest ET receives its first instruction packet ten cycles and its last packet 17 cycles after the GT issues the first fetch command. While the latency appears high, the pipelining enables a high-fetch bandwidth of 16 instructions per cycle in steady state, with multiple fetched blocks in flight at once, producing one instruction per ET per cycle.

On an I-cache miss, the GT instigates a distributed I-cache refill, using the global refill network (GRN) to transmit the refill block's physical address to all of the ITs. Each IT processes the misses for its own chunk independently, and can simultaneously support one outstanding miss for each executing thread (across four threads).

3.3.3.2 Distributed Execution

Dataflow execution of a block begins with the injection of block inputs by the RTs. An RT may begin to process an arriving read instruction even if the entire block has not yet been fetched. Each RT first searches the write queues of all older in-flight blocks. If no matching, in-flight write to that register is found, the RT simply reads that register from the architectural register file and forwards it to the consumers in the block via the OPN. If a matching write is found, the RT takes one of two actions: if the write instruction has received its value, the RT forwards that value to the read instruction's consumers. If the write instruction is still awaiting its value, the RT buffers the read instruction, which will be woken up by a tag broadcast when the pertinent write's value arrives. While CAMs are typically expensive in terms of energy, they are relatively small, and are accessed less often than the number of instructions, due to the block model.

Arriving OPN operands wake up instructions within an ET, which selects and executes enabled instructions. The ET uses the target fields of the selected instruction to determine where to send the resulting operand. Arithmetic operands traverse the OPN to other ETs, while load and store instructions' addresses and data are sent on the OPN to the DTs. Branch instructions deliver their next block addresses to the GT via the OPN.

An issuing instruction may target its own ET or a remote ET. If it targets its local ET, the dependent instruction can be woken up and executed in the next cycle using a local bypass path to permit back-to-back issue of dependent instructions. If the target is a remote ET, a control packet is formed in the cycle before the operation will complete execution, and is then sent to wake up the dependent instruction early. The OPN is tightly integrated with the wakeup and select logic. When a control packet arrives from the OPN, the targeted instruction is accessed and may be speculatively woken up. The instruction may begin execution in the following cycle as the OPN router injects the arriving operand directly into the ALU. Thus, for each OPN hop between dependent instructions, there will be a one-cycle bubble before the consuming instruction executes.

Figure 3.6 shows an example of how a code sequence is executed on the RTs, ETs, and DTs. The leftmost labels (e.g., N[1]) are included for readability only and

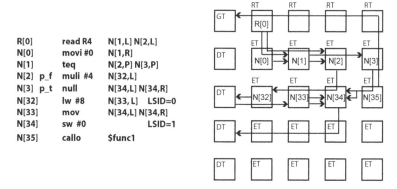

R[0]   read R4   N[1,L] N[2,L]
N[0]   movi #0   N[1,R]
N[1]   teq   N[2,P] N[3,P]
N[2] p_f   muli #4   N[32,L]
N[3] p_t   null   N[34,L] N[34,R]
N[32]   lw #8   N[33,L]   LSID=0
N[33]   mov   N[34,L] N[34,R]
N[34]   sw #0   LSID=1
N[35]   callo $func1

Fig. 3.6 TRIPS program and program execution

are not part of the instructions. Block execution begins when the read instruction R[0] is issued to RT0, triggering delivery of R4 via the OPN to the left operand of two instructions, teq (N[1]) and muli (N[2]). When the test instruction receives the register value and the immediate "0" value from the movi instruction, it fires and produces a predicate which is routed to the predicate field of N[2]. Since N[2] is predicated on false, if the routed operand has a value of 0, the muli will fire, multiply the arriving left operand by four, and send the result to the address field of the lw (load word). If the load fires, it sends a request to the pertinent DT, which responds by routing the loaded data to N[33]. The DT uses the load/store IDs (0 for the load and 1 for the store, in this example) to ensure that they execute in the proper program order if they share the same address. The result of the load is fanned out by the mov instruction to the address and data fields of the store.

If the predicate's value is 1, N[2] will not inject a result into the OPN, thus suppressing execution of the dependent load. Instead, the null instruction fires, targeting the address and data fields of the sw (store word). Note that although two instructions are targeting each operand of the store, only one will fire, due to the predicate. When the store is sent to the pertinent DT and the block-ending call instruction is routed to the GT, the block has produced all of its outputs and is ready to commit. Note that if the store is nullified, it does not affect memory, but simply signals the DT that the store has issued. Nullified register writes and stores are used to ensure that the block always produces the same number of outputs for completion detection.

3.3.3.3 Block/Pipeline Flush Protocol

Because TRIPS executes blocks speculatively, a branch misprediction or a load/store ordering violation causes periodic pipeline flushes. These flushes are implemented using a distributed protocol. The GT is first notified when a mis-speculation occurs, either by detecting a branch misprediction itself or via a GSN message from a DT indicating a memory-ordering violation. The GT then initiates a flush wave on the

GCN which propagates to all of the ETs, DTs, and RTs. The GCN includes an eight-bit block identifier mask indicating which block or blocks must be flushed. The processor must support multi-block flushing because all speculative blocks after the one that caused the misspeculation must also be flushed. This wave propagates at one hop per cycle across the array, holding the flush command high for three cycles at each point to eliminate mispredicted in-flight state. As soon as it issues the flush command on the GCN, the GT may issue a new dispatch command to start a new block. Because both the GCN and GDN guarantee in-order delivery with no contention, the instruction fetch/dispatch command can never catch up with or pass the flush command.

3.3.3.4 Block Commit Protocol

Block commit is the most complex of the microarchitectural protocols in TRIPS, and involves three phases: completion, commit, and commit acknowledgment. In phase one, a block is complete when it has produced all of its outputs, the number of which is determined at compile time and consists of up to 32 register writes, up to 32 stores, and exactly one branch. After the RTs and DTs receive all of the register writes or stores for a given block, they inform the GT using the global status network (GSN). When an RT detects that all block writes have arrived, it informs its west neighbor. The RT completion message is daisy-chained westward across the RTs, until it reaches the GT indicating that all of the register writes for that block have been received. The DTs use a similar protocol for notifying the GT that all stores have completed, but require some additional communication among them because the number of stores that will arrive at each DT depends on the store addresses computed during program execution.

During the second phase (block commit), the GT broadcasts a commit command on the global control network and updates the block predictor. The commit command informs all RTs and DTs that they should commit their register writes and stores to architectural state. To prevent this distributed commit from becoming a bottleneck, the logic supports pipelined commit commands. The GT can legally send a commit command on the GCN for a block in the next cycle after a commit command has been sent for the previous older in-flight block, even if the commit commands for older blocks are still in flight. The pipelined commits are safe because each tile is guaranteed to receive and process them in order. The commit command on the GCN also flushes any speculative in-flight state in the ETs and DTs for that block.

The third phase acknowledges the completion of commit. When an RT or DT has finished committing its architectural state for a given block and has received a commit completion signal from its neighbor on the GSN (similar to block completion detection), it signals commit completion on the GSN. When the GT has received commit completion signals from both the RTs and DTs, it knows that the block is safe to deallocate, because all of the block's outputs have been written to architectural state. When the oldest block has acknowledged commit, the GT initiates a block fetch and dispatch sequence for that block slot.

3.3.4 Discussion

The tiled nature of the TRIPS processor made it easy to implement, even with a design team consisting of a small number of graduate students. Although the TRIPS chip consists of 11 different types of tiles, the reuse (106 instances) and the design style enforced via well-engineered interconnection networks resulted in a straight-forward verification strategy and fully functional first silicon.

Our experimental results show that even with a nascent compiler, one TRIPS processor can be competitive with commercial microprocessors on serial programs. Table 3.2 summarizes the performance of the TRIPS prototype relative to two commercial systems; a full performance evaluation of the TRIPS prototype can be found in [6]. Because of the different process technologies and design methodologies, we use cycle count as the performance metric, thus normalizing out the differences in clock rate. The ratio of processor speed to memory speed is approximately the same across all of the systems. We also use the GNU C compiler gcc on the Intel platforms as it has similar scalar optimizations as the TRIPS compiler. In general, TRIPS performs well on computationally intensive applications, approaching the performance of the Core 2 on EEMBC and SpecFP. On less regular applications, TRIPS compiled code performs less well relative to the Core 2 because of more frequent and less predictable branches as well as less inherent instruction-level parallelism in the application.

Table 3.2 Performance relative to Intel Core 2

Benchmark suite	Optimization	Intel Core 2 (gcc)	Intel Pentium 4 (gcc)	TRIPS
Simple	TRIPS hand optimized	1.00	0.51	2.77
EEMBC	Compiled	1.00	0.53	0.73
SpecINT 2000	Compiled	1.00	0.49	0.35
SpecFP 2000	Compiled	1.00	0.47	1.00
Process		65 nm	90 nm	130 nm
Power		65 W (TDP)	84 W (TDP)	24 W (actual)
Scaled power				19 W (scaled)
Clock rate		1.6 GHz	3.6 GHz	366 MHz

However, the TRIPS compiler is still relatively immature and has many opportunities for improvement. To determine the underlying capabilities of the architecture, we extracted key functions from 15 different benchmarks and performed source-code and some assembly code optimizations to enable better block formation. On these programs (labeled *Simple* in Table 3.2), TRIPS often performs three to four times better than the Core 2, and averages 2.8 times better. For these applications, exploiting instruction-level parallelism is effective. We also found that while some of the applications we examined used the abundance of execution and data resources effectively, others had serial program phases that resulted in inefficient use of the TRIPS tiles. Because the fixed TRIPS design was a poor match for these applications, we sought to devise architectures that could be adapted to the parallelism granularity of the running application: applications with serial threads would use narrow-issue processors while applications with more abundant ILP would employ wider-issue processors. The configurable processor design described in the next section is the result of this investigation.

Because it does not require the large associative structures found in conventional processors, TRIPS has the potential for greater power efficiency. This efficiency is not obvious in the TRIPS prototype, as it lacks any power management such as clock gating or voltage/frequency scaling. A rudimentary scaling model that accounts for power management, technology scaling, clock rate scaling, and additional leakage power shows that TRIPS should consume about 20 W at the same frequency and technology as the Core 2. While 20 W cannot be compared to the thermal design point (TDP) of 65 W for the Core 2, achieving comparable or better power consumption at three times the performance on general purpose codes is within the potential of TRIPS architectures. Further, power efficiency can be achieved by adapting the granularity of the processors to the concurrency of the application.

3.4 Composable Lightweight Processor Architectures

TFlex is a second-generation EDGE microarchitecture that provides the capability of dynamically composing cores to form variable-granularity processors. Figure 3.7 shows an array of 32 TFlex tiles and the internals of one of the cores. The L2 cache banks and the peripheral circuitry are not shown in this diagram. In the figure, each TFlex core includes two execution units with a dual-issue pipeline, an

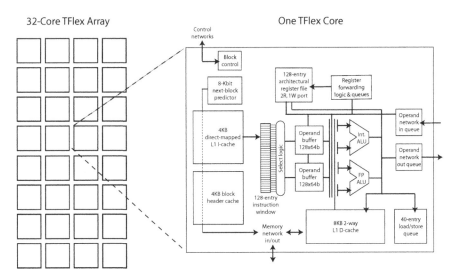

Fig. 3.7 TFlex core microarchitecture

8 Kbit next-block predictor, a 4 KB instruction cache, an 8 KB data cache, a 128-instruction window, a 40-entry load/store queue, and ports to networks that connect to other cores. A TFlex area model estimates that a single TFlex core would consume 2.4 mm^2 in a 65 nm technology, which would make it smaller than two of the termite cores in Table 3.1.

A truly composable processor has microarchitectural structures that linearly increase (or decrease) in capacity as participating cores are added (or removed). For example, doubling the number of cores should double the number of useful load/store queue entries, the usable state in the branch predictors, and the cache capacities. The key mechanisms for dynamically composing cores into differently sized logical processors are (1) composing multiple microarchitecture structures into a single logical but distributed microarchitecture structure, (2) interleaving the addresses to the structure across its different banks, and (3) employing a programmable hashing function that changes the interleaving depending on the number of cores that compose the logical processor. For example, when two CLP cores form a single logical processor, the level-1 data cache addresses are interleaved across the banks using the low-order bits of the index to select the core containing the right cache bank. If two more CLP cores are added, the hashing function is changed to interleave the cache-line addresses across four banks. Concretely, the TFlex microarchitecture (Fig. 3.8) uses three distinct hash functions for interleaving across three classes of structures.

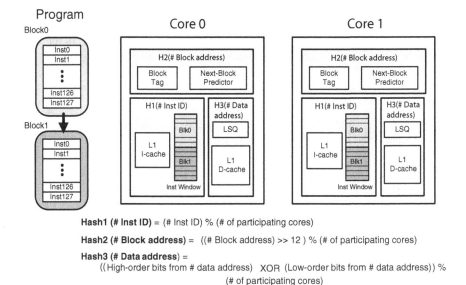

Hash1 (# Inst ID) = (# Inst ID) % (# of participating cores)

Hash2 (# Block address) = ((# Block address) >> 12) % (# of participating cores)

Hash3 (# Data address) =
((High-order bits from # data address) XOR (Low-order bits from # data address)) %
(# of participating cores)

Fig. 3.8 Interleaving of different microarchitectural structures for a two-core processor

Block starting address: The next-block predictor resources (e.g., BTBs and local history tables) and the blog I-cache tag structures are partitioned based on the starting virtual address of a particular block, which corresponds to the program counter in a conventional architecture. Predicting control flow and fetching instructions in TFlex occur at the granularity of a block, rather than individual instructions.

Instruction ID within a block: A block contains up to 128 instructions, which are numbered in order. Instructions are interleaved across the partitioned instruction windows and instruction caches based on the instruction ID, theoretically permitting up to 128 cores each holding one instruction from each block.

Data address: The load–store queue (LSQ) and data caches are partitioned by the low-order index bits of load/store addresses, and registers are interleaved based on the low-order bits of the register number.

The register names are interleaved across the register files. However, because a single core must have 128 registers to support single-block execution, register file capacity goes unused when multiple cores are aggregated. Because interleaving is controlled by bit-level hash functions, the number of cores that can be aggregated to form a logical processor must be a power of two.

In addition, composability requires distributed protocols to accomplish traditionally centralized microarchitecture operations. We adapted the distributed protocols of the TRIPS microarchitecture to a system consisting of homogeneous cores which are described in the subsequent sections.

3.4.1 Overview of TFlex Operation

Since TFlex employs an EDGE instruction set, its block execution model is similar to TRIPS. However, when cores are composed, TFlex uses a distributed fetch protocol in which each in-flight block is assigned a single *owner core*, based on a hash of the block address. The owner core is responsible for initiating fetch of that block and predicting the next block. Once the next-block address is predicted, the owner core sends that address to the core that owns the next predicted block. For each block that it owns, an owner core is also responsible for launching pipeline flushes, detecting block completion, and committing the block.

Figure 3.9 provides an overview of TFlex execution for the lifetime of one block. It shows two threads running on non-overlapping sets of eight cores each. In the block fetch stage, the block owner accesses the I-cache tag for the current block and broadcasts the fetch command to the I-cache banks in the participating cores (Fig. 3.9a). Upon receiving a fetch command from a block owner, each core fetches its portion of the block from its local I-cache, and dispatches fetched instructions into the issue window. In parallel, the owner core predicts the next block address and sends a control message – plus the next block address and global history register – to the next block owner to initiate fetch of the subsequent block (Fig. 3.9b). Up to eight blocks (one per core) may be in flight for eight participating cores. Instructions are executed in dataflow order when they are ready (Fig. 3.9c). When a block completes, the owner detects completion, and when it is notified that it holds the oldest block, it launches the block commit protocol, shown in Fig. 3.9d. While Fig. 3.9e–h shows the same four stages of execution for the next block controlled by

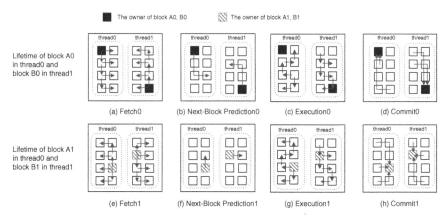

Fig. 3.9 TFlex execution stages: execution of two successive blocks (A0, A1) and (B0,B1) from two different threads executing simultaneously on a 16-core TFlex CLP with each thread running on eight cores

a different owner, fetch, execution, and commit of the blocks are actually pipelined and overlapped.

3.4.2 Composable Instruction Fetch

In the TFlex microarchitecture, instructions are distributed across the private I-caches of all participating cores, but fetches are initiated by each of the instruction block's owner cores. The owner core manages the tags on a per-block basis, and functions as a central registry for all operations of the blocks it owns. The cache tags for a block are held exclusively in the block owner core. In a 32-core configuration, each core caches four instructions from a 128-instruction block.

In TFlex, the fetch bandwidth and the I-cache capacity of the participating cores scale linearly as more cores are used. While this partitioned fetch model amplifies fetch bandwidth, conventional microarchitectures cannot use it because current designs require a centralized point to analyze the fetch stream (for register renaming and instruction number assignment) to preserve correct sequential semantics. This constraint poses the largest challenge to composable processors built with conventional ISAs. In contrast, EDGE ISAs overcome these deficiencies by explicitly and statically encoding the dependences among the instructions within a block; the dynamic total order among all executing instructions is obtained by concatenating the block order and the statically encoded order within a block. Given large block sizes, distributed fetching is feasible because intra-block instruction dependence relations are known a priori.

3.4.3 Composable Control-Flow Prediction

Control-flow predictors are some of the most challenging structures to partition for composability since the predictor state has traditionally been physically central-ized to facilitate few cycles between successive predictions. The TFlex compos-able predictor treats the distributed predictors in each composed core as a single logical predictor, exploiting the block-atomic nature of the TRIPS ISA to make this distributed approach tenable. Similar to the TRIPS prototype microarchitec-ture, the TFlex control-flow predictor issues one next-block prediction for each 128-instruction hyperblock – a predicated single entry, multiple exit region – instead of one prediction per basic block.

Figure 3.10 illustrates the distributed next-block predictor which consists of eight main structures. Each core has a fully functional block predictor which is identical to all others across the cores. The block predictor is divided into two main components: the exit predictor, which predicts which exit will be taken out of the current block; and the target predictor, which predicts the address of the next block. Each branch in a block contains three *exit* bits in its instruction, which are used to form histories without traditional taken/not taken bits.

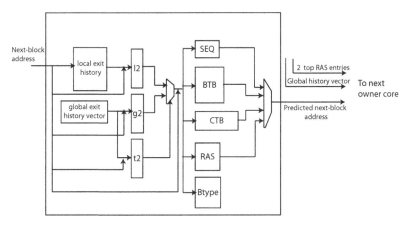

Fig. 3.10   TFlex next-block predictor

The exit predictor is an Alpha 21264-like tournament-style hybrid predictor [13] and is composed of traditional two-level local, global, and choice predictors that use local and global block exit histories. Speculative history updates are maintained and used to make predictions, updating local and global histories when blocks commit. The target address is predicted by first predicting the type of the exit branch – a call, a return, a next sequential block, or a regular branch – using the branch type (Btype) predictor. The Btype predictor result selects one of those four possible next-block targets, which are provided by a next-block adder (SEQ), a branch target buffer (BTB) to predict branch targets, a call target buffer (CTB) to predict call targets, and a return address stack (RAS) to predict return targets.

The predictor organization described above is naturally composable. The local histories are composable since, for a fixed composition, the same block address will always map to the same core. Local predictions do not lose any information even though they are physically distributed. Similarly, the Btype, BTB, and CTB tables hold only the target addresses of the blocks owned by that core. With this organization, the capacity of the predictor increases as more cores are added, assuming that block addresses tend to be distributed among cores equally.

To enable prediction with current global histories, the predicted next-block address is sent from the previous blocks' owner to the owning core of the predicted block along with the global exit history. This forwarding enables each prediction to be made with the current global history, without additional latency beyond the already incurred point-to-point latency to transmit the predicted next-block address from core to core.

The return address stack (RAS) is maintained as a single logical stack across all the cores, since it represents the program call stack which is necessarily logically centralized. The TFlex microarchitecture uses the RAS entries across different tiles to enable a deeper composed RAS. Instead of using address interleaving, the RAS is sequentially partitioned across all the cores (e.g., a 32-entry stack for two cores would have entries 0–15 in core 0 and 16–31 in core 1). The stacks from the participating cores form a logically centralized but physically distributed global stack. If the exit branch type is predicted as a call, the corresponding return address is pushed on to the RAS by sending a message to the core holding the current RAS top. If the branch is predicted as a return, the address on the stack is popped off by sending a pop-request to the core holding the RAS top. Since prediction is a time-critical operation, the latency of predicting return addresses is optimized by passing the two top RAS values from owner core to owner core. Recovery upon a misprediction is the responsibility of the mispredicting owner, which rolls back the mis speculated state and sends the updated histories and RAS pointers to the next owner core, as well as the corrected top-of-stack RAS information to the core that will hold the new RAS top.

3.4.4 Composable Instruction Execution

Each instruction in an EDGE block is statically assigned an identifier between 0 and 127 by the TRIPS compiler. A block's instructions are interleaved across the cores of a composable processor by the hash function described above. Changing the mapping function when cores are added (or removed) achieves composability of instruction execution. Figure 3.11 shows the mechanism that the TFlex design uses to support dynamic issue across a variable number of composed cores. Each instruction in an EDGE ISA block contains at least one nine-bit *target* field, which specifies the location of the dependent instruction that will consume the produced operand. Two of the nine bits specify which operand of the destination instruction is targeted (left, right, or predicate), and the other seven bits specify which of the

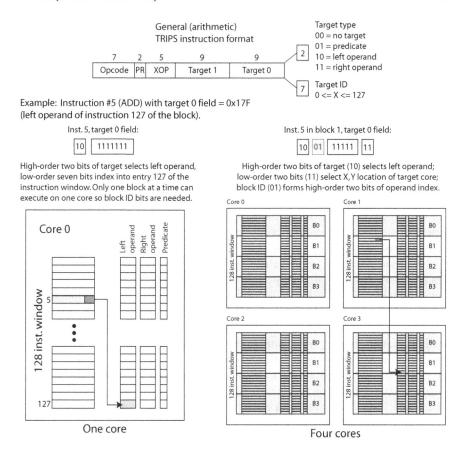

Fig. 3.11 Block mapping for one-core and four-core processors

128 instructions is targeted. Figure 3.11 shows how the target bits are interpreted in single-core mode, with instruction five targeting the left operand of instruction 127. In single-core mode, all seven target identifier bits are used to index into a single 128-instruction buffer.

Figure 3.11 also shows how the microarchitecture interprets the target bits when running in a four-core configuration. The four cores can hold a total of four instruction blocks, but each block is striped across the four participating cores. Each core holds 32 instructions from each of the four in-flight blocks. In this configuration, the microarchitecture uses the two low-order bits from the target to determine which core is holding the target instruction, and the remaining five bits to select one of the 32 instructions on that core. The explicit dataflow target semantics of EDGE ISAs make this operation simple compared to what would be required in a RISC or CISC ISA, where the dependences are unknown until an instruction is fetched and analyzed.

3.4.5 Composable Memory System

As with clustered microarchitectures [19, 26], L1 data caches in a composable processor can be treated as a single logical cache by treating the individual caches as address-partitioned banks. When running in one-core mode, each thread accesses only its bank. When multiple cores are composed, the L1 cache becomes a cache-line interleaved aggregate of all the participating L1 caches. With each additional core, a thread obtains more L1 D-cache capacity and an additional memory port. The cache bank accessed by a memory instruction is determined by the low-order bits of the set index; the number of bits is determined by the number of participating cores. All addresses within a cache line will always map to the same bank in a given configuration. When a core computes the effective address of a load, the address and the target(s) of the load are routed to the appropriate cache bank, the lookup is performed, and the result is directly forwarded to the core containing the target instruction of the load.

One of the microarchitectural challenges to support a composable memory system is to handle memory disambiguation efficiently on a distributed substrate with a variable number of cores. The TFlex microarchitecture relies on load–store queues (LSQs) to disambiguate memory accesses dynamically. As more cores are aggregated to construct a larger window, more entries in the LSQ are required to track all in-flight memory instructions. Partitioning LSQ banks by address and interleaving them with the same hashing function as the data caches is a natural way to build a large, distributed LSQ. However, unless each LSQ bank is maximally sized for the worst case (the instruction window size), a system must handle the situation when a particular LSQ bank is full, and cannot slot an incoming memory request (called "LSQ overflow"). TFlex uses a low-overhead NACK mechanism to handle LSQ overflows efficiently [31].

3.4.6 Composable Instruction Commit

Block commits in TFlex are staged across four phases. First, the block owner detects that a block is complete when participating cores inform the owner core that the block has emitted all of its outputs – stores, register writes, and one branch. Register files send a message to the owner core when they have received all of their register writes, and each completing store sends a control message to the owner core, which can then determine when all outputs have been received. Second, when the block becomes the oldest one, the block owner sends out a *commit* command on the control network to the participating cores. Third, all distributed cores write their outputs to architectural state, and when finished, respond with *commit acknowledgment* signals. Finally, the owner core broadcasts a resource deallocation signal, indicating that the youngest block owner can initiate its own fetch and overwrite the committed block with a new block.

3.4.7 Coherence Management

For a level-two (L2) cache organization, we consider a 4 MB shared design (shown in Fig. 3.1) that contains 32 cache banks connected by a switched mesh network. To maintain coherence among private L1 caches, the shared L2 cache uses a standard on-chip directory-based cache coherence protocol with L1 sharing vectors stored in the L2 tag arrays. The composition of the cores on the chip does not affect how the coherence protocol is implemented. A sharing status vector in the L2 tag keeps track of L1 coherence by treating each L1 cache as an independent coherence unit. When a composition changes – adding cores to some composed processors and removing them from others – the L1 caches need not be flushed; the new interleaved mapping for the composed L1 D-cache banks will result in misses, at which point the underlying coherence mechanism forwards the request to the lines stored in the old L1 cache banks, invalidating them or forwarding a modified line.

3.5 Experimental Evaluation

To explore the benefits of flexibility and composability, we compare the performance of the composable TFlex architecture to a configuration similar to the TRIPS architecture on single-threaded applications. Then we employ a set of multiprogrammed workloads to measure the benefits of composability on throughput-oriented workloads. These analyses use a validated cycle-accurate simulator with a set of both compiler-generated and hand-optimized benchmarks shown in Table 3.3. For the TFlex configurations, the programs are scheduled assuming a 32-core composable processor.

Table 3.3 Single-core TFlex microarchitecture parameters

Parameter	Configuration
Instruction supply	Partitioned 8 KB I-cache (1-cycle hit); Local/Gshare Tournament predictor (8 K+256 bits, 3 cycle latency) with speculative updates; Num. entries: Local: 64(L1) + 128(L2), Global: 512, Choice: 512, RAS: 16, CTB: 16, BTB: 128, Btype: 256.
Execution	Out-of-order execution, RAM structured 128-entry issue window, dual-issue (up to two INT and one FP).
Data supply	Partitioned 8 KB D-cache (2-cycle hit, 2-way set-associative, 1-read port and 1-write port); 44-entry LSQ bank; 4 MB decoupled S-NUCA L2 cache [15] (8-way set-associative, LRU-replacement); L2-hit latency varies from 5 cycles to 27 cycles depending on memory address; average (unloaded) main memory latency is 150 cycles.
Simulation	Execution-driven simulator validated within 7% of TRIPS hardware
Hand-optimized	3 kernels (conv, ct, genalg), 7 EEMBC benchmarks (a2time, autocor, basefp, bezier, dither, rspeed, tblook), 2 Versabench (802.11b, 8b10b) [25]
Compiled	14 SPEC CPU benchmarks currently supported (8 Integer, 6 FP), simulated with single simpoints of 100 million cycles [32].

The sizes of the structures in the core ensure that one core can execute and atomically commit one EDGE block. Each core has a 128-entry instruction window to accommodate all instructions in a block, 128 registers, and an LSQ large enough to hold at least 32 loads/stores. The sizes of the remaining structures, including the I-cache, D-cache, and the branch predictor, were sized to balance area overhead and performance. The simulated baseline TRIPS microarchitecture matches that described by Sankaralingam et al. [29], with the exception that the L2 capacity of the simulated TRIPS processor is 4 MB to support a more direct comparison with TFlex on the multiprogramming workloads.

The TFlex architecture also includes two microarchitectural optimizations that could be applied to improve the baseline performance of the TRIPS microarchitecture. First, the bandwidth of the operand network is doubled to reduce inter-ALU contention. Second, TFlex cores support limited dual issue – two integer instructions but only one floating-point instruction issued per cycle – as opposed to the single-issue execution tiles in TRIPS.

3.5.1 Serial Programs

Figures 3.12 and 3.13 show the performance of the TRIPS prototype architecture and that of TFlex compositions on low and high ILP benchmarks, for 2–32 cores. The performance of each benchmark is normalized to the performance of a single TFlex core. On low-ILP benchmarks, TFlex compositions with fewer cores typically meet the performance of more cores as the exposed ILP does not justify wide issue. On some benchmarks (such as sixtrack), four-core performance exceeds that of eight or more since the protocol overheads executing across a larger number of

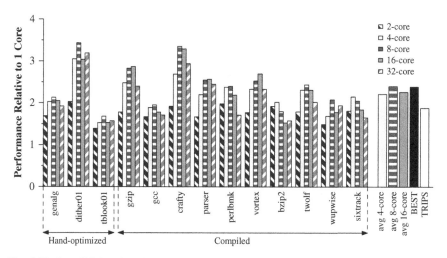

Fig. 3.12 Low ILP benchmarks: normalized performance of applications running on 2–32 CLP cores

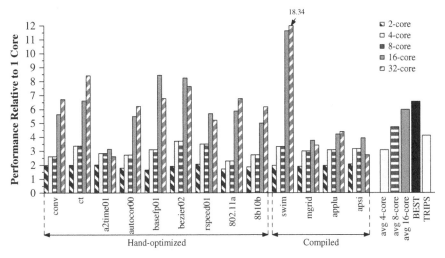

Fig. 3.13 High ILP benchmarks: normalized performance of applications running on 2–32 CLP cores

distributed processors can hinder performance. In general, eight cores shows the highest average performance on the low-ILP benchmarks.

On high-ILP benchmarks, the best single composition ranges from 16 to 32 cores, depending on the application. Averaged across all of the low and high ILP benchmarks, the 16-core TFlex configuration performs best and shows a factor of 3.5 speedup over a single TFlex core. Additional performance can be obtained by adjusting the number of TFlex cores per processor on a per-application basis. This configuration, represented by the bar "BEST," provides an additional 13% performance increase and an overall speedup over a single TFlex core of a factor of four. These results indicate that a combination of scalable processor mechanisms and the capability to dynamically adjust the processor granularity provides a system that can run sequential applications effectively across multiple cores to achieve substantial speedups.

On average, an eight-core TFlex processor, which has area and issue width comparable to the TRIPS processor, outperforms TRIPS by 19%, reflecting the benefits of additional operand network bandwidth as well as twice the L1 cache bandwidth stemming from fine-grained distribution of the cache. The best per-application TFlex configuration outperforms TRIPS by 42%, demonstrating that adapting the processor granularity to the application granularity provides significant improvements in performance.

3.5.2 Multitasking Workloads

To quantify throughput gains provided by the increased flexibility of CLPs, we measured the performance of multi-programming workloads on both the TFlex design and fixed-granularity CMPs. To isolate the contributions of composability to flexibility, we modeled fixed-granularity CMPs by configuring a 32-core TFlex system to use fixed numbers of logical processors composed of equal numbers of cores. For example, an experiment labeled "CMP-4" is a TFlex configuration that has eight composed processors with four cores each.

The TFlex simulator models inter-processor contention for the shared L2 cache and main memory. The 12 benchmarks used in this experiment were selected from the hand-optimized benchmark suite, which exhibit considerable diversity in ILP. We varied the workload size (i.e., degree of multiprogramming) from two to 16 applications running simultaneously.

We measured throughput performance for TFlex on multiple workload sizes (number of threads) and compared them to different fixed-core configurations using a *weighted speedup* (WS) metric [35]. Weighted speedup results are normalized to a single-processor TFlex configuration that produces the best performance. We used an optimal dynamic programming algorithm to find the core assignments that maximize WS on TFlex, rather than employing an on-line algorithm [10]. The dynamic programming algorithm uses the results from Fig. 3.12 and 3.13 to compute individual benchmark speedup as a function of the number of cores.

For each workload size, Fig. 3.14 reports the average weighted speedup on TFlex and on the set of fixed-granularity CMPs. The best CMP granularity changes depending on the workload size (i.e., the number of threads). CMP-16 is the best for workloads with two threads, CMP-8 is the best for four-thread workloads, CMP-4 is the best for 6–8 threads, and CMP-2 is the best for 12–16 threads. Because of the capability of selecting the best granularity for each workload, the TFlex array consistently outperforms the best fixed-granularity CMP. The set of bars labeled AVG show the average WS across all workloads. While CMP-4 is the best granularity for fixed CMPs, the TFlex design produces an average of 26% higher WS and a maximum of 47% higher WS.

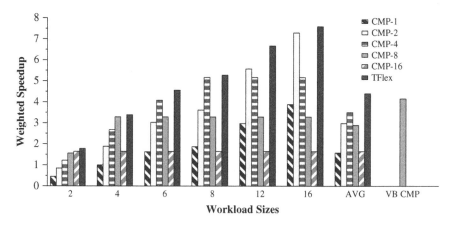

Fig. 3.14 Weighted speedup comparison between TFlex and fixed-granularity CMPs

Figure 3.14 also demonstrates the value of asymmetric processor composition in TFlex by comparing to a hypothetical flexible CMP that can change its granularity dynamically but is symmetric, requiring each processor to be composed from equal number of cores. The results for this hypothetical system, labeled VB CMP (Variable Best CMP), show that the capability of composing processors of different granularities for different simultaneously executing programs results in a 6% improvement in throughput. For a given granularity, even when the number of threads exactly matches the number of processors, some threads may under-utilize their own processor's resources. In such a scenario, TFlex is able to allocate a portion of these under-utilized resources to applications which can make better use of them. Table 3.4 lists the fraction of applications at each workload size that are assigned to a given processor granularity and demonstrates that the optimal processor granularity varies even within a given workload size. For workloads of size 8, TFlex allocates four cores to the majority of the applications, but also allocates two cores to 28% of them and eight cores to 14% of them. A CMP-4 configuration, conversely, would allocate only four cores regardless of the application characteristics. The difference between the VB CMP bar and the TFlex average bar represents the performance benefit of allocating different numbers of cores to applications running simultaneously.

Table 3.4 Percentage of applications mapped to each composed processor in TFlex for each workload size

	Cores (%)					
Workload size	1	2	4	8	16	32
2	0	0	13	35	53	0
4	0	0	31	54	15	0
6	0	18	43	37	2	0
8	0	28	59	14	0	0
12	9	82	9	0	0	0
16	32	52	16	0	0	0

3.5.3 Composability Overheads and Benefits

Since the TRIPS prototype processor and the proposed TFlex implementation share the same ISA and and several microarchitectural structures, a comparison of TFlex and TRIPS implementations can reveal the area overhead for providing composability. Table 3.5 presents the area of different microarchitectural components in an eight-core TFlex processor and the 30 tiles in a single TRIPS processor. The area of each microarchitectural component in a single TFlex core was estimated from the post-synthesis netlist (before final place-and-route) of the 130 nm ASIC implementation of the TRIPS prototype.

The TRIPS and TFlex designs differ in a number of features that are unrelated to composability. For example, the techniques in TFlex for reducing the LSQ capacity per tile could be applied to TRIPS. The difference in cache and branch predictor capacities are also independent of composability. The only area costs for compos-

Table 3.5 Area (mm^2) estimate for eight-core TFlex and area of a single TRIPS core

Structures	eight TFlex cores (16-issue)		Single TRIPS processor (16-issue)	
	Size (8 tiles)	Area	Size (30 tiles)	Area
Fetch (block predictor, I-cache)	66 K-bit predictor, 64 KB I-cache	10.9	80 K-bit predictor, 80 KB I-cache	7.7
Register files	1024 entries	6.5	512 entries	3.0
Execution resources (issue window, ALUs)	1 K-entry issue window 16-INT ALU, 8-FP ALU	23.6	1 K-entry issue window 16-INT ALU, 16-FP ALU	39.4
L1 D-cache subsystem	64 KB D-cache		32 KB D-cache	
(D-cache, LSQ, MSHR)	352-entry LSQ	27.8	1024-entry LSQ	33.4
Routers		7.0		11.0
Total		75.8		94.5

ability are (1) larger register file, (2) some additional control logic to adjust the hashing functions, and (3) some additional bits in control networks to manage the distributed protocols. We conservatively estimate that this area overhead comes to less than 10% in the processors, and less than that when the on-chip memory system is included.

Despite the register file overhead, TFlex has better performance/area because the features used for supporting composability simultaneously improve performance (by aggregating resources) and reduce area (by distribution). In addition, we estimate that the total power consumption of eight TFlex cores is 15% less than a single TRIPS processor, indicating that composability imposes little if any power penalty. Results also show that TFlex achieves twice the power efficiency (performance^2/Watt) of TRIPS and 25% better efficiency than the best fixed-core design, because of its ability to adapt the processor granularity to the application's concurrency [16].

3.6 Summary

Since clock rate scaling can no longer sustain computer system performance scaling, future systems will need to mine concurrency at multiple levels of granularity. Depending on the workload mix and number of available threads, the balance between instruction-level and thread-level parallelism will likely change frequently for general-purpose systems, rendering the design-time freezing of processor granularity (bulldozers, chainsaws, or termites) in traditional CMPs an undesirable option. Composable lightweight processors (CLPs) provide the flexibility to allocate resources dynamically to different types of concurrency, ranging from running a single thread on a logical processor composed of many distributed cores to running many threads on separate physical cores. The small size of an individual TFlex CLP core places it near the regime of termites so that it can exploit substantial threaded parallelism. When the application or environment demands, CLP cores can be aggregated to form bulldozer-sized processors. In addition to pure performance, the system can also use energy and/or area efficiency to choose the best-suited processor configuration. We envision multiple methods of controlling the allocation of cores to threads. At one end of the spectrum, the operating system could make such decisions based on the number of threads to run and their criticality. The OS could even monitor how each thread uses its allocated resources and reallocate them among the threads as necessary. At the other end of the spectrum, the hardware could potentially adjust the number of cores per thread in an automated fashion. This capability would further blur the distinction between conventional uniprocessors and multiprocessors, which we view as a promising direction.

Acknowledgments The authors wish to thank the entire TRIPS hardware team for their contributions to the design and implementation of the TRIPS and TFlex architectures. The team members include: Rajagopalan Desikan, Saurabh Drolia, Mark Gebhart, M.S. Govindan, Paul Gratz, Divya Gulati, Heather Hanson, Changkyu Kim, Haiming Liu, Robert McDonald, Ramadass Nagarajan, Nitya Ranganathan, Karthikeyan Sankaralingam, Simha Sethumadhavan, Sadia Sharif, Premkishore Shivakumar, Stephen W. Keckler, and Doug Burger. The TRIPS project was supported by the Defense Advanced Research Projects Agency under contract F33615-01-C-4106 and by NSF CISE Research Infrastructure grant EIA-0303609.

References

1. M. Annavaram, E. Grochowski, and J. P. Shen. Mitigating Amdahl's law through EPI throttling. In *International Symposium on Computer Architecture*, pages 298–309, June 2005.
2. S. Bell, B. Edwards, J. Amann, R. Conlin, K. Joyce, V. Leung, J. MacKay, M. Reif, L. Bao, J. Brown, M. Mattina, C.-C. Miao, C. Ramey, D. Wentzlaff, W. Anderson, E. Berger, N. Fairbanks, D. Khan, F. Montenegro, J. Stickney, and J. Zook. TILE64 processor: A 64-core SoC with mesh interconnect. In *International Solid-State Circuits Conference*, pages 88–89, Feb 2008.
3. D. Burger, S. Keckler, K. McKinley, M. Dahlin, L. John, C. Lin, C. Moore, J. Burrill, R. McDonald, and W. Yoder. Scaling to the end of silicon with EDGE architectures. *IEEE Computer*, 37(7), 2004.
4. J. Dorsey, S. Searles, M. Ciraula, S. Johnson, N. Bujanos, D. Wu, M. Braganza, S. Meyers, E. Fang, and R. Kumar. An integrated quad-core opteron processor. In *International Solid-State Circuits Conference*, pages 102–103, Feb 2007.
5. J. Friedrich, B. McCredie, N. James, B. Huott, B. Curran, E. Fluhr, G. Mittal, E. Chan, Y. Chan, D. Plass, S. Chu, H. Le, L. Clark, J. Ripley, S. Taylor, J. Dilullo, and M. Lanzerotti. Design of the Power6 microprocessor. In *International Solid-State Circuits Conference*, pages 102–103, Feb 2007.
6. M. Gebhart, B. Maher, P. Gratz, N. Ranganathan, J. Diamond, B. Robatmili, S. Govindan, S. W. Keckler, D. Burger, and K. McKinley. TRIPS System Evaluation. Technical Report TR-08-31, The University of Texas at Austin, Department of Computer Sciences, Sep 2008.
7. S. Ghiasi and D. Grunwald. Aide de Camp: Asymmetric Dual Core Design for Power and Energy Reduction. Technical Report CU-CS-964-03, The University of Colorado, Department of Computer Science, 2003.
8. E. Grochowski, R. Ronen, J. Shen, and H. Wang. Best of both latency and throughput. In *International Conference on Computer Design*, pages 236–243, Oct 2004.

9. T. R. Halfhill. Floating point buoys clearspeed. *Microprocessor Report*, 17(11):19–24, Nov 2003.

10. T. Ibaraki and N. Katoh. *Resource Allocation Problems: Algorithmic Approaches*. MIT Press, 1988.

11. E. Ipek, M. Kirman, N. Kirman, and J. F. Martínez. Core fusion: Accommodating software diversity in chip multiprocessors. In *International Symposium on Computer Architecture*, pages 186–197, June 2007.

12. J. A. Kahle, M. N. Day, H. P. Hofstee, C. R. Johns, T. R. Maeurer, and D. Shippy. Introduction to the cell multiprocessor. *IBM Journal of Research and Development*, 49(4/5), July 2005.

13. R. Kessler. The Alpha 21264 microprocessor. *IEEE Micro*, 19(2):24–36, March/April 1999.

14. B. Khailany, T. Williams, J. Lin, E. Long, M. Rygh, D. Tovey, and W. J. Dally. A Programmable 512 GOPS stream processor for signal, image, and video processing. In *International Solid-State Circuits Conference*, pages 272–273, Feb 2007.

15. C. Kim, D. Burger, and S. W. Keckler. An adaptive, non-uniform cache structure for wire-delay dominated on-chip caches. In *International Conference on Architectural Support for Programming Languages and Operating Systems*, pages 211–222, Oct 2002.

16. C. Kim, S. Sethumadhavan, M. Govindan, N. Ranganathan, D. Gulati, D. Burger, and S. W. Keckler. Composable lightweight processors. In *International Symposium on Microarchitecture*, pages 381–394, Dec 2007.

17. R. Kumar, K. I. Farkas, N. P. Jouppi, P. Ranganathan, and D. M. Tullsen. Single-ISA heterogeneous multi-core architectures: The potential for processor power reduction. In *International Symposium on Microarchitecture*, pages 81–92, Dec 2003.

18. R. Kumar, D. M. Tullsen, P. Ranganathan, N. P. Jouppi, and K. I. Farkas. Single-ISA heterogeneous multi-core architectures for multithreaded workload performance. In *International Symposium on Computer Architecture*, pages 64–75, June 2004.

19. F. Latorre, J. González, and A. González. Back-end assignment schemes for clustered multithreaded processors. In *International Conference on Supercomputing*, pages 316–325, June 2004.

20. E. Lindholm and S. Oberman. The NVIDIA GeForce 8800 GPU. In *Proceedings of the Symposium on High Performance Chips (HotChips-19)*, August 2007.

21. B. A. Maher, A. Smith, D. Burger, and K. S. McKinley. Merging head and tail duplication for convergent hyperblock formation. In *International Symposium on Microarchitecture*, pages 480–491, Dec 2006.

22. M. R. Marty and M. D. Hill. Virtual hierarchies to support server consolidation. In *International Symposium on Computer Architecture*, pages 46–56, June 2007.

23. S. Melvin, M. Shebanow, and Y. Patt. Hardware support for large atomic units in dynamically scheduled machines. In *Workshop on Microprogramming and Microarchitecture*, pages 60–63, Nov 1988.

24. U. G. Nawathe, M. Hassan, K. C. Yen, A. Kumar, A. Ramachandran, and D. Greenhill. Implementation of an 8-core, 64-thread, power-efficient SPARC server on a chip. *IEEE Journal of Solid-State Circuits*, 43(1), Jan 2008.

25. R. M. Rabbah, I. Bratt, K. Asanović, and A. Agarwal. Versatility and VersaBench: A New Metric and a Benchmark Suite for Flexible Architectures. Technical Report MIT-LCS-TM-646, Massachusetts Institute of Technology, Computer Science and Artificial Intelligence Laboratory, June 2004.

26. P. Racunas and Y. N. Patt. Partitioned first-level cache design for clustered microarchitectures. In *International Conference on Supercomputing*, pages 22–31, June 2003.

27. P. Salverda and C. Zilles. Fundamental performance challenges in horizontal fusion of In-order cores. In *International Symposium on High-Performance Computer Architecture*, pages 252–263, Feb 2008.

28. K. Sankaralingam, R. Nagarajan, H. Liu, C. Kim, J. Huh, D. Burger, S. W. Keckler, and C. R. Moore. Exploiting ILP, TLP, and DLP with the polymorphous TRIPS architecture. In *International Symposium on Computer Architecture*, pages 422–433, June 2003.

29. K. Sankaralingam, R. Nagarajan, R. McDonald, R. Desikan, S. Drolia, M. S. S. Govindan, P. Gratz, D. Gulati, H. Hanson, C. Kim, H. Liu, N. Ranganathan, S. Sethumadhavan, S. Sharif, P. Shivakumar, S. W. Keckler, and D. Burger. Distributed microarchitectural protocols in the TRIPS prototype processor. In *International Symposium on Microarchitecture*, pages 480–491, Dec 2006.
30. S. Sethumadhavan, R. McDonald, R. Desikan, D. Burger, and S. W. Keckler. Design and implementation of the TRIPS primary memory system. In *International Conference on Computer Design*, pages 470–476, Oct 2006.
31. S. Sethumadhavan, F. Roesner, J. S. Emer, D. Burger, and S. W. Keckler. Late-binding: Enabling unordered load – store queues. In *International Symposium on Computer Architecture*, pages 347–357, June 2007.
32. T. Sherwood, E. Perelman, and B. Calder. Basic block distribution analysis to find periodic behavior and simulation points in applications. In *International Conference on Parallel Architectures and Compilation Techniques*, pages 3–14, Sep 2001.
33. A. Smith, J. Burrill, J. Gibson, B. Maher, N. Nethercote, B. Yoder, D. Burger, and K. S. McKinley. Compiling for EDGE architectures. In *International Symposium on Code Generation and Optimization*, pages 185–195, Mar 2006.
34. A. Smith, R. Nagarajan, K. Sankaralingam, R. McDonald, D. Burger, S. W. Keckler, and K. S. McKinley. Dataflow predication. In *International Symposium on Microarchitecture*, pages 89–100, Dec 2006.
35. A. Snavely and D. M. Tullsen. Symbiotic jobscheduling for a simultaneous multithreaded processor. In *International Conference on Architectural Support for Programming Languages and Operating Systems*, pages 234–244, Nov 2000.
36. B. Stackhouse, B. Cherkauer, M. Gowan, P. Gronowski, and C. Lyles. A 65 nm 2-billion-transistor quad-core Itanium processor. In *International Solid-State Circuits Conference*, pages 92–93, Feb 2008.
37. D. Tarjan, M. Boyer, and K. Skadron. Federation: Out-of-Order Execution using Simple In-Order Cores. Technical Report CS-2007-11, University of Virginia, Department of Computer Science, Aug 2007.
38. M. B. Taylor, W. Lee, J. Miller, D. Wentzlaff, I. Bratt, B. Greenwald, H. Hoffmann, P. Johnson, J. Kim, J. Psota, A. Saraf, N. Shnidman, V. Strumpen, M. Frank, S. P. Amarasinghe, and A. Agarwal. Evaluation of the raw microprocessor: An exposed-wire-delay architecture for ILP and streams. In *International Symposium on Computer Architecture*, pages 2–13, June 2004.
39. M. Tremblay and S. Chaudhry. A third-generation 65 nm 16-core 32-thread plus 32-scout-thread CMT SPARC Processor. In *International Solid-State Circuits Conference*, pages 82–83, Feb 2008.
40. S. Vangal, J. Howard, G. Ruhl, S. Dighe, H. Wilson, J. Tschanz, D. Finan, A. Singh, T. Jacob, S. Jain, C. Roberts, Y. Hoskote, N. Borkar, and S. Borkar. An 80-tile sub-100 W TeraFLOPS processor in 65-nm CMOS. *IEEE Journal of Solid-State Circuits*, 43(1):29–41, Jan 2008.
41. E. Waingold, M. Taylor, D. Srikrishna, V. Sarkar, W. Lee, V. Lee, J. Kim, M. Frank, P. Finch, R. Barua, J. Babb, S. Amarasinghe, and A. Agarwal. Baring it all to software: Raw machines. *IEEE Computer*, 30(9):86–93, 1997.

Chapter 4
Speculatively Multithreaded Architectures

Gurindar S. Sohi and T.N. Vijaykumar

Abstract Using the increasing number of transistors to build larger dynamic-issue superscalar processors for the purposes of exposing more parallelism has run into problems of diminishing returns, great design complexity, and high power dissipation. While chip multiprocessors (CMPs) alleviate these problems by employing multiple smaller, power-efficient cores to utilize the available transistors, CMPs require parallel programming which is significantly harder than sequential programming. *Speculatively multithreaded* architectures address both the programmability issues of CMPs and the power–complexity–performance problems of superscalar processors. Speculatively multithreaded architectures partition a sequential program into contiguous program fragments called *tasks* which are executed in parallel on multiple cores. The architectures execute the tasks in parallel by speculating that the tasks are independent, though the tasks are not guaranteed to be independent. The architecture provides hardware support to detect dependencies and roll back misspeculations. This chapter addresses the key questions of how programs are partitioned into tasks while maximizing inter-task parallelism and how inter-task control-flow and data dependencies (register and memory dependencies) are maintained especially in the distributed multicore organization employed by the speculatively multithreaded architectures.

4.1 Introduction

Computer architects are continually looking for novel ways to make productive use of the increasing number of transistors made available by improvements in semiconductor technology. Conventional dynamic (a.k.a. out-of-order issue) superscalar processors have utilized the transistors to improve performance by exploiting instruction-level parallelism. The processors execute multiple instructions in

G.S. Sohi (✉)
Computer Sciences Department
University of Wisconsi, Madison, WI, USA
e-mail: sohi@cs.wisc.edu

S.W. Keckler et al. (eds.), *Multicore Processors and Systems*, Integrated Circuits and Systems, DOI 10.1007/978-1-4419-0263-4_4, © Springer Science+Business Media, LLC 2009

parallel and overlap functional unit and memory latencies. However, brute-force scaling of the processors to exploit more transistors of successive semiconductor technology generations by executing more instructions in parallel has run into two problems: (1) The processors establish a single *window* of instructions to search for independent instructions. Due to limited instruction-level parallelism in nearby instructions, enlarging the window does not yield enough parallelism to continue to improve performance. (2) Moreover, many of the key circuits in the processors (e.g., issue logic, register bypass, register file) scale super-linearly with the number of instructions executed in parallel and with the window size [21]. Combined with faster clocks, this super-linear scaling results in great design complexity and high power dissipation.

One option to alleviate the above problems is chip multiprocessors (CMPs), which employ multiple cores to exploit the ever-increasing transistor counts. CMPs avoid the above scaling problem by employing many small cores instead of one large core, where each smaller core's circuits handle fewer instructions and provide smaller window sizes. Thus, the smaller cores alleviate both issues of design complexity and power dissipation. CMPs address the limited parallelism problem by requiring the programmer to write parallel programs where the burden of identifying and providing parallelism falls on the programmer. Unfortunately, parallel programming is significantly harder than sequential programming. Parallel programs require data management (data distribution and movement) and synchronization, both of which represent a considerable burden for the programmer well beyond programming the desired functionality. The burden is considerable because data management can be difficult and can have a significant impact on performance, while poor synchronization can lead to correctness problems such as deadlocks and livelocks.

While conventional multiprocessors have been successful despite the need for parallel programming, a key difference between multiprocessors and CMPs is that while multiprocessors in the past have targeted niche market segments (e.g., database servers), CMPs are expected to be programmed by non-expert programmers. Such ubiquitous parallel programming for all applications and programs may be a hard goal to achieve. The alternate approach of writing sequential programs to be parallelized automatically by compilers has, to date, achieved success only for a limited class of numerical applications.

Speculatively multithreaded architectures address both the programmability issues of CMPs and the scaling and limited-parallelism problems of dynamic superscalar processors. Speculatively multithreaded processors partition a sequential program into contiguous program fragments called *tasks* which are executed in parallel on multiple cores. Though the tasks are not guaranteed to be independent, execution proceeds by speculating that the tasks are independent. The architecture provides hardware support to detect and roll back misspeculations. By employing multiple smaller cores instead of one large core (as do CMPs), speculatively multithreaded processors avoid the scaling limitations of dynamic superscalar processors. Because parallelism among farther-away instructions is significantly larger than that in nearby instructions [3], these processors avoid the limited-parallelism problem of superscalars. Yet these processors rely on the time-tested and familiar sequential programming model, avoiding the programmability problems of CMPs. Furthermore, by employing multicore hardware, this approach provides an easy growth path from the era of sequential programming to the era of parallel programming, if parallel programming becomes widespread; and if the latter era were never to be realized, then speculative multithreading would become a prominent multicore architecture.

Wisconsin Multiscalar [26] (formerly called Expandable Split-Window Paradigm [8]) was the first project to propose the idea and an implementation of speculatively multithreaded architectures. Inspired by Multiscalar, many other projects proposed and pursued a number of alternative implementations and extensions of the idea (e.g., Stanford Hydra [12], UIUC IA-COMA [7], CMU STAMPede [27], Dynamic Multithreading [1], and Core Fusion [15]). While the idea of speculative multithreading and its advantages are appealing, there are many challenges:

- What choice of tasks would result in the most thread-level parallelism?
- How are control-flow dependencies from one task to another handled?
- How are data dependencies – register and memory – among tasks handled?
- How does the architecture retain its multicore nature (i.e., distributed organization) while handling these dependencies?

The rest of the chapter addresses the above challenges. We begin by describing the reasoning behind and the goals targeted by speculatively multithreaded architectures. Next, we describe the high-level execution model, the hardware organization, and the split between hardware and software to address the above challenges. Then we explain the details of how the hardware and compiler address the above challenges. We then provide an analysis of performance, complexity, and scalability of the architectures. Finally, we connect speculative multithreading to recent proposals of architectural support for transactional memory, a parallel programming model.

4.2 The Vision of Speculatively Multithreaded Architectures

4.2.1 Where is the Parallelism?

As mentioned in the previous section, a previous study of the limits of instruction-level parallelism [3] shows that though sufficient parallelism exists in sequential programs, the parallelism was not evenly distributed in the dynamic instruction stream. That is, instructions in close proximity, such as those *within* a sub-window, are likely to be dependent, thus limiting the parallelism in a small instruction window, but instructions that are far apart are likely to be independent. Consequently, speculatively multithreaded architectures establish a large instruction window by stringing together many smaller *sub-windows* to exploit the parallelism *across* the sub-windows; each sub-window is mapped to and executed on a core in a speculatively multithreaded processor.

Another study [9] supports this idea of building a large window from smaller sub-windows by showing that on average fewer than two instructions use the result of another instruction (i.e., a dynamic register value) and that most instruction results are used only by one other instruction. Because the consumer instruction is usually near the producer instruction in the instruction stream, both instructions are in the same sub-window. Consequently, only a few values need to go from one sub-window to another (i.e., from one core to another), keeping inter-core communication bandwidth and power within manageable limits.

4.2.2 Goals

To be competitive against CMPs and dynamic superscalar processors, speculatively multithreaded architectures need to meet some goals. We list the goals for the hardware and application software separately.

The hardware should meet the following requirements:

- Use of simple, regular hardware structures
- speeds comparable to single-issue processors
- Easy growth path from one generation to the next

  - reuse existing processing cores to the extent possible
  - no centralized bottlenecks

- Exploit available parallelism
- Place few demands on software (compilers and applications):

  - require few guarantees.

The architecture should employ simple, regular hardware structures so that the clock speed is not negatively impacted. Moreover, an ideal growth path from one generation to the next, as semiconductor resources increase, is one that simply uses more processing cores (e.g., four 2-wide cores), rather than one that requires a redesign of the core (e.g., to go from a single 4-wide to an 8-wide core). Increasing the number of cores without impacting the clock speed is possible only if there are no centralized structures whose complexity increases super-linearly with the number of cores. Finally, the ensemble of cores should collectively be able to exploit the available parallelism in a power-efficient and low-complexity manner. Finally, the architecture should place few demands on the software. In particular, the architecture should not require the compiler to provide independence guarantees for exploiting parallelism that would require compiler analysis that was either impossible or would be complex and time consuming.

The application software has the following requirements:

- Write programs in ordinary languages (e.g., C, C++, and Java)
- Target uniform hardware–software interface

  - facilitate software independence and growth path

- Maintain uniform hardware–software interface (i.e., do not tailor for specific architecture)

  * minimal OS impact
  * facilitate hardware independence and growth path

While computer scientists have proposed numerous programming languages where the expression of parallelism may be easier (e.g., functional languages), most programs in the real world are written in imperative languages such as C and C++ which make it hard for the compiler to extract parallelism automatically. Moreover, it is unrealistic to expect millions of (non-expert) programmers to adopt entirely new programming styles. Software should also target a uniform hardware–software interface (i.e., instruction set architecture) so that recompilation (either statically or dynamically) is not required when a program is run on a different hardware implementation. A uniform interface is especially important when considering the operating system, as it is not practical to have multiple versions of the executable code of an operating system. This uniformity is also important to allow hardware design to proceed independently of software.

4.2.3 Putting Things Together

Based on the above requirements, a speculatively multithreaded processor is an ensemble of *processing units* or cores which work collectively in parallel along with additional hardware/software.[1] The hardware and software complexity of the processor is similar to that of simple, non-ILP cores. A high-level sequencer sequences through the static program in "task-sized" steps, assigning a *task* of instructions to a processing unit. A *task* is a sub-graph of the program's *control-flow graph (CFG)*, which results in a contiguous portion of the dynamic instruction stream when processed (e.g., loop iterations or function invocations).[2] Each processing unit processes its task just as any simple processor would process a sequential program, albeit one comprising of only a few tens or hundreds of dynamic instructions. The processing units are ordered to maintain the appearance of sequential execution, and collectively the large instruction window is split into multiple, ordered, smaller instruction windows.

Figures 4.1 and 4.2 depict the instruction windows of a conventional superscalar processor and of a speculatively multithreaded processor. The key sources of power consumption and complexity in a conventional, centralized wide window superscalar processor are wide-issue wake-up logic, wide register bypasses, and large, multiported register rename tables and register files [21]. All of these structures are distributed over multiple cores in a speculatively multithreaded processor such that the per-core structures are small enough to be manageable.

---

[1]Throughout this chapter, we use the terms "processing units" and "cores" interchangeably.

[2]Throughout this chapter, we use the terms "task" and "thread" interchangeably.

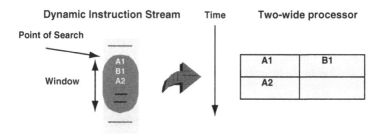

Fig. 4.1 Instruction window of conventional uniprocessor (superscalar). The window in the dynamic instruction stream starts at the point of search and includes instructions A1, B1, and A2. A2 is dependent on A1 but B1 is independent. Thus, A1 and B1 execute in parallel on a two-wide processor. A2 executes later in time

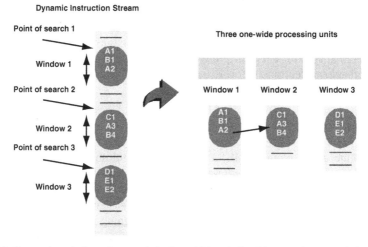

Fig. 4.2 Instruction window of a speculatively multithreaded architecture. The speculatively multithreaded architecture splits the dynamic instruction stream into three small windows, each of which starts at a point of search and executes on a processing unit. The three windows consist respectively of instructions A1, B1, and A2, and C1, A3, and B4, and D1, E1, and E2. All the instructions starting with the same letter are in a dependence chain where each instruction depends on the previous. Each window executes on a processing unit and in parallel with the other windows. The A3–A2 dependence is handled to maintain sequential semantics

The collection of the sub-windows being handled by each processing core effectively constructs a large instruction window that is required to expose parallelism. The distributed implementation of the architecture allows physical scalability as we increase the number of cores (e.g., up to a few tens of cores). By avoiding inter-core communication as much as possible and wherever unavoidable ensuring the communication to be as near-neighbor as possible, the architecture can achieve good performance scalability. Some components in the memory sub-system are difficult to scale to much larger numbers of cores, especially in terms of performance scalability. We later discuss performance and physical and performance scalability in greater detail.

While the above discussion assumes a multicore organization, some speculatively multithreaded architectures such as dynamic multithreading (DMT) [1] and implicitly multithreaded architecture [22] run the tasks on a centralized simultaneously multithreaded processor (SMT) [28]

4.3 High-Level Execution Model

With the above big picture in mind, we now describe the execution model. Speculatively multithreaded architectures employ the compiler or hardware to partition a sequential program into speculative tasks. Though the tasks are *not* necessarily data- or control-flow independent of each other, the hardware speculates that the tasks are independent and executes them in parallel. Tasks are spawned via prediction, from one task to the next, and the tasks proceed in parallel assuming there is no data dependence among the tasks. Each task is mapped to a (virtual or physical) core. Figure 4.3 illustrates the execution model using a simple example. The program state of the tasks in registers and memory is committed in *sequential program order* after confirming the speculation to be correct (i.e., no violations of control-flow or data dependencies). Instructions *within* a task commit in program order *locally* within the core, just as they would in a uniprocessor, and tasks themselves commit in overall program order *globally* across the cores, giving rise to a two-level hierarchical commit.

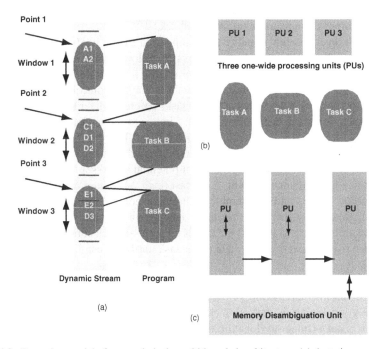

Fig. 4.3 Execution model of a speculatively multithreaded architecture. (**a**) A static program partitioned into three tasks which correspond to the three windows. (**b**) Tasks assigned to processing units (cores). (**c**) Hardware organization showing the processing units connected for communicating inter-task data dependence values. Memory disambiguation unit is for handling memory dependencies (figure taken from [29] with permission)

Inter-task control-flow dependencies arise when branch decisions inside a task determine which task should be executed next, and inter-task data dependencies arise when a register or memory value produced by one task is consumed by another task. Similar to conventional processors, only true data dependencies across tasks matter as all false dependencies can be removed by renaming, as we explain later.

If the compiler or the hardware detects control-flow or data dependence across tasks, then there are hardware and compiler mechanisms to enforce the correct order among the dependent instructions within the tasks. Inter-task register dependencies are detected and synchronized by the compiler and memory dependencies can be detected and synchronized by the hardware, as an optional optimization. However, if execution violates a control-flow or data dependence (e.g., either due to misprediction in task spawning or due to a consumer executing before its producer and getting incorrect data), hardware detects such violations and rolls back the offending task and all tasks later in program order. While tasks earlier than the offending task in program order are not affected by the incorrectness of the offending task, later tasks may have consumed incorrect data from the offending task. A simple scheme is to roll back all later tasks irrespective of whether a later task is incorrect because the task either has consumed incorrect data or is simply incorrect due to control-flow misprediction but a more efficient scheme (perhaps requiring more hardware support) could determine which later tasks, or even instructions within a task, are unaffected and save them from rollback.

At a high level, the functionality to be supported by the hardware and software to implement the above execution model is as follows: First, the system should provide a scheme to partition sequential programs into suitable speculative tasks to be executed in parallel. Second, because there may be control-flow dependence among tasks, we need a way to launch later tasks before the later task's control-flow dependencies on earlier tasks are resolved. Similar to uniprocessors using branch prediction to fetch future instructions before control-flow dependencies on earlier instructions are resolved, speculatively multithreaded architectures use inter-task prediction (i.e., predict one among the set of potential successor tasks); because the predictor can be implemented to be fairly fast (e.g., 2–3 cycles), a single predictor would suffice for the entire ensemble of multiple cores. Figure 4.4 shows inter-task prediction. Third, because the state written by a speculative task should not update the architectural state until the task is committed, each core should buffer the speculative register and memory state of the task until they can be committed. This buffering

- provides automatic renaming of the register and memory locations that are speculatively written by the task so that multiple concurrent tasks may write to the same register or memory location and require the multiple versions to be held separately,
- attempts to direct each register read or memory load to the correct version of register or memory location from among the multiple speculative versions,

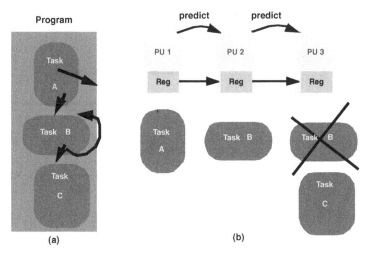

Fig. 4.4 Inter-task prediction in a speculatively multithreaded architecture. (**a**) Control flow of a program partitioned into three tasks. Note that Task B is a loop body where each iteration is an instance of the task. (**b**) Task prediction: One instance of Task B is predicted and assigned to PU 2 and another instance to PU 3. The second prediction is incorrect and rolled back and the correct task (Task C) is assigned on PU 3

- detects violation of data dependence when an incorrect version is read (e.g., because the correct version has not been produced by the time the read occurs, as illustrated in Fig. 4.5), and
- allows easy rollback by discarding the buffer contents should the task be rolled back.

As part of the third requirement above, the system should provide a mechanism to enforce true register and memory data dependencies between tasks when those dependencies are known to the system before execution. Because register dependencies are known at compile time, the compiler provides enough synchronization information so that hardware can enforce the dependencies. Figure 4.6 illustrates this idea through an example. For memory dependencies not known to the compiler, we will describe a hardware mechanism that learns memory dependencies hinted by tracking repeated rollbacks and synchronizes the dependencies in hardware.

A key common requirement to support the above functionality is tracking of program order among the tasks. This ordering is critical in determining inter-task control-flow mispredictions. When one task completes, determining whether the next task is predicted correctly can be done only if the identity of the next task in program order is known. Similarly, this ordering is critical in enforcing true register dependencies and detecting violations of true memory dependencies. Program order among the tasks is tracked by placing the cores in a (logical or physical) ring and assigning successive threads in program order to successive cores in ring order.

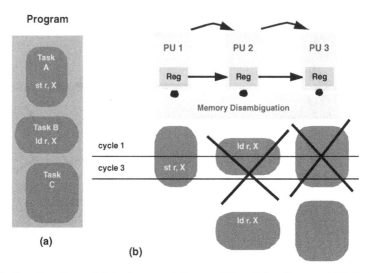

Fig. 4.5 Data dependence violation in a speculatively multithreaded architecture. (**a**) Task A stores into address X from which Task B should load as per sequential semantics. (**b**) During execution, however, Task B's load is executed ahead of Task A's store, resulting in a dependence violation which is detected by the memory disambiguation hardware. Task B is rolled back and restarted

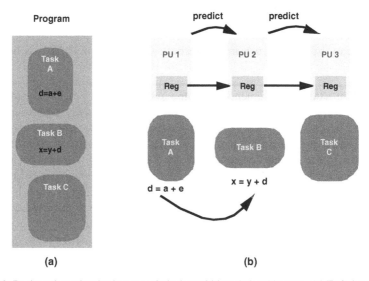

Fig. 4.6 Register dependencies in a speculatively multithreaded architecture. (**a**) Task A produces the register d which is consumed by Task B. The compiler detects this dependence and inserts some synchronization information. (**b**) During execution, the consumer instruction waits for the producer instruction

Because the program order is maintained by the hardware, supporting precise interrupts falls out naturally. If a speculative task's instruction encounters an exception, execution is frozen until all previous tasks have committed (i.e., the current task reaches the head of the program-order ring) and all instructions prior to the exception instruction in the task have committed. In conventional and speculatively multithreaded processors, interrupts trigger an immediate insertion of an exception into the instruction stream to reduce interrupt latency, and thus interrupts are handled similar to exceptions. Thus, exceptions and interrupts can be handled in a precise manner.

4.3.1 Organization and Hardware–Software Split

Each core has its own register file and depending on the specific implementation, each core may have its own private cache or there may be a single cache shared among the cores. Though the latter would not scale to a large number of cores, the latter is simpler which may make it a better choice for a small number of cores. The task predictor predicts and assigns tasks to cores. Figure 4.3(c) shows the organization of the hardware and the ring order among the cores.

To implement the execution model, the responsibility for control-flow, register values, and memory values is split between the hardware and the compiler. It is important to understand this split before going into implementation details. One caveat is that while we describe one such split, some shift of the responsibilities between the compiler and hardware is possible and has been explored. For example, while our description uses the compiler for partitioning a sequential program into tasks, dynamic multithreading (DMT) [1] uses hardware for task partitioning (e.g., spawn a new task after a function call or a loop).

For inter-task control flow, each task needs to predict the next task to be executed. While conventional branch prediction has only two options to choose between, tasks may potentially lead to more than two successor tasks and thus task prediction typically has more than two options. Similar to BTBs which learn branch targets, task prediction needs to learn the possible successor tasks. Two choices to learn the successors are as follows:

1. the compiler explicitly embeds the list of the successors for every task (up to some maximum) and the hardware prediction simply chooses one and
2. a BTB-like hardware table simply learns the successors from past instances of a task such that the table can be restricted to some maximum.

We describe the former while the latter is a simple variant. Figure 4.7 shows a hardware implementation. The details of the figure are explained in the rest of this section.

Though the tasks share a single logical register space, the physical register files are private to each core. The private register files provide register renaming *across*

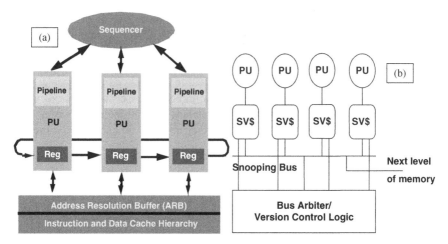

Fig. 4.7 Hardware implementation of a speculatively multithreaded architecture. (**a**) The sequencer performs inter-task prediction, and each processing unit (core) has its own pipeline and register file. Memory accesses go to the address resolution buffer (ARB) for memory disambiguation. The ARB is backed up by a conventional memory hierarchy. (**b**) A distributed alternative to the centralized ARB called the speculative versioning cache (Fig. 4.7(b) taken from [11] with permission)

tasks, while renaming *within* tasks is done by each core's conventional register renaming. This across-task renaming allows each core's register file to buffer the tasks' speculative register values. The values are used to update architectural state upon commit and discarded upon rollback. To honor inter-task register dependencies the compiler annotates the producer task's instructions to send their register values to the next task in program order over a special point-to-point ring dedicated for register values (other topologies for the interconnect would work by super-imposing the cores' logical ring order among the interconnects). If the next task does not re-define the register then the task propagates the register value to its successor task; otherwise, propagation of the previous value is stopped and the new value is sent.

Two key optimizations reduce the bandwidth demand on the register ring. First, register values that have gone around the entire ring and have updated all the register files are stopped from unnecessarily continuing further. Second, register values that are dead beyond a task, as determined by the compiler, are stopped from being propagated. Because register values are not long-lived, this optimization significantly reduces bandwidth demand.

Because memory dependencies are harder for the compiler to determine statically, especially for irregular programs with dynamic data structures, memory values are handled entirely in hardware, using a shared cache-like structure called the address resolution buffer (ARB) [10] or a distributed coherence-like scheme called the speculative versioning cache (SVC) [11] (another design similar to the SVC is proposed in [12]). The ARB and SVC provide per-core buffering so that store values from multiple tasks for a single memory location can be held separately (i.e., memory renaming). The buffered values update the architectural memory state upon task commit or are discarded upon rollback. To provide the correct value to loads that occur out of program order with respect to stores from different tasks, the ARB and SVC ensure that a load from an address receives its value from the closest preceding store to the same address across tasks in program order among the stores that have occurred before the load. However, if a consumer load from one task executes before the producer store in another task, then the load would receive an incorrect value. The ARB and SVC detect this incorrect ordering when the producer store executes, and trigger a rollback of the task containing the incorrect consumer load. While the above discussion is for memory dependencies *across* tasks, dependencies *within* a task are handled by conventional load–store queues present in the core. Other works [7, 27] have extended the SVC idea to directory-based schemes to scale the SVC beyond snoopy bus-based coherence. These extensions will be helpful when the number of cores in a CMP exceeds what can be supported by a snoopy bus. While these previous designs did not allow speculative blocks to be replaced from caches, an impediment to long threads, a recent design called Bulk [5] uses Bloom filters to decouple memory disambiguation from the caches allowing speculative blocks to be replaced.

Repeated rollbacks due to violation of memory dependencies cause significant performance degradation. To address this problem the key idea is to leverage the recurrent nature of the misspeculations and learn the identities of the offending instructions from past instances of rollbacks and synchronize them in the future (i.e., force the consumer load to wait for the producer store).

4.4 Implementation

The important components of the implementation follow from the high-level execution model and organization and include task selection, control flow, register file, and memory disambiguation. To understand the design decisions involved in the implementation, we first discuss the key factors affecting performance in multiscalar architectures.

4.4.1 Performance Factors

Because the choice of tasks fundamentally and drastically affects the amount of parallelism exploited, task selection is central to achieving performance in speculatively multithreaded architectures. Task partitioning must address the following key performance overheads in such processors:

- inter-task control dependencies,
- inter-task data dependencies (register and memory),
- load imbalance, and
- serialization due to large tasks.

Because of the complex nature of this goal, task selection is perhaps best done by the compiler though some proposals advocate using the hardware (as mentioned before), so that dusty-deck code can be run on such a processor.

Rollbacks due to task misprediction are the main way in which control flow causes performance overhead. Mispredictions occur either because of inherent unpredictability in control flow (e.g., data-dependent control flow) or because of tasks having too many successor tasks. The amount of loss depends on how late the misprediction is detected. Because task prediction is verified at the end of each task, longer tasks result in later detection of misprediction and larger performance overhead. However, shorter tasks are not desirable because they reduce the amount of parallelism.

Inter-task register dependencies are detected and enforced by the compiler by synchronizing producers and consumers. The synchronization causes consumers to wait for producers to send their register values resulting in performance overhead. The amount of waiting depends on the relative position or execution timing of the dependent instructions in their respective tasks. Producers positioned late in their tasks with consumers positioned early in their tasks result in long wait. In contrast, if execution reaches a producer and consumer in their respective tasks at about the same time then the amount of waiting is less. Further, if the producer is early and consumer is late, then the dependence does not cause any waiting at all. Compared to register dependencies, memory dependence may cause waiting or rollbacks depending upon whether the dependent instructions are synchronized by dependence prediction and synchronization hardware or not. Note that any compiler-enforced synchronization would also result in waiting. As with register dependencies, the relative positions of the producer instructions (stores) and consumer instructions (loads) determine the amount of waiting or the amount of computation that is rolled back, which is a measure of the amount of time lost due to rollback.

Because tasks commit in program order, tasks following much longer tasks may complete execution before the preceding task but have to wait till the preceding task commits. Such waiting is a form of load imbalance, as illustrated in Fig. 4.8. Reducing load imbalance does not require that all tasks be of similar size, but that

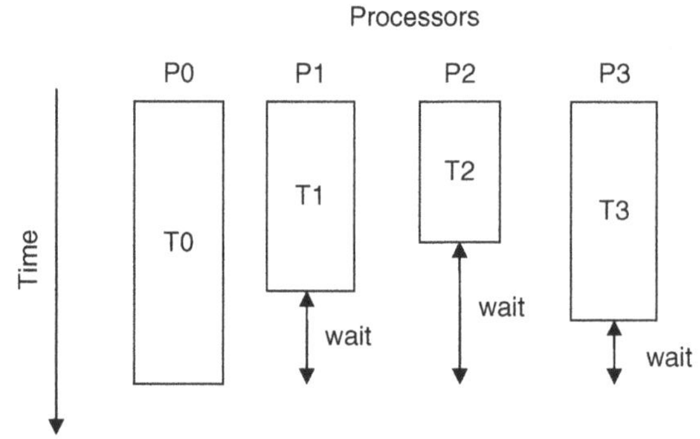

Fig. 4.8 Load imbalance. Tasks T1, T2, and T3 have to wait for Task T0 to finish and commit before they can commit and free the processing units for other tasks

adjacent tasks do not vary significantly in size. Tasks may vary in size gradually such that tasks far apart in program order can vary in size without incurring much load imbalance.

Finally, large tasks may lead to serialization and thus reduce the exploited parallelism. They may also run out of space to buffer speculative values and have to stall until such space is available.

4.4.2 Task Selection

While task selection can be done to address in isolation each of the performance overheads listed in Sect. 4.4.1, the key challenge is the fact that the above overheads do not occur in isolation but rather interact with each other in complex and often opposing ways. For instance, while a particular choice of tasks may avoid inter-task data dependencies, the tasks may have severe inter-task control-flow dependencies. Similarly, one choice of tasks may avoid inter-task control-flow dependencies but may incur load imbalance. We first discuss strategies to address each of the overheads in isolation and then describe how to combine the strategies into one task selection scheme.

Control-flow mispredictions can be reduced by exploiting control-flow independence and easy-to-predict branches. Task selection can exploit control independence by selecting task boundaries at the control-independent points in the control-flow graph. Such selection amounts to including *within* the task the entire *if-else* paths until they re-converge (i.e., avoid task boundaries that cut a control-dependent path before reaching a control-independent point). Figure 4.9 illustrates this point. In addition to exploiting control-flow independence, task boundaries can be placed at easy-to-predict branches, such as loop branches (one could think of control-independent points as the trivially easy-to-predict task boundaries). Task boundaries at loop branches result in the loop body being selected as a task and any branches within a loop body do not affect the next-task prediction.

Performance overhead due to data dependence can be reduced by avoiding task boundaries that cut data dependencies. Consequently, including both the producer and consumers of a dependence *within* the same task avoids this problem, similar to control-flow dependence. A simple case is a loop-body task with independent loop iterations where each iteration includes data dependencies, as shown in Fig. 4.10(a–d). One difficulty is that, unlike register dependencies which are known at compile time, memory dependencies may not be known due to ambiguities caused by pointers and control flow. This is a classic problem that automatic parallelization has faced for a long time. While better pointer analysis mitigates this problem, there is one key difference between automatic parallelization and our task selection problem. In automatic parallelization, the analysis cannot incur false negatives (i.e., stipulate that a dependence does not exist when in reality it does, which would result in incorrect execution). False positives, which assert that dependence exists when it does not, may result in loss of parallelization opportunities but do not affect

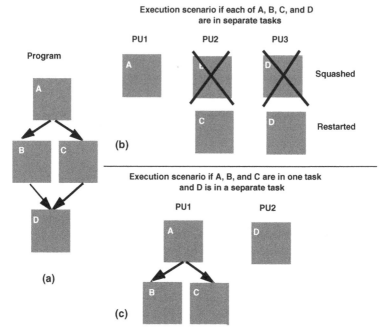

Fig. 4.9 Control-flow considerations in task selection. (**a**) Control flow in a program where A, B, C, and D are some code blocks. (**b**) One task selection option where A, B, C, and D are separate tasks which incur control-flow misprediction. (**c**) A better task selection option where A, B, and C are one task and D is another task. Because D is control independent of A, B, and C, this selection does not incur any misprediction

correctness. Thus, automatic parallelization requires conservative analysis which must guarantee that all dependencies are identified. In contrast, in a speculatively multithreaded architecture, false negatives would be caught by the hardware; they may result in performance loss but not incorrect execution. Consequently, our task selection need not be conservative and need not identify all dependencies. Our strategy uses profiling to identify frequently occurring memory dependencies to ensure that such dependencies are not cut by task boundaries.

For cases where it is hard to include all data dependencies within tasks, one idea is to schedule the producer early and the consumer late in their respective tasks to reduce the waiting or the potential rollback penalty. This idea is shown in Fig. 4.10(e). This scheduling involves moving producer instructions up and consumer instructions down their respective tasks. The code motion part is similar to code motion in compiler instruction scheduling to hide memory and operational latencies in uniprocessor pipelines. We describe the special and important case of loop induction variables later.

Reducing load imbalance amounts to estimating execution times and selecting tasks such that adjacent tasks do not vary significantly in size. Including several nested function calls within a task can make the task large and cause load imbalance

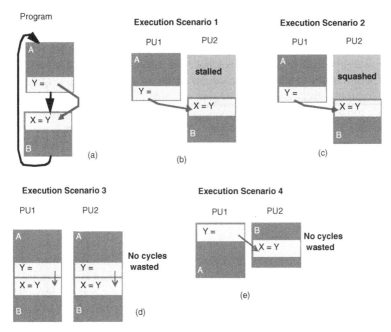

Fig. 4.10 Data dependence considerations in task selection. (**a**) A loop-body task with two code blocks A and B where each iteration includes the data dependence between X and Y. A task selection option where A and B are separate tasks. This selection incurs either (**b**) waiting due to synchronization or (**c**) rollbacks due to dependence violation. (**d**) A better task selection where A and B (the entire loop body) is one task so that the dependence is contained within the task. (**e**) A and B are separate tasks where Y is scheduled early in its task and X is scheduled late in its task to reduce waiting or rollback

and serialization problems. However, including a function call within a task would be a good choice if the call is present in a loop and all loop iterations make the call. In this case the load is balanced and there is likely to be abundant parallelism.

As previously mentioned, the key issue with task selection is combining strategies which target different overheads of task selection. While the general problem of optimum decomposition of a program into tasks is NP-complete [24], one approach to combine the individual strategies is by prioritizing the overheads in some order so that the strategy for the higher-priority overhead is preferred over that for a lower-priority overhead (e.g., [29] gives control-flow strategies the highest priority). However, such prioritization is static and does not compare the performance resulting from the individual strategies for a given segment of code to pick the best strategy for the code segment. To that end, another paper [16] uses min-cut max-flow as a heuristic to give equal consideration to all the overheads. The heuristic starts with the program's control-flow graph, where the nodes are basic blocks and the edges are control-flow paths, annotated with data-dependence edges. The idea is to assign weight to a control-flow or to a data-dependence edge based on the overhead incurred if the edge were to be cut. That is, control-flow edges weights correspond to misprediction penalty and data-dependence edge weights correspond to either wait time for register dependencies or misspeculation penalty for memory dependencies. While the overheads due to data dependence and thread prediction are represented as edge weights, load imbalance does not lend itself to being represented as edge weight. Consequently, the heuristic adapts balanced min-cut [30], originally proposed for circuit placement and routing, to achieve *balance* by modifying a min-cut to reduce the difference between the estimated runtimes of the tasks resulting from the min-cut. Thus, the heuristic accounts for all the overheads in estimating the overall execution time and trade-offs one overhead for another to pick those balanced cuts that minimize the execution time. The details of the heuristic can be found in [16].

4.4.3 Control Flow

Speculatively multithreaded execution proceeds by predicting the next task and assigning it to the next processing core. Next-task prediction is similar to branch prediction except that instead of predicting the outcome of a branch, next-task prediction predicts the successor of the current task. Thus, the granularity of prediction is larger and also the number of choices is larger.

Similar to branch prediction [25, 31] next-task prediction needs to *choose* one of the successors to be spawned and *provide* the successor's starting address. While task selection helps this prediction by including hard-to-predict branches within tasks, we need a mechanism to provide good prediction accuracy. Similar to branch prediction, the next-task prediction can also benefit from two-level schemes where the first level encodes the preceding tasks in a shift register and the second level provides a prediction table of saturating counters, one per potential successor up to some number of successors (e.g., 4). Thus, the first level provides a history pattern and the second level provides the most likely successor corresponding to the counter that is most saturated. When the successor is known upon task completion, the corresponding counter is incremented and the others are decremented. Like branch prediction, there are many design variations possible: per-task or global history and per-task or global prediction table. A detailed analysis of these variations can be found in [4].

The starting address of the predicted successor can be provided in one of two ways:

1. using a hardware table, like the branch target buffer, where the starting addresses of the successors of a task can be learnt from past instances of the task, or
2. the compiler can provide a list of a task's successors' start addresses as part of the task code, as illustrated in Fig. 4.11.

Because the start addresses are used in prediction, the addresses need not be always correct and any incorrect addresses are corrected when the prediction is verified.

Fig. 4.11 Control-flow information. Task 0's successors include Tasks 1, 2, and 3. The list of starting addresses of the successors is provided above Task 0's code. The stop bits are shown at the appropriate instructions

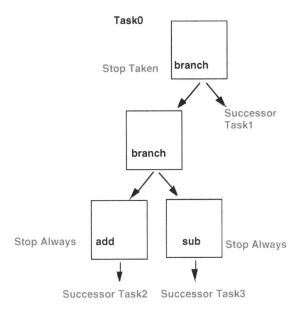

Thus, the former scheme's learning may be inaccurate and the latter scheme need not provide all the successors' addresses if some are not known statically, such as targets of indirect calls or indirect jumps. Both schemes take up some storage, either in the form of a table or in the instruction cache as part of the task code. The advantage of the former is that statically unknown successors can be handled seamlessly with the rest of the statically known successors, whereas the unknown case is problematic for the latter and may require profiling. The advantage of the latter is somewhat simpler hardware.

While the list of successors can be built in hardware or software, the hardware needs to know when a task has ended. To that end, the compiler annotates the instructions with extra bits called *stop bits*. Because a task may end on the taken path of a branch but not on the not-taken path (e.g., the first branch in Fig. 4.11) or vice versa, or may end on both paths, the stop bits include the following flavors: stop always, stop taken, or stop not-taken. The stop bits are illustrated in Fig. 4.11.

4.4.4 Register Values

Recall from Sect. 4.3 that while the tasks share a global register space, the cores provide private register files for each task. Consequently, the register values consumed by the instructions of a task can come from other instructions in the task (*intra-task* case) or from previous in-flight tasks which send their values to later tasks via the register ring described earlier. The inter-task dependence in turn gives rise to two more cases in addition to the intra-task case: either the value has not been sent by the time the consumer instruction undergoes its conventional register renaming (*unarrived inter-task* case) or the value has already been received (either from an in-flight predecessor task or from a previously committed task) and placed in the local register file by that time (*arrived inter-task* case). Recall from Sect. 4.3.1 that a register value received from a previous task is propagated to the next task if the receiving task does not produce a new value for the register; otherwise, the new value is sent to the next task.

The following list specifies whether and what the consumer should wait for in the above three cases:

- In the intra-task case, a given instruction should wait for its register operand to come from within the core as controlled by the core's conventional register renaming.
- In the unarrived case, a given instruction should wait for the operand to come from the previous task.
- In the arrived case, the instruction should obtain the operand from the local register file.

The key issue is how does the core distinguish among these cases to obtain the correct register operand? The intra-task case can be identified locally at runtime by conventional register renaming hardware within a processing core which identifies instructions that are dependent on earlier instructions within the task. Similarly, any register that arrives from an in-flight predecessor task can be identified as such. However, the unarrived inter-task case and that arrived from a previously committed task case (well before the current task starts) cannot be distinguished *locally* because there is no way to know whether a register value will arrive in the future and therefore the consumer instruction should wait, or whether it has already arrived and therefore the consumer instruction need not wait. We need information from the currently in-flight previous tasks identifying the set of registers produced by those tasks. To this end, the compiler determines the set of registers produced by a task and provides the set to the hardware as a bit mask, called the *create mask*. The mask has a "1" for each register that is written in the task and "0" otherwise. One complication is ambiguity due to control flow within a task where some registers may be written only in some of the control-flow paths but not all. For such cases, we conservatively include all the registers that are written in *some* control-flow path within the task. Figure 4.12 shows an example of the create mask. Upon start, each task receives from its in-flight predecessor task an aggregate create mask, which is a logical OR of all the in-flight preceding tasks' create masks, adds its own create mask to the aggregate, and sends the new aggregate to the successor task. Consumer instructions in each task should wait for the registers in this aggregate mask that have not arrived by the time the instruction is renamed. Such registers correspond to the unarrived inter-task case and are distinguished from the register already arrived from a previously committed task by the fact that the latter is not present in the

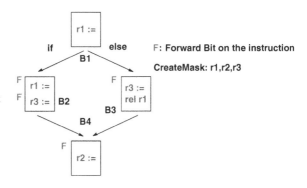

Fig. 4.12 Register information for an example task. This task includes four basic blocks B1–B4 and writes to registers r1, r2, and r3 which are in the create mask. The forward bits are set on the last write of a register (indicated by "F"). Because r1 is not written on the *else* path, there is a release for r1 (indicated as "rel r1")

aggregate mask (the issue raised in the previous paragraph). In addition, tasks use their create masks to decide whether to propagate a register value received from a previous task (if the register is absent in the mask) or not (if the register is present in the mask). Because the create mask is conservative, it is possible that a register in a task's create mask is not written by the task. If care is not taken, a side-effect of the combination of this possibility and the propagation policy would be that a task would not send the register value to later tasks, resulting in their being blocked indefinitely. Avoiding this problem is related to when register values are sent to the next task which we describe next.

In addition to the set of registers produced by a task, we also need to know at what point in a task it is safe to send a register value to successor tasks. In general, the point is the last write of the register within the task because the last write value, and not earlier write values, should be consumed by later tasks. The compiler annotates the last write with a "forward" bit on the instruction so that the value produced by the instruction is sent to later tasks. However, identifying this point is non-trivial in the cases where a task writes to a register multiple times in the presence of control flow within the task. For instance, it is possible that there is a register write within a task and a later write to the same register in the *if* path of an *if-else* control-flow structure but not the *else* path (e.g., r1 in Fig. 4.12). Then, the earlier write is the last write in the *else* path but the later write is the last write in the *if* path. Consequently, setting the forward bit of the earlier write would result in an incorrect send for the *if* path but a correct send for the *else* path. To address this problem, we set the forward bit of the later write in the *if path* and introduce a "release" instruction in the beginning of the *else* path to send the register value if execution enters the else path (e.g., rel r1 in Fig. 4.12). Thus, register values whose propagation from a previous task to the next task is stopped by a task which does not write to the registers due to control flow are explicitly released by the task.

As part of commit, the task ensures that all the register values that need to be sent to the successor task have been sent before the task is committed. Upon a rollback, because *all* the tasks from the offending task onwards are rolled back, there is no need to restore registers that were written incorrectly. After a roll-back, all incorrect values will be overwritten by correct values. To ensure that the consumers will be forced to wait for the producers, the aggregate mask from the last task before the rolled back task is re-sent to the new task launched after the rollback.

It is possible that a task may exit on one of the two paths of a branch but not on the other. In such a case, the continuing path may write to many registers that are not written in the exiting path. Consequently, the exiting path will need to release those registers. Because these releases occur at an exit of a task, they are called *end-releases*. There is an option of not explicitly specifying end-releases and have the hardware implicitly end-release those registers that are on the create mask but have not yet been forwarded or released. This option saves on the code space that the explicit end-releases would have taken.

The create masks and forward and release information can be learnt by the hardware instead of being provided by the compiler. While the advantage of such hardware is the ability to run dusty-deck code that cannot be recompiled, such hardware incur some complexity especially for the forward and release instructions which require some dataflow analysis.

Recall from Sect. 4.3.1 the register ring bandwidth optimization based on the observation that register values that are dead need not be propagated further. A register is considered dead at the end of a task only if it is dead at all exits. The compiler provides register death information for each task. One option to provide this information is to overload the create mask by adding the dead registers to it without any corresponding forwards or releases, as the registers are dead and therefore need not be sent to successor tasks. This overloading does not incur any additional code space overhead. However, this option clashes with implicit end-releases where the hardware would mistake the dead registers for those that need to be end released and it would implicitly end release the dead registers, undoing the bandwidth optimization. Instead, the compiler provides register death information as an optional bit mask called the *dead mask*, in addition to the create mask. The dead mask has a "1" for each register that is dead at the exit of the task and "0" otherwise. If a dead register is on the create mask, meaning the register is written in the task and the value is dead at the task's exit, then the register will have a forward or an explicit or implicit release. The hardware does not send the dead register values to successor tasks even if the registers have forwards or explicit or implicit releases. For instance in Fig. 4.12, assume that r1 is dead beyond the task and is in the dead mask (not shown). The hardware would ignore the forward of r1 in the left path; and the compiler would not insert the release of r1 in the right path nor would the hardware implicitly end-release r1. While the dead mask imposes some code space overhead, the dead mask allows implicit end-releases which saves the space overhead of explicit releases. Overall, dead mask overhead is less than that of explicit releases.

The successor start addresses, create and dead masks are bundled into a *task descriptor* and placed above the task code.

Because loops are a key source of parallelism and task selection often selects loop iterations as tasks, a key performance optimization involving register values involves loop induction variables. In such tasks, even if the loop iterations are independent the loop induction variables introduce a loop-carried dependence. Modern optimizing compilers place the increment at the bottom of the loop and not on the top. Consequently, the increment occurs at the *end* of the previous iteration and is needed usually near the *beginning* of the next iteration. This dependence timing results in completely serializing the iterations though they are independent. A simple optimization to avoid this problem is to move the loop induction increments to the top of the loop and adjust the offsets of the memory accesses that use the induction variable for address calculation. In addition to the loop induction increments, the loop branch may need to be moved up requiring adding an extra jump at the bottom of the loop. Because there is no advantage for conventional processors to move induction increments to loop tops and because the extra jump is an overhead, compilers for conventional processors do not perform this code motion. However, for speculatively multithreaded architectures the benefits of removing the serialization far exceeds the cost of the extra jump.

4.4.5 Memory Values

Tasks share a single memory address space similar to the shared register space, and there can be intra-task or inter-task memory dependence. One key difference from register values is that memory dependencies are harder for the compiler to detect. Accordingly, the compiler does not do anything special about memory dependencies; the hardware carries the full burden for ensuring that they are respected. Since memory dependencies are much less frequent than register dependencies, the memory system is designed assuming that dependencies are the uncommon case and dependencies that exist can be learnt by the hardware and synchronized on the fly if necessary.

Intra-task memory dependence is handled by the processing core's conventional load–store queues. Inter-task dependence needs special support to be built into the memory system. As mentioned in Sect. 4.4, two choices for designing this support are the banked cache-like ARB and the SVC, which resembles a snoopy-based cache coherence scheme.

Recall from Sect. 4.3 that the following functionality is required, irrespective of ARB or SVC: First, because the tasks are speculative, store values are buffered either in the ARB or in the SVC and committed to the architectural memory state in program order. Specifically, stores from multiple speculative tasks to the same address which create speculative versions should be buffered separately to provide versioning via memory renaming and should be committed to the architectural memory state in program order. It may seem that versioning may be unnecessary because stores to the same address from multiple tasks would be uncommon. However, for loop-iteration tasks containing function calls, the stack frames of the callees of the different iterations are located at the same addresses, and stores in the stack frames need to be renamed for the loop iterations to execute in parallel. Second, because tasks execute asynchronously with respect to each other, dependent loads and stores may execute in any order, some of which result in incorrect execution. Handling inter-task memory dependence involves two cases: (a) ensuring that execution obtains the correct values when the dependent instructions happen to execute in the correct order and (b) detecting the cases when the dependent instructions happen to execute in an incorrect order and allowing for safe rollback after misspeculations. Case (a) implies that if there are uncommitted stores to an address in multiple tasks, a load from the same address should obtain the value of the closest preceding store to the same address in some task preceding the load's task. Case (b) implies that each store to an address should check if a later load to the same address executed prematurely.

We discuss the ARB first because it is simpler than the SVC.

4.4.5.1 Address Resolution Buffer (ARB)

The ARB is a shared cache-like structure that serves memory accesses from all the cores. The ARB can be thought of as a replacement for the conventional L1 cache and is backed up by a regular L2 cache. The ARB is indexed like a conventional direct-mapped or set-associative cache where each associative way of the set holds as many blocks as there are cores instead of one block and each set holds as many blocks as the associativity times the number of cores. By providing space for multiple blocks per set, the ARB ensures that even if a location is written by *all* the in-flight tasks there is enough renaming space, satisfying the first requirement above for speculative versioning. Within each way of a set, the blocks are ordered based on the program order among the corresponding tasks (or cores). When a task commits, its dirty blocks in the ARB are written back; and upon a rollback, the task's blocks are simply invalidated. To achieve high bandwidth, the ARB can be banked similar to caches. Figure 4.13 shows the ARB's structure.

For the second requirement above of inter-task dependence, each load searches among blocks in its way that are older than the load for matching stores that have already occurred to the same address. Loads and stores are distinguished via Load and Store bits per block, as shown in Fig. 4.13. Upon one or more matches, the load obtains the value of the older store closest to the load. This search allows loads to get values from address-matching uncommitted stores, satisfying case (a) above. Each store searches among blocks in its way that are younger than the store for matching loads that have already executed. Any matching load is a prematurely executed load which should have obtained its value from the current store. Such premature loads cause violation of inter-task memory dependence and trigger a rollback of the premature load's task, satisfying case (b) above.

One issue with the ARB is that commits of large tasks with many dirty blocks hold up the core by waiting for all the dirty blocks to be written back. This wait may delay the launch of a later task on the same core. Another issue is that there can be no replacement of blocks that belong to an in-flight task. Replacing the block implies that we would lose the information that the task has made an access to it in the past, and thus the ability to detect memory dependence violations at

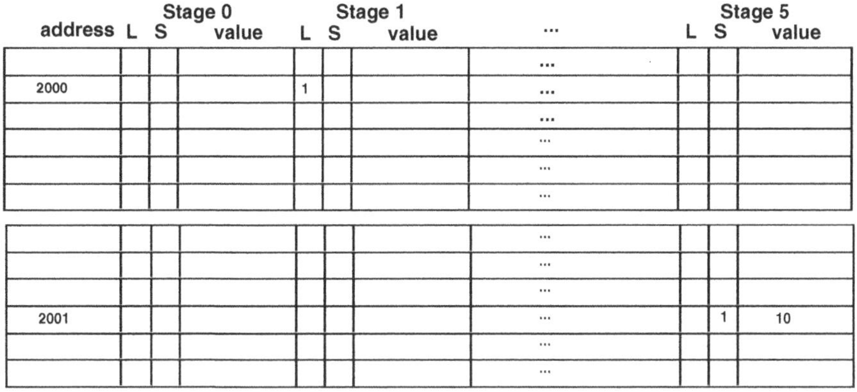

address	Stage 0			Stage 1			···	Stage 5		
	L	S	value	L	S	value		L	S	value
							···			
2000				1			···			
							···			
							···			
							···			
							···			
							···			
							···			
							···			
2001							···	1		10
							···			
							···			

Fig. 4.13 Address resolution buffer. This example has two banks and uses a direct-mapped structure where each set has space for six tasks (stages 0–5). The active ARB window includes four stages for four current tasks while the other stages are for previously committed tasks whose data are being written back and for to-be-assigned tasks. The extra stages serve the purpose of extra buffering in the ARB for committed data to be written back. Each stage holds a load and store bit (denoted by L and S) for a given address to track if the core performed a load or a store to the address. The figure shows a load from the address 2000 in the second task (stage 1) and a store to the address 2001 in the sixth task (stage 5)

a later time. Therefore, if a replacement is called for, the task simply suspends execution until all previous tasks have committed so that the task is no longer speculative, resolves all intra-task speculations, writes back its blocks from the ARB, and enters a mode where accesses bypass the ARB and directly go to the next-level cache.

If inter-task memory dependence is tracked at the ARB block granularity and not at the word granularity (i.e., read and write bits are per block and not per word), then there could be many false rollbacks similar to false sharing in conventional multiprocessors because the ARB cannot determine that a load and a store from different tasks accessed different words in the same block. To reduce such false rollbacks, the ARB could employ read/write bits per word of the block. These bits provide finer resolution of read/write information allowing the ARB to avoid false rollbacks. At the extreme, the read/write bits could be maintained per byte, but doing so would significantly increase the space overhead of the bits. Thus, there is a trade-off between frequency of false rollbacks and space overhead. This trade-off is quantified in detail in [10].

4.4.5.2 Speculative Versioning Cache (SVC)

SVC extends a snoopy-based coherent cache where each core has its own private L1 cache. SVC extends the standard coherence protocol to enable speculative multithreading. Similar to conventional cache coherence, the high-level goal is to allow cache hits to proceed quickly without requiring checks involving the caches of other cores, while cache misses can snoop into the other caches to determine the correct action. The idea is that because coherence controls access to cache blocks on an individual access basis, we can extend coherence to detect when loads or stores in one thread access the same address as stores in another thread, potentially causing a memory dependence violation. At a high level, the SVC uses Load and Store bits per block (similar to the ARB) which are set by loads and stores so that future accesses can check for dependence violation. We describe the details of these bits later. Figure 4.14 shows the high-level structure of the SVC.

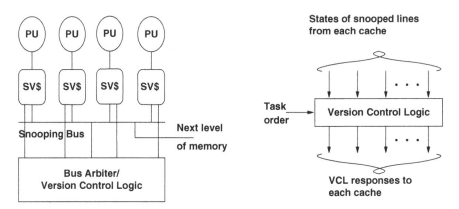

Fig. 4.14 Speculative versioning cache. The private caches provide versioning by leveraging snoopy coherence. The version control logic manages the versions of a cache block (figure taken from [11] with permission)

The private caches allow separate buffering of each task's memory state allowing speculative versioning, which is the first requirement in Sect. 4.4.4. However, unlike the ARB which writes back all the blocks upon commit, the SVC allows *in-cache* commits of speculative data simply by changing the state of the blocks as committed via a "flash" operation in the cache which can be made fast in CMOS. This commit strategy allows fast commits by setting the Commit bits of the blocks whose Load and/or Store bits are set, without requiring any data movement, and eliminates ARB's write-back bottleneck at commit. Because commits are more common than rollbacks, the SVC eliminates this key bottleneck.

There are two key implications of SVC's commit strategy. First, multiple uncommitted or committed versions of a block may exist in the caches where uncommitted versions occur as accesses happen and committed versions occur as tasks commit. While it is clear that the committed versions are before the uncommitted versions in program order, we need to maintain program order among the committed versions and among the uncommitted versions. The ordering among the committed versions allows SVC to identify the last version to supply to later loads. Similarly the ordering among the uncommitted versions allows SVC to identify the correct version to supply to a given load from the closest preceding store, similar to the ARB, as we describe below. The uncommitted versions can be ordered easily by a snoop based on the program order among the corresponding tasks. However, for the tasks corresponding to the committed versions, there is no information available on the relative order among the committed tasks. To address this issue, SVC places the versions in a list ordered by the program order among the versions' tasks at the time of the store that creates each of the versions. This list is called the *version ordering list (VOL)*. More than one copy of a version may exist in the VOL due to later loads making copies of an earlier version (the copies would be contiguous in the ordering). There is no extra snoop overhead for insertion into the VOL which is managed by the *version control logic (VCL)* as shown in Fig. 4.14. Because a store to an invalid or shared block performs a snoop in conventional coherence, the snoop allows the newly created version to be inserted in the VOL. A store to a modified or exclusive block does not perform a snoop, but then only one version of the block exists which does not require any ordering. At the time of the insertion, all previous committed versions are cleaned up by writing back the last committed version and invalidating the other committed versions.

Second, because we cannot lose the information captured in the Load bit of the block, stores cannot simply invalidate the versions in the VOL that have Load bits set; otherwise, an older store that occurs later in time cannot detect the dependence violation by the loads that set the Load bits. However, not invalidating can cause a different problem: a load younger than the store should read the store value and not some uninvalidated older version. To handle this problem, stores mark younger blocks in the VOL with a possibly *Stale bit* without invalidating the blocks. The younger blocks are only possibly stale and not definitely stale because such a block is not stale for a load that is younger than the block but older than the other versions in the VOL. For such a load, the stale block continues to provide data. An access to a possibly stale block triggers a miss to consult the VOL via a snoop and confirm whether the block is valid or stale for the access. The SVC state bits are shown in Fig. 4.15.

Fig. 4.15 SVC state bits. Bits to record load, store, commit, and potentially stale blocks. The pointer field is for implementing the VOL by pointing to the next cache block in the VOL (figure taken from [11] with permission)

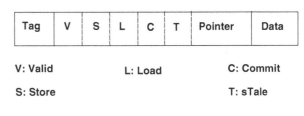

Based on the above two implications of SVC commit which is part of the first requirement of Sect. 4.4.4 (i.e., versioning), we now address the second requirement of inter-task memory dependence. Case (a) of this requirement is that loads should receive the closest previous store value. Accordingly, load misses consult the VOL via a snoop and return the closest previous version. This version could be an uncommitted version, the last committed version, or a clean copy from L2 or memory. The first possibility is handled by the ordering among the uncommitted tasks, the second possibility is handled by the ordering among the committed versions, and the third possibility is straightforward. Case (b) of the requirement is that stores should detect premature loads and trigger rollbacks. To allow stores to detect premature loads, all loads set the Read bit, and store misses consult the VOL via snoops to check for younger loads with the Read bit set.

The only remaining event is a task rollback where all uncommitted blocks (Commit bit reset) are simply invalidated. One side-effect of this invalidation is that the VOL may have invalid entries. However, the invalid entries are cleaned up on the next access that consults the VOL. The details of the SVC states and their transitions can be found in [11].

4.4.5.3 Memory Dependence Prediction and Synchronization

While the ARB and SVC can detect memory (i.e., store-to-load) dependence violations and trigger rollbacks, repeated rollbacks hurt performance. Such repeated rollbacks occur due to memory dependencies present in code that are repeatedly executed and violated. A key point to note is that while memory dependencies are far fewer than register dependencies, repeated rollbacks cause significant performance loss due to the high penalty of rollbacks. Fortunately, such repeated rollbacks can be avoided by having the hardware learn about the dependence and force the corresponding load and store instructions to be synchronized (i.e., the load should wait for the store). This observation was first made in [11].

To that end, the following issues need to be addressed:

- we need to identify the stores and loads that are involved in dependencies,
- we need to isolate the persistently problematic dependencies from the non-problematic ones, and
- we need to force loads to wait for the appropriate stores.

For the first issue of identifying dependencies, we could use either the data address or the instruction addresses. However, because *most* accesses to an address are not involved in a dependence, we use the instruction (load and store) addresses to identify dependencies. Isolating dependencies is straightforward because memory dependence violations are detected by the ARB or the SVC. A violation is detected when a store finds a prematurely executed load. To ensure that the dependence is a persistent one (the second issue above), the offending instructions are recorded in a prediction table which uses saturating counters to learn about repeated occurrences of violations. A synchronization table is used to perform the synchronization where the load–store pair's instruction addresses are recorded upon repeated violations. If the load occurs first then the load records its existence in the synchronization table entry and waits. The store locates the entry with its instruction address and signals the load to proceed. If the store occurs first then the store marks the table entry so loads occurring later see the marked entry and proceed without waiting. More details about memory dependence prediction and synchronization and its use can be found in [4, 19, 20].

4.5 Performance, Complexity, and Scalability

We now discuss how speculatively multithreaded architectures meet the hardware and software criteria listed in Sect. 4.2.1. The criteria are

- to use simple, low-complexity hardware amenable to high clock speed and easy growth path,
- to exploit available parallelism,
- to place as few demands on software (compilers and applications) and to target programs in ordinary, imperative languages, and
- to provide uniform hardware–software interface.

First, by using simple cores each of which is comparable to a dual or single-issue uniprocessor, speculatively multithreaded architectures can achieve high clock speed. The architectures can be extended easily with technology generations by adding more cores, like any multicore, without making each core more complex. While the register file and prediction hardware are different from conventional implementations, the differences are unlikely to add significantly to the implementation complexity. Similarly, the ARB is likely to be only as complex as a modern load–store queue in a conventional uniprocessor. However, the SVC is a key source of complexity though it can leverage snoopy coherence. Because of its multiwriter nature, the SVC protocol is more complex than conventional snoopy coherence. SVC's increased number of protocol states leads to increased possibilities for complicated interactions among memory accesses, which can be challenging. On the compiler side, though register forwarding and task successors need to be specified by the compiler, the complexity of these specifications is comparable to conventional register allocation, and hence manageable. On the other hand, because task selection is a hard problem that has many opposing factors which need to be balanced for good performance, task selection may increase compiler complexity.

Second, by establishing a wider window to exploit distant parallelism via thread-level parallelism as opposed to instruction-level parallelism, speculatively multithreaded architectures can achieve high performance. Previous limit studies show window sizes exceeding tens of thousands of instructions to achieve commit rates of a few hundreds of instructions per cycle [3]. Though these studies are optimistic due to their idealistic assumptions about control-flow prediction and memory latencies, the wide window of speculative multithreading can achieve significant performance improvement over uniprocessors.

To explain the parallelism exploited by speculative multithreading, we can categorize applications into three categories:

1. Automatically compiler-parallelizable, loop-intensive programs (e.g., many numeric applications such as SPECFP 2000) making up the *automatic* category;
2. loop-intensive programs that are not amenable to compiler analysis due to ambiguous control-flow and memory dependencies (e.g., numeric applications with call through pointers and memory accesses through pointers in

a type-unsafe manner such as sparse matrix computations) making up the *ambiguous-loop* category; and
3. non-loop intensive programs with ambiguous dependencies (e.g., non-numeric applications such as SPECint 2000) making up the *non-loop* category.

While all three categories exhibit more parallelism at wider windows, the automatic and ambiguous-loop categories have more parallelism than the non-loop category. While only the automatic category can be sped up by CMPs and a parallelizing compiler, speculatively multithreaded architectures achieve good performance on all three categories. While significant speedups can be achieved on the non-loop category (e.g., an eight-core processor, where each core is a two-issue superscalar, can achieve 2–3× speedups over a single core), even greater speedups of 4–5× are possible in the ambiguous-loop category. The wide window of speculative multithreading exposes the parallelism across the iterations of the loops in the programs in the ambiguous-loop category. In this category, speedups are limited by inter-iteration memory dependencies as control flow and load imbalance are not major factors because of the loop-intensive nature of the programs. In the non-loop category, all the three factors are important though memory dependencies more so than the other two.

In addition to performance, the question of scalability over the number of cores is important to ensure that the architecture can both increase its compute power by adding more cores and also be able to utilize the ever-increasing number of transistors afforded by future technology generations. Two aspects of scalability include physical scalability and performance scalability. The distributed nature of the implementation where each core is similar to a uniprocessor of the recent past is amenable to good physical scalability. While ARB-like centralized disambiguation can scale up to a few cores, SVC-like distributed disambiguation can scale up to a few tens of cores, similar to snoopy-based coherence. While further physical scaling may be possible using directory-based disambiguation which avoids the bus bottleneck, performance scalability becomes an issue. In the distributed implementation of a speculatively multithreaded architecture, most of a given program's computation and data are local to the processing units (cores) and their caches, requiring inter-core data movement only infrequently. Thus, the implementation achieves good performance scalability up to a few tens of cores. This reason is the same for the good performance scalability of CMPs. However, scaling to a much larger number of cores may be difficult because of inter-task memory dependencies which require inter-core communication through the SVC cache hierarchy. The key point here is that while some of the conventional multithreaded code may scale to such large number of cores by employing aggressive programming to avoid communication, the speculative tasks from a sequential program have dependencies that cannot be removed (i.e., avoided or scheduled away) easily by the compiler. These dependencies lead to communication through the memory system which makes performance scaling to such large number of cores hard.

Third, the speculative nature of the hardware frees the compiler and application from providing strict guarantees of independence and enables continued use of imperative languages for programming. And finally, the hardware–software interface in terms of the instruction set, system calls, interrupts/exceptions, and virtual memory is identical to that of a conventional uniprocessor without including any details of speculative multithreading. The interface naturally allows independent growth paths for software and hardware.

4.6 Speculative Multithreading and Transactional Memory

We conclude this chapter by pointing out a key connection between speculatively multithreaded architectures and hardware transactional memory (HTM) architectures which was originally proposed in [14] and significantly improved by others (e.g., UTM [2], TCC [13, 6], VTM [23], and LogTM [17, 18]). Transactional memory (TM) has been proposed to address some of the programmability problems with lock-based synchronization in shared memory (e.g., deadlocks and races). Instead of specifying critical sections using locks, transactional programmers specify atomic regions of code. Locks are named and incorrect lock order can lead to deadlocks and incorrect lock names can lead to races. In contrast, atomic regions are unnamed and the implementation (hardware or software) ensures atomicity of the regions.

Most current TM implementations concurrently execute transactions under the optimistic assumption that the transactions do not have conflicts which would violate the transaction's atomicity (i.e., a read or write to an address from one transaction and a write to the same address from another transaction constitutes a conflict). TMs detect conflicts and roll back all but one of the offending transactions. Thus, just as speculatively multithreaded architectures support speculative parallelism among potentially dependent tasks, TMs support speculative concurrency among potentially conflicting transactions. Just as speculatively multithreaded architectures detect dependence violation among speculative tasks and roll back the offending tasks, HTMs detect conflicts and roll back all but one of the offending transactions.

TM allows any *serializable order* among transactions whereas sequential programming enforces *sequential order* among tasks. Though the two architectures' programming models are completely different, the architectures have significant similarities in their hardware functionality in that both architectures support speculation in the memory hierarchy. Consequently, HTMs employ some of the key ideas from the memory systems of speculatively multithreaded architectures, such as detecting conflicts via Load/Store bits in the caches, buffering speculative versions in the caches (while some HTMs buffer the versions in hardware caches others use software), and committing speculative data when speculation succeeds. Hence, much of the same hardware can be used to support both programming models.

4.7 Concluding Remarks

Speculatively multithreaded architectures establish instruction windows that are much larger than those of conventional uniprocessors. The large windows enable the architectures to exploit distant thread-level parallelism which is more abundant than the nearby instruction-level parallelism exploited by conventional uniprocessors. By building the large window as a collection of many smaller windows, the architectures are amenable to a distributed implementation of multiple cores, where each small window maps to a core. The architectures can increase their instruction window size to improve performance simply by adding more cores without making the cores more complex or more power-hungry. Thus, speculatively multithreaded architectures provide a scalable, power-efficient solution to continue to exploit the ever-increasing number of transistors on a chip for higher performance. These attributes match well with the multicores' original goals of scalability, power efficiency, and high performance. Furthermore, speculatively multithreaded architectures achieve these goals while maintaining the programmability advantage of sequential programming over conventional multicores which require parallel programming.

References

1. H. Akkary and M. A. Driscoll. A dynamic multithreading processor. In *Proceedings of the 31st Annual ACM/IEEE International Symposium on Microarchitecture*, pages 226–236, 1998.
2. C. S. Ananian, et al. Unbounded transactional memory. In *HPCA '05: Proceedings of the 11th International Symposium on High-Performance Computer Architecture*, pages 316–327, 2005.
3. T. M. Austin and G. S. Sohi. Dynamic dependency analysis of ordinary programs. In *Proceedings of the 19th Annual International Symposium on Computer Architecture*, pages 342–351, May 1992.
4. S. E. Breach. *Design and Evaluation of a Multiscalar Processor*. Ph.D. thesis, University of Wisconsin-Madison, 1998.
5. L. Ceze, J. Tuck, J. Torrellas, and C. Cascaval. Bulk Disambiguation of Speculative Threads in Multiprocessors. In *Proceedings of the 33rd Annual International Symposium on Computer Architecture*, pages 227–238, 2006
6. J. Chung, et al. Tradeoffs in transactional memory virtualization. In *Proceedings of the 12th International Conference on Architectural Support for Programming Languages and Operating Systems*, pages 371–381, 2006.
7. M. Cintra, J. F. Martnez, and J. Torrellas. Architectural support for scalable speculative parallelization in shared-memory multiprocessors. In *Proceedings of 27th Annual International Symposium on Computer Architecture*, June 2000.
8. M. Franklin and G. S. Sohi. The expandable split window paradigm for exploiting fine-grained parallelism. In *Proceedings of the 19th Annual International Symposium on Computer Architecture*, pages 58–67, May 19–21, 1992.
9. M. Franklin and G. S. Sohi. Analysis for Streamlining Inter-operation Communication in Fine-Grain Parallel Processors. In *Proceedings of the 27th Annual ACM/IEEE International Symposium on Microarchitecture*, pages 226–236, Dec 1992.
10. M. Franklin and G. S. Sohi. ARB: A hardware mechanism for dynamic reordering of memory references. *IEEE Transactions on Computers*, 45(6):552–571, May 1996.
11. S. Gopal, T. N. Vijaykumar, J. E. Smith, and G. S. Sohi. Speculative versioning cache. In *Proceedings of the 4th International Symposium on High-Performance Computer Architecture*, Feb 1998.
12. L. Hammond, M. Willey, and K. Olukotun. Data speculation support for a chip multiprocessor. In *Proceedings of the 8th International Conference on Architectural Support for Programming Languages and Operating Systems*, pages 58–69, Oct 02–07, 1998.

13. L. Hammond, et al. Transactional memory coherence and consistency. In *ISCA '04: Proceedings of the 31st Annual International Symposium on Computer Architecture*, page 102, 2004.
14. M. Herlihy and J. E. B. Moss. Transactional memory: architectural support for lock-free data structures. In *ISCA '93: Proceedings of the 20th Annual International Symposium on Computer Architecture*, pages 289–300, 1993.
15. E. Ipek, M. Kirman, N. Kirman, and J. F. Martinez. Core fusion: accommodating software diversity in chip multiprocessors. In *Proceedings of the 34th Annual International Symposium on Computer Architecture*, pages 186–197, 2007.
16. T. A. Johnson, R. Eigenmann, and T. N. Vijaykumar. Min-cut program decomposition for thread-level speculation. In *Proceedings of the ACM SIGPLAN 2004 Conference on Programming Language Design and Implementation*, pages 59–70, 2004.
17. K. E. Moore, J. Bobba, M. J. Morovan, M. D. Hill, and D. A. Wood. LogTM: log-based transactional memory. In *HPCA '06: Proceedings of the 12th International Symposium on High-Performance Computer Architecture*, pages 254–265, 2006.
18. M. J. Moravan, et al. Supporting nested transactional memory in LogTM. In *Proceedings of the 12th International Conference on Architectural Support for Programming Languages and Operating Systems*, pages 359–370, 2006.
19. A. Moshovos, S. E. Breach, T. N. Vijaykumar, and G. S. Sohi. Dynamic speculation and synchronization of data dependences. In *Proceedings of the 24th Annual International Symposium on Computer Architecture*, pages 181–193, June 2–4, 1997.
20. A. Moshovos. Memory Dependence Prediction, Ph.D. thesis, University of Wisconsin, Dec 1998.
21. S. Palacharla, N. P. Jouppi, and J. E. Smith. Complexity-effective superscalar processors. In *Proceedings of the 24th International Symposium on Computer Architecture*, pages 206–218, June 1997.
22. Il Park, Babak Falsafi, and T. N. Vijayakumar. Implicitly-Multithreaded Processors. In *Proceedings of the 30th Annual International Symposium on Computer Architecture (ISCA'03)*, page 39, 2003.
23. R. Rajwar, M. Herlihy, and K. Lai. Virtualizing transactional memory. In *ISCA '05: Proceedings of the 32nd Annual International Symposium on Computer Architecture*, pages 494–505, 2005.
24. V. Sarkar and J. Hennessy. Partitioning parallel programs for macro-dataflow. In *Proceedings of the Conference on LISP and Functional Programming*, pages 202–211, 1986.
25. J. E. Smith. A study of branch prediction strategies. In *Proceedings of the 8th Annual International Symposium on Computer Architecture*, pages 135–148, May 1981.
26. G. S. Sohi, S. E. Breach, and T. N. Vijaykumar. Multiscalar processors. In *Proceedings of the 22nd Annual International Symposium on Computer Architecture*, pages 414–425, June 22–24, 1995.
27. J. G. Steffan, C. B. Colohan, A. Zhai, and T. C. Mowry. A scalable approach to thread-level speculation. In *Proceedings of 27th Annual International Symposium on Computer Architecture*, pages 1–12, June 2000.
28. D. M. Tullsen, S. J. Eggers, and Levy, H. M. Simultaneous multithreading: Maximizing on-chip parallelism. In *Proceedings of the 22nd Annual International Symposium on Computer Architecture*, pages 392–403, June 1995.
29. T. N. Vijaykumar and G. S. Sohi. Task selection for a multiscalar processor. *The 31st International Symposium on Microarchitecture (MICRO-31)*, Dec 1998.
30. H. H. Yang and D. F. Wong. Efficient network flow based min-cut balanced partitioning. *IEEE Transactions on Computer-Aided Design of Integrated Circuits and Systems*, 15(12), Dec 1996.
31. T. Yeh and Y. N. Patt. Alternative implementations of two-level adaptive branch prediction. In *Proceedings of the 19th Annual international Symposium on Computer Architecture*, pages 124–134. May 1992.

Chapter 5
Optimizing Memory Transactions for Multicore Systems

Ali-Reza Adl-Tabatabai, Christos Kozyrakis, and Bratin Saha

Abstract The shift to multicore architectures will require new programming technologies that enable mainstream developers to write parallel programs that can safely take advantage of the parallelism offered by multicore processors. One challenging aspect of shared memory parallel programming is synchronization. Programmers have traditionally used locks for synchronization, but lock-based synchronization has well-known pitfalls that make it hard to use for building thread-safe and scalable software components. Memory transactions have emerged as a promising alternative to lock-based synchronization because they promise to eliminate many of the problems associated with locks. Transactional programming constructs, however, have overheads and require optimizations to make them practical. Transactions can also benefit significantly from hardware support, and multicore processors with their large transistor budgets and on-chip memory hierarchies have the opportunity to provide this support.

In this chapter, we discuss the design of transactional memory systems, focusing on software optimizations and hardware support to accelerate their performance. We show how programming languages, compilers, and language runtimes can support transactional memory. We describe optimization opportunities for reducing the overheads of transactional memory and show how the compiler and runtime can perform these optimizations. We describe a range of transactional memory hardware acceleration mechanisms spanning techniques that execute transactions completely in hardware to techniques that provide hardware acceleration for transactions executed mainly in software.

"Unlocking Concurrency" [2] ©[2006] ACM, Inc. Reprinted with permission

A.-R. Adl-Tabatabai (✉)
Intel Corporation, Hillsboro, OR, USA
e-mail: ali-reza.adl-tabatabai@intel.com

5.1 Introduction

Multicore architectures are an inflection point in mainstream software development because they force developers to write parallel programs. Unlike before, where a developer could expect an application's performance to improve with each successive processor generation, in this new multicore world, developers must write explicitly parallel applications that can take advantage of the increasing number of cores that each successive multicore generation will provide.

Multicore processors will require new programming technologies that enable mainstream developers to write scalable and robust parallel applications. Today's mainstream programming languages were not designed with parallelism in mind and force the use of low-level constructs for parallel programming. Multicore processors will have an opportunity to provide hardware support that reduces the overheads associated with these new programming technologies.

Parallel programming poses many new challenges to the mainstream developer, one of which is the challenge of synchronizing concurrent access to memory shared by multiple threads. Programmers have traditionally used locks for synchronization, but lock-based synchronization has well-known pitfalls. Simplistic coarse-grained locking does not scale well, while more sophisticated fine-grained locking risks introducing deadlocks and data races. Furthermore, scalable libraries written using fine-grained locks cannot be easily composed in a way that retains scalability and avoids deadlock and data races.

Transactional memory (TM) provides a new concurrency control construct that avoids the pitfalls of locks and significantly eases concurrent programming. Transactional memory brings to mainstream parallel programming proven concurrency control concepts used for decades by the database community. Transactional language constructs are easy to use and can lead to programs that scale. By avoiding deadlocks and automatically allowing fine-grained concurrency, transactional language constructs enable the programmer to compose scalable applications safely out of thread-safe libraries.

Transactional memory can benefit greatly from hardware support, and multicore processors with their large transistor budgets and on-chip memory hierarchies have the opportunity to provide such hardware support. This hardware support ranges from executing transactions completely in hardware to providing hardware that accelerates a software implementation of TM. Some hardware vendors have already announced products that will include hardware transactional memory.

In this chapter, we describe how transactions ease some of the challenges programmers face using locks and describe the challenges system designers face implementing transactions in programming languages. One of the major challenges of implementing TM is its performance, so we concentrate on software optimizations and hardware support for improving TM performance. We show how the programming system can implement transactional memory language support, and we describe compiler and runtime optimizations for improving its performance. Finally, we present various hardware mechanisms for accelerating TM in a multicore processor.

5.2 Multicore Programming with Transactions

In this section, we describe the programmer's view of transactional memory. In later sections, we describe implementation and optimization of transactional memory using software and hardware techniques.

A memory transaction is a sequence of memory operations that execute *atomically* and *in isolation*. Transactions execute atomically, meaning they are an all or nothing sequence of operations. If a transaction commits, then its memory operations appear to take effect as a unit. If a transaction aborts, then none of its stores appear to take effect, as if the transaction never happened.

A transaction runs in isolation, meaning it executes as if it is the only transaction running on the system. This means that the effects of a memory transaction's stores are not visible to other transactions until the transaction commits, and that no other transaction modifies its state while it runs.

Transactions appear to execute in some serial order (*serializability*), and thus give the illusion that they execute as a single atomic step with respect to other concurrent transactions in the system. The programmer can reason serially inside a transaction because no other transaction will perform any conflicting operation on its memory state.

Of course a transactional memory system does not really execute transactions serially; otherwise, it would defeat the purpose of parallel programming. Instead, the system "under the hood" allows multiple transactions to execute concurrently as long as it can still provide atomicity and isolation for each transaction. Further, unlike in databases where all operations are performed inside transactions, some memory operations may execute outside of transactions; as we see later, this makes possible a range of semantics for the isolation and ordering guarantees between transactions and non-transactional memory accesses.

5.2.1 Language Support for Transactions

The best way to provide the benefits of transactional memory to the programmer is to replace locks with a new language construct such as atomic {B} that executes the statements in block B as a transaction [18, 3, 10, 29]. A first-class language construct not only provides syntactic convenience for the programmer, but also enables static analyses that provide compile-time safety guarantees and enables compiler optimizations to improve performance. It is also possible to define a programming model that provides transactions implicitly [1] or a programming model that provide new language constructs for speculative parallelism leveraging a transactional memory infrastructure [6, 39].

Figure 5.1 illustrates how an atomic statement could be introduced and used in an object-oriented language such as Java. This figure shows two different implementations of a thread-safe Map data structure. The code in Fig. 5.1(a) shows a lock-based Map using Java's synchronized statement. The get() method simply delegates the

```
class LockBasedMap                    class AtomicMap
   implements Map                        implements Map
{                                     {
  Object mutex;                         Map m;
  Map m;                                AtomicMap(Map m) {
  LockBasedMap(Map m) {                   this.m = m;
    this.m = m;                         }
    mutex = new Object();             public Object get() {
  }                                     atomic {
  public Object get() {                   return m.get();
    synchronized (mutex) {              }
      return m.get();                 }
    }                                 // other Map methods
  }                                   . . .
  // other Map methods              }
  . . .
}
            (a)                                   (b)
```

Fig. 5.1 Lock-based versus transactional Map data structure

call to an underlying non-thread-safe Map implementation but first acquires a lock that guards all the other calls to this hash map.

Using locks, the programmer has explicitly forced all threads to execute any call through this synchronized wrapper serially. Only one thread at a time can call any method on this hash map. This is an example of coarse-grained locking. It is easy to write thread-safe programs in this way – you simply guard all calls through an interface with a single lock, forcing threads to execute inside the interface one at a time.

The code in Fig. 5.1(b) shows the same code but using a new atomic statement instead of locks. This atomic statement declares that the call to get() should be done atomically, as if it was done in a single execution step with respect to other threads. Like coarse-grained locking, it is easy for the programmer to make an interface thread-safe by simply wrapping all the calls through the interface with an atomic statement. Rather than explicitly forcing one thread at a time to execute any call to this hash map, however, the programmer has instead declared to the system that the call should execute atomically. The system now assumes responsibility for guaranteeing atomicity and implements concurrency control under the hood.

5.2.2 Scalability of Transactional Memory

Unlike coarse-grained locking, transactions can provide scalability as long as the data access patterns allow transactions to execute concurrently. Transactions can be implemented in such a way that they allow concurrent read accesses as well as concurrent accesses to disjoint, fine-grained data elements (for example, different objects or different array elements). Using transactions, the programmer gets these forms of concurrency automatically without having to code them explicitly in the program.

It is possible to write a concurrent hash map data structure using locks so that you get both concurrent read accesses and concurrent accesses to disjoint data. In fact, the recent Java 5 libraries provide a version of hash map called concurrent hash map that does exactly this. The code for concurrent hash map, however, is significantly longer and more complicated than the version using coarse-grained locking. In general, writing highly concurrent lock-based code such as concurrent hash map is very complicated and bug prone, and thereby introduces additional complexity to the software development process.

Figure 5.2 compares the performance of the three different versions of hash map. It plots the time it takes to complete a fixed set of insert, delete, and update operations on a 16-way SMP machine [3]. Lower execution time is better. As the numbers show, the performance of coarse-grained locking does not improve as the number of processors increase, so coarse-grained locking does not scale. The performance of fine-grained locking and transactional memory, however, improve as the number of processors increase. So for this data structure, transactions give the same scalability and performance as fine-grained locking but with significantly less programming effort. Transactions transfer the complexities of scalable concurrency management from the application programmers to the system designers. Using a combination of software and hardware, a TM runtime guarantees that concurrent transactions from multiple threads execute atomically and in isolation.

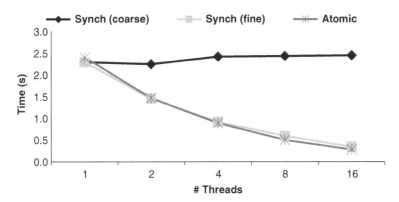

Fig. 5.2 Performance of transactions versus locks for hash map

5.2.3 Composing Software Using Transactions

Although highly concurrent libraries built using fine-grained locking can scale well, a developer does not necessarily retain scalability after composing larger

applications out of these libraries. As an example, assume the programmer wants
to perform a composite operation that moves a value from one concurrent hash map
to another while maintaining the invariant that threads always see a key in either
one hash map or the other, but never in neither and never in both. Implementing
this requires that the programmer resort to coarse-grained locking, thus losing the
scalability benefits of a concurrent hash map (Fig. 5.3(a)). To implement a scalable solution to this problem, the programmer must somehow reuse the fine-grained
locking code hidden inside the implementation of concurrent hash map. Even if the
programmer had access to this implementation, building a composite move operation out of it risks introducing deadlock and data races, especially in the presence of
other composite operations.

```
move(Object key) {                    move(Object key) {
  synchronized(mutex) {                 atomic {
    map2.put(key,map1.remove(key));       map2.put(key,map1.remove(key));
  }                                     }
}                                     }
              (a)                                   (b)
```

Fig. 5.3 Thread-safe composite operation

Transactions, on the other hand, allow the programmer to compose applications
out of libraries safely and still achieve scalability. The programmer can simply wrap
a transaction around the composite move operation (Fig. 5.3(b)). The underlying
transactional memory system will allow two threads to perform a move operation
concurrently as long as the two threads access different hash table buckets in both
underlying hash map structures. So transactions allow a programmer to take separately authored scalable software components and compose them together into
larger components, in a way that still provides as much concurrency as possible
but without risking deadlocks due to concurrency control.

5.2.4 Transaction Semantics

Atomic blocks can provide much richer semantics than locks. By providing a mechanism to roll back side effects, transactions enable a language to provide *failure
atomicity*. In lock-based code, programmers must make sure that exception handlers properly restore invariants before releasing locks, often leading to complicated
exception handling. In a transaction-based language, the atomic statement can roll
back all the side effects of the transaction (automatically restoring invariants) if an
uncaught exception propagates out of its block. This reduces the amount of exception handling code significantly and improves robustness as uncaught exceptions
inside a transaction will not compromise a program's invariants. It is also possible

to provide failure atomicity via an abort construct that allows the programmer to explicitly rollback the effects of an atomic block.

Atomic blocks must provide a way for threads to co-ordinate via condition synchronization. To use lock-based condition synchronization correctly, programmers must carefully avoid lost wakeup bugs and handle spurious wakeups when using condition variables. Transactions can support condition synchronization via a *retry* construct [19] that rolls back a transaction and automatically re-executes it when a value that was read by the transaction changes, avoiding the pitfalls of lock-based condition synchronization.

The semantics of atomic blocks must define a memory model that considers the interaction between transactions and memory operations that execute outside a transaction. A range of memory models is possible: *Strong atomicity* [7, 36] (also known as *strong isolation*) isolates atomic blocks from non-transactional accesses so that non-transactional accesses appear ordered with respect to atomic blocks. Weak *atomicity* (*isolation*) leaves the semantics undefined if the same memory location is accessed both inside and outside an atomic block, requiring the programmer to segregate transactional and non-transactional data. Specialized type systems can enforce this segregation statically [19], though such type systems have so far not been developed for mainstream languages such as C/C++ and Java. Between these weak and strong extremes, other memory models are possible that define atomic block semantics in terms of locks, and thus define the memory model in terms of the underlying language's memory model [26]. The simplest such model, single global lock atomicity, defines the semantics of an atomic block to be the same as a critical section that acquires a single system-wide global lock.

Atomic blocks also introduce several semantic challenges. While locks allow arbitrary operations inside critical sections, some operations (such as certain I/O operations) are not supported naturally inside atomic blocks because their side effects cannot be delayed until commit or rolled back on abort (imagine a transaction that displays a window to the user and waits for input). It is possible to allow arbitrary I/O operations inside transactions by allowing transactions to become irrevocable [41] (or inevitable [38]) so that they are guaranteed to commit without rollback; but irrevocable transactions do not compose with failure atomicity or retry mechanisms, both of which require transaction rollback. Similarly, locks allow parallelism nested inside critical sections, but the semantics of parallelism inside atomic blocks is unclear. These semantic issues are open research topics.

5.3 Implementing TM in a Multicore Environment

The key mechanisms used by a TM implementation for enforcing atomicity and isolation are *data versioning* and *conflict detection*.

As transactions execute, the TM implementation must simultaneously manage multiple versions of data. With *eager versioning*, a write access within a transaction immediately writes to memory the new data version. The old version is buffered in an undo log. This requires the acquisition of locks (or an equivalent hardware mechanism such as exclusive ownership) prior to a transactional write to prevent other code from observing the uncommitted new versions. If the transaction aborts, the old versions must be restored from the undo log. *Lazy versioning* stores all new data versions in a write buffer until the transaction completes whereby the new versions are copied to the actual memory addresses. Locks (or equivalent hardware assist) only need to be acquired during the copy phase. However, on every transactional read, the write buffer must be searched to access the most recent data version.

A TM system must also prevent conflicting transactional operations – that is, two or more transactions operating concurrently on the same data with at least one transaction doing a write. For this, the TM system maintains the *read set* and *write set* for each transaction, which respectively include the addresses it read from and wrote to during its execution. Under *pessimistic conflict detection*, the system checks for conflicts progressively as transactions read and write data. Thus, conflicts are detected early and can be handled either by stalling one of the transactions or by aborting one transaction and retrying it later. The alternative is *optimistic conflict detection* that assumes conflicts are rare and postpones all checks until the end of each transaction. Before committing, a transaction validates that no other transaction is reading the data it wrote or writing the data it read. Since conflicts are detected late a transaction may end up being inconsistent. The TM implementation must then ensure that an inconsistent transaction does not generate any visible effect such as raising exceptions. Once a conflict is detected the TM system can use different policies to resolve it, which are typically referred to as *contention management* [34].

Table 5.1 shows how a number of different TM implementations handle conflict detection and version management. On the one hand, there is still no convergence on the best strategy. On the other hand, the best strategy will also depend on factors such as application characteristic and implementation complexity. Note that TM systems usually do not employ eager versioning and optimistic conflict detection. This combination can result in an inconsistent transaction writing inconsistent values to memory, which get subsequently read by a different agent (for example, thread or process) leading to application failure.

The *granularity of conflict detection* is also an important design parameter. Some TM systems detect conflicts at granularities that are programmer visible entities. For example, *object level* detection is often used in object-oriented environments, wherein two transactions conflict on accessing the same object. Some TM systems

Table 5.1 TM implementation examples

		Version Management	
		Eager	Lazy
Conflict Detection	*Pessimistic*	HW: LogTM [27] SW: McRT [32]	HW: LTM [4] SW: DSTM [21]
	Optimistic		HW: TCC [17] SW: TL2 [15]

also use *element level* detection for arrays – that is, transactions conflict if they access the same array element. In contrast, some TM systems use address-based conflict detection; for example, *cache-line* detection signals a conflict only if two transactions access the same cache line.

Finally, some TM implementations use virtual addresses for detecting conflicts, while most hardware TM implementations use physical addresses for conflict detection. Using virtual addresses makes it difficult to extend transactions across different processes or to ensure consistency against device or DMA accesses. On the other hand, using virtual addresses allows the system to provide better conflict diagnostics.

5.4 Software Transactional Memory (STM)

A TM system can be implemented entirely in software (STM). Indeed an efficient STM that runs on stock hardware is critical to enabling transactional programming for two reasons: (1) Even if some processor introduces hardware support for transactional memory at some point, the vast majority of the processors deployed in the market at that time will not have hardware TM support. Effectively, software vendors will still have to write their programs to work on processors without the hardware support. For transactional memory to take root, software vendors must be able to use the same programming model for both new (TM-enabled) and existing (non-TM-enabled) hardware. In other words, we need a high-performance STM at least for existing hardware. (2) Most hardware TM proposals rely on a software mechanism to handle transactions that are large (that is, not cache resident), have I/O, require debugging, throw exceptions, and have other features not directly supported in hardware. We need an efficient STM to handle these cases efficiently.

5.4.1 Compiling Transactions

A STM implementation uses read and write barriers (that is, inserts instrumentation) for all shared memory reads and writes inside transactional code blocks. This allows the runtime system to maintain the metadata that is required for data versioning and conflict detection. The instrumentation is best inserted by a compiler because it allows the instrumentation to be optimized such as through redundancy elimination, detecting local variable access, and the like; we will discuss compiler optimizations more in the next section.

Figure 5.4 shows an example of how a compiler can translate the atomic construct in a STM implementation [40]. Figure 5.4(a) shows an atomic code block written by the programmer, and Fig. 5.4(b) shows the compiler translation of the atomic block. We use a simplified control flow to ease the presentation. The **setjmp** function checkpoints the current execution context so that the transaction can be restarted on an abort. The **stmStart** function initializes the runtime data structures.

int foo (int arg)	int foo (int arg)
{	{
...	jmpbuf env;
atomic	...
{	do {
b = a + 5;	if (setjmp(&env) == 0) {
}	**stmStart();**
...	temp = **stmRead**(&a);
}	temp1 = temp + 5;
	stmWrite(&b, temp1);
	stmCommit();
	break;
	}
	} while (1);
	...
(a) User Code	(b) Compiled Code

Fig. 5.4 Translating an atomic construct for a STM

Accesses to the global variables **a** and **b** are mediated through the barrier functions **stmRead** and **stmWrite**. The **stmCommit** function completes the transaction and makes its changes visible to other threads. The transaction's consistency gets validated periodically during its execution (validation checks that no other transaction has written to any of the data read by the transaction), and if a conflict is detected, the transaction is aborted. The periodic validation is usually performed by the barrier functions, but some systems can also have the compiler insert explicit calls to a runtime validation function. On an abort, the STM library rolls back all the updates performed by the transaction, uses a longjmp to restore the context saved at the beginning of the transaction, and re-executes the transaction.

Since transactional memory accesses need to be instrumented, a compiler needs to generate an extra copy of any function that may be called from inside a transaction. This copy contains instrumented accesses and is invoked when the function is called from within a transaction. A JIT (just in time) compiler can produce the instrumented version on demand [3], but a static compiler must either produce dual versions of all code or rely on programmer annotations to specify which functions can be called from inside transactions. The compiler can retarget direct function calls or virtual method calls to the instrumented clone, but indirect function calls through function pointers need to look up the transactional version of the call target dynamically.

An unoptimized STM using the mechanisms outlined in this section incurs an overhead of 60–80% compared to a corresponding lock-based program on realistic workloads. However, the STM mechanisms can be heavily optimized. For example, an atomic section does not need multiple read barriers if the same variable is read multiple times. Section 5.5 describes a number of optimizations that are used by modern high-performance STMs (for example, the Intel C/C++ STM released on whatif.intel.com [29]) to reduce this overhead.

The need for STM instrumentation makes it difficult to support calling pre-compiled uninstrumented binaries inside transactions. One option that has been used by some TM systems is to make a transaction irrevocable before calling pre-compiled binaries and to disallow any other transaction in the system while the irrevocable transaction executes [29]. Alternatively, the system can use dynamic binary translation to insert read and write barriers into binaries that are called from inside transactions.

Creating transactional clones of functions facilitates a *pay-as-you-go* policy in which a program pays the overheads of instrumentation only when executing inside a transaction. A pay-as-you-go policy is possible if the language provides weak atomicity or single global lock atomicity semantics [26]. But providing strong atomicity semantics in a system that allows the same data to be accessed both inside and outside transactions requires instrumentation of accesses in non-transactional code [36]. In a JIT-compiled environment such as Java, the JIT compiler can instrument non-transactional code dynamically and optimize the instrumentation overheads [36, 35]. In a statically compiled environment, strong atomicity requires recompiling the entire application, which is impractical in many cases due to pre-compiled binaries.

5.4.2 Runtime Data Structures

The read and write barriers operate on *transaction records*, pointer-sized metadata associated with every datum that a transaction may access. For example, to implement eager versioning, the write barrier acquires an exclusive lock on the transaction record corresponding to the memory location being updated prior to performing the store. Similarly, the read barrier may acquire a read lock on the corresponding transaction record before reading a data item. In other cases, the transaction record may store a timestamp to denote the last transaction to have modified the associated data. The read barrier then records the timestamp associated with the read data and uses it for periodically validating the transaction's consistency.

It is interesting to note that multicore processors are friendlier to STMs than multiprocessors (multisocket systems) because the locking overhead in a multicore processor is much lower than in a multiprocessor system – a few tens of cycles as opposed to hundreds of cycles. This benefits STM implementations because they perform interlocked operations on the transaction records.

The runtime system also maintains a *transaction descriptor* for each transaction. The descriptor contains the transaction's state such as the read set, the write set, and the undo log for eager versioning (or the write buffer for lazy versioning). The STM runtime exports an API that allows other components of the language runtime (such as the garbage collector or a debugger) to inspect and modify the contents of the descriptor such as the read set, write set, or the undo log. The garbage collector (GC), for example, must be cognizant of the TM implementation because a transactions' read and write sets may contain pointers to objects, and the GC must keep those objects alive and update their pointers inside the read and write sets if it moves them. The TM system must therefore enumerate its internal data structures to the garbage collector [3]. If a TM system uses address-based conflict detection (such as cache-line-based detection) then it must take into account that objects may change their addresses in the middle of a transaction as a result of a GC copy.

5.4.3 Conflict Detection

STM implementations detect conflict in two cases: (1) the read or write barrier finds that a transaction record is locked by some other transaction resulting in pessimistic conflict detection; and (2) the transaction finds during validation that an element in its read set has been changed by another transaction, resulting in optimistic conflict detection. On detecting conflict, the STM uses a contention resolution scheme such as causing transactions to back off in a randomized manner, or aborting and restarting some set of conflicting transactions.

The granularity of conflict detection in an STM system depends on the language environment. Managed type-safe languages can use object or element granularity conflict detection whereas unmanaged languages such as C or C++ typically use address-based conflict detection at the cache-line or word granularity. Implementing object-level conflict detection in C++ is difficult because it requires a custom memory allocator and because pointer manipulations make it difficult to associate an object with any arbitrary pointer [23].

STM systems that use optimistic conflict detection have additional requirements when used inside a language environment. Some optimistic STMs allow a transaction to see an inconsistent view of memory temporarily. It is important to guarantee, however, that any error or exception that occurs inside a transaction is based on a consistent view of memory. Managed languages provide type-safety and raise well-defined exceptions if a program encounters errors such as illegal memory access, division by zero, and the like. If an inconsistent transaction encounters an error, the TM implementation can check consistency in the managed language runtime's exception handler. This is not the case in a language like C, where an illegal memory access can crash the application. In such a language, the TM implementation must make sure that a transaction always remains consistent.

5.5 Optimizing Software Transactional Memory

Optimizations can significantly improve the performance of transactional memory by eliminating unnecessary STM operations and by reducing their costs. These optimizations can be done either at compile time or dynamically at runtime. The optimization algorithms and the representation of TM operations in the compiler depend on the language (for example, Java versus C or C++) and on the TM system's version management and conflict detection policy. In this section, we concentrate on optimizations for optimistic, eager-versioning STMs in a managed, object-oriented language such as Java. Optimizations for other types of STM systems or for unmanaged languages such as C or C++ would operate similarly. Many of these optimizations also apply in TM systems that use hardware support for transactions described in Sect. 5.6.

5.5.1 Optimization Opportunities

The compiler and runtime have several opportunities for eliminating redundant operations across multiple barriers. These redundancies arise when a transaction accesses the same datum multiple times and consequently performs redundant barrier operations such as logging the same address into the read set multiple times or checking a datum for conflicts with other transactions when it already has exclusive ownership on that datum. The redundancies depend on whether the STM detects conflicts at the field level or the object level.

The following redundancy elimination opportunities exist:

- Once a write barrier acquires write ownership of a field x.f and logs its previous value in the undo log, the transaction does not need to repeat these write barrier operations on subsequent writes to x.f, and the transaction does not need to perform read barrier operations on subsequent reads of x.f. If the STM detects conflicts at the object granularity, then once a write barrier acquires write ownership of an object, the transaction does not need to acquire ownership again on subsequent writes to any field of x, and it does not need to perform a read barrier on subsequent reads from any field of x.
- Once a read barrier has logged the address of a field x.f into the read set and checked for conflicting writes to x.f, the transaction does not need to log x.f into the read set again on subsequent reads of x.f. In an STM that does not eagerly maintain read set consistency, the transaction does not need to check for conflicts again either. If the STM detects conflicts at the object granularity, then these redundancies arise on subsequent reads from any field of x.

Optimizations can also eliminate barriers on transaction-private or immutable data since such data can never be accessed concurrently in a conflicting way. Transaction-private data, which is data visible to only one transaction, can only be reached by a single transaction and thus can never be accessed concurrently. Examples of transaction-private data include local stack-allocated variables, objects freshly allocated inside a transaction, and objects referenced only by local variables or other private heap objects.

Accesses to immutable data do not require read barriers since such data will never be written after initialization. Examples of immutable data include the virtual method pointer field of each object in a Java or C++ system, the virtual method dispatch table in Java or C++, and final fields in Java. Some built-in classes (such as the String class in Java) are known to be immutable and the compiler can hard code this knowledge [36].

5.5.2 Runtime Optimizations

The STM runtime can detect and optimize barrier redundancies using runtime filtering techniques. *Read set filtering* avoids redundantly logging a datum into the read set if it already exists in the read set, and *undo log filtering* avoids logging the same location more than once in the undo log [20]. Eliminating redundant logging reduces the footprint of the logs and, in the case of read set filtering, reduces the time it takes to validate the read set on commit. Runtime filtering can reduce the execution time of long-running transactions by $10\times$ or more [20]. Filtering optimizations require a fast set structure (such as a hash table or bit vector) that allows the read or write barrier to check membership quickly. This extra lookup adds to the barrier overheads, so it should only be done when the gains from avoiding redundant logging outweigh its costs. Filtering can also be done during the garbage collection phases, which enumerates the read set and undo log anyway. As we discuss in a later section on hardware support, hardware support can also accelerate filtering [33].

The runtime can detect thread-private objects using *dynamic escape analysis* [36], which tracks whether an object is reachable by only the thread that allocated it. Dynamic escape analysis detects such objects at runtime by tracking whether more than one thread can reach an object. A newly allocated object starts as private, meaning that only the thread that created it can reach it. An object is marked as public, meaning that multiple threads can reach it, when a pointer to it is stored into a static variable or into another public object. Read and write barriers must check at runtime whether an object is private, and skip over the barrier operations if an accessed data is private.

5.5.3 Compiler Optimizations

The compiler can identify many of the optimization opportunities using standard or new compiler analyses. Standard data flow analyses and optimizations such as common sub-expression elimination and partial redundancy elimination can eliminate many of the barrier redundancies [3, 20, 40]. Escape analyses [5], which detect whether an object escapes from a method or a thread, can detect transaction-private objects at compile time (compared to dynamic escape analysis, which is performed at runtime). The compiler can also detect immutable data by analyzing a program and reasoning that a particular object, field, or array is never written after it becomes shared by multiple threads.

Compiler optimizations can improve transaction performance significantly – in some programs, compiler optimizations can reduce the single-thread overhead of

STM from 40 to 120% to less than 20% [3]. Compilation scope, control flow ambiguities, and aliasing constrain such compiler optimizations, so many redundancies remain that the STM runtime can pick up dynamically via runtime filtering techniques.

Even when a transaction accesses different data it may access runtime metadata (such as the transaction descriptor or the read and write set pointers) multiple times redundantly. The transaction descriptor typically resides in thread-local storage (TLS), and the read and write set pointers reside in this descriptor, so the compiler can eliminate redundant accesses to TLS by keeping these values in registers. Similarly, two read barriers to different memory locations will both have to access the pointer to the read set from the transaction descriptor, and the compiler can eliminate this redundant access by keeping the read set pointer in a register. How aggressively the compiler performs such optimizations depends on the target architecture as a classic tradeoff exists between redundancy elimination and register pressure.

Besides eliminating redundancies across barriers, optimizations can hoist read and write barrier operations using partial redundancy elimination or loop invariant code motion optimizations. For example, the compiler can hoist the ownership acquisition and undo log operations if a transaction writes to the same field x.f repeatedly inside a loop; similarly for read barrier operations. If a field x.f is read and then written, the compiler can hoist the ownership acquisition operation before the read so that it is not necessary to log in the read set.

Code motion and redundancy elimination optimizations are subject to several constraints for correctness. First, barrier operations make sense only in the scope of their enclosing top-level (that is, non-nested) transaction, so an optimization can hoist a barrier operation only within its atomic block. Second, a barrier operation in one top-level does not make a barrier in a different top-level transaction redundant, so an optimization can eliminate barrier operations due to other barriers within the same top-level transaction only. Third, to support partial rollback of nested transactions (required for failure atomicity), optimizations cannot hoist undo log operations across different nesting levels and cannot eliminate undo log operations due to redundancy from a different nesting level.

5.5.3.1 Representing Barrier Operations

To allow optimization of the TM barrier operations, the compiler's intermediate representation (IR) must model them as first-class operations (rather than calls to library functions) so that optimizations can analyze their effects. With proper IR support, standard redundancy elimination optimizations, such as partial redundancy elimination, code motion, and common sub-expression elimination require few modifications to optimize TM operations.

To maximize optimization opportunities, the IR must expose the operations that comprise the read and write barriers to the optimization phases. Representing each read or write barrier as a single coarse-grain operation that reads or writes transactional memory hides many optimization opportunities from the compiler. Instead, a

"RISC-like" representation that decomposes each barrier into a finer-grain sequence of operations uncovers many optimizations to the compiler; optimizations then result in not only fewer but also cheaper STM operations at runtime. All STM operations, for example, access the transaction descriptor. Exposing the load of the descriptor in the compiler IR allows common sub-expression elimination and code motion to optimize this access.

Note that optimization of STM operations is orthogonal to optimization of the read and write operations. Aliasing may constrain code motion or redundancy elimination of read or write operations, but aliasing does not constrain the same optimizations for the read and write barrier operations. For example, the STM operations of two read operations might be redundant but the two reads might not be redundant due to an intervening write through an ambiguous pointer. So the compiler IR should decouple the STM operations from the memory accesses.

The example in Fig. 5.5 illustrates the optimization opportunities uncovered by a compiler representation that exposes the underlying STM operations. Figure 5.5(a) shows a simple method that assigns the two incoming parameters a and b to two fields x and y of the *this* object. The value assigned to field y depends on the value

foo(int a,int b) {	foo(int a, int b) {
this.x = a;	**stmWrite**(&this.x, a);
this.y = b;	**stmWrite**(&this.y, b);
if (this.z == 0)	t= **stmRead**(&this.z);
this.y = a;	if (t==0)
}	**stmWrite**(&this.y, a);
	}
(a)	(b)
foo(int a, int b) {	foo(int a, int b) {
txn = **stmGetDesc**();	txn = stmGetDesc();
stmOpenWrite(txn,this);	**stmOpenWrite**(txn,this);
stmLog(txn,&this.x)	**stmLog**(txn,&this.x)
this.x = a;	this.x = a;
stmOpenWrite(txn,this);	**stmLog**(txn, &this.y)
stmLog(txn, &this.y)	this.y = b;
this.y = b;	if (this.z == 0) {
stmOpenRead(txn,this);	this.y = a;
if (this.z == 0) {	}
stmOpenWrite(txn,this);	}
stmLog(txn,&this.y);	
this.y = a;	
}	
}	
(c)	(d)

Fig. 5.5 Example illustrating optimizations uncovered by a compiler intermediate representation that exposes underlying STM operations: (**a**) original function, (**b**) representation using coarse-grain read and write barriers, (**c**) representation using finer-grain STM operations, (**d**) representation after optimizations of STM operations

in a third field z of the *this* object. Figure 5.5(b) shows the representation of this method using simple read and write barriers. This representation is generic in that it can be used to generate code for any STM algorithm. Note that the compiler has not used read barriers when accessing the incoming parameters since these are private to the transaction and thus don't require conflict detection or version management by the STM. In Fig. 5.5(c) the compiler has exposed the operations that constitute the read and write barriers of the STM algorithm, which uses eager versioning and object-granularity conflict detection. The compiler first loads the transaction descriptor into a temporary so that it can be reused by the STM operations inside the method. The write barriers have been decomposed into open-for-write and undo logging operations (stmOpenWrite and stmLog, respectively) followed by the write operation. The read barrier has been decomposed into an open-for-read operation (stmOpenRead) followed by the read operation. Exposing the STM operations allows the compiler to detect redundant STM operations. In Fig. 5.5(d) the compiler has eliminated the redundant open-for-write, open-for-read, and undo logging operations. The resulting code uses much cheaper STM operations compared to the code in Fig. 5.5(b).

While we have assumed optimizations for an eager-versioning STM in a managed setting, the IR for lazy-versioning STM or for an unmanaged language such as C would look different, but the optimization opportunities would be similar. Not surprisingly, the more we couple the IR to the STM algorithm, the more optimization opportunities we uncover.

Coupling the IR to the STM algorithm binds the generated executable to the STM algorithm at compile time. This is undesirable if the STM runtime wants to use multiple different STM algorithms at runtime for the same program [29] – for example, based on the program's execution profile, the runtime may want to switch from optimistic to pessimistic concurrency control or to use different hardware acceleration algorithms. In a JIT-compiled environment, binding may be less of an issue as it occurs at runtime and the JIT compiler can regenerate the code if necessary.

5.5.3.2 Code Generation Optimizations

In addition to the global optimizations we have discussed so far, the compiler can also optimize the code generated for the barrier operations. To remove the call-and-return overheads of the STM barriers, the compiler can inline the common fast-path of each barrier operation. This fast-path includes the case in which no contention exists with other transactions and no buffer overflow occurs while adding to the read or write sets, or the undo log. The read barrier fast-path can be accomplished with as few as around ten instructions [3]. Partial inlining saves not only the overheads of the call and return branches, but also the overheads of saving and restoring registers. Partial inlining, however, does increase code size so it should be applied only to the hot parts of an application to reduce its impact on memory system performance.

5.6 Hardware Support for Transactions

STM frameworks support transactional execution on shared memory multiprocessors of any type, including symmetric multiprocessors (SMPs) such as dual-socket or quad-socket servers. The recent trend towards multicore chips or single-chip multiprocessors (CMPs) does not change the basic functionality of an STM. Moreover, the large, on-chip caches available in CMPs reduce the overhead of STM tasks. Instrumentation functions execute faster as auxiliary data structures such as read sets, write sets, and undo logs typically fit in on-chip caches. The caches also support low-latency, high-bandwidth communication between threads for shared data structures such as transactional records and descriptors. As a result, CMPs exhibit good scalability across a wide range of transactional applications, including those that make frequent use of short transactions. On the other hand, the increasing number of cores in CMPs poses a challenge for STM developers as the algorithms and data structures used to orchestrate transactional execution must be highly concurrent.

Even though STM performance scales well with the number of cores in a CMP, the instrumentation overhead is still a significant bottleneck in absolute terms. Compared to the execution time of sequential code without instrumentation, applications executing under STM slow down by a factor of up to $7\times$ [9]. In other words, the instrumented code requires up to eight cores before it becomes faster than the sequential code running on a single core. To address the inefficiency, there are several proposals for hardware support for transactional execution. The proposals vary in their focus, hardware complexity, and performance.

5.6.1 Hardware-Accelerated STM

Hardware-accelerated STM (HASTM) implements a small set of hardware primitives that target the major bottleneck in STM execution: tracking and validating the transaction read set and write set in order to implement conflict detection [33, 9]. To capture the read set and write set, the STM uses at least one barrier per unique address accessed within each transaction. The barrier checks if the corresponding record is locked and inserts its timestamp in a software structure. The validation process traverses the software structures in order to check the validity of records in the read set, and subsequently releases the locks for records in the write set. While write barriers also implement data versioning, the cost of conflict detection is significantly higher as most transactions read more data than they write.

To accelerate the common case of read set and write set management in STM, HASTM uses hardware to implement a generic filtering mechanism [33]. This mechanism allows software to mark fine-grain memory blocks using per-thread mark bits. There are two key capabilities enabled with mark bits. First, software can query if the mark bit has been previously set for a single given block of memory and that there have been no writes to the memory by other threads since the block was marked. Second, software can query if there has potentially been any writes by other threads to any of the set of memory blocks the software has marked.

The HASTM hardware implements mark bits using additional metadata for each line in private caches. The bits are set and reset using software. The hardware tracks if any marked cache line has been invalidated due to an eviction (conflict or capacity miss) or a write access by another thread (coherence miss). Mark bits are used in STM barriers in the following way. The read barrier checks and sets the mark bit for the address of the transaction record or data. If the mark bit was previously set, the record is not inserted into the software structure for the read set. To validate the transaction, the STM quickly queries the hardware if any marked cache lines have been invalidated. If this is not the case, the hardware guarantees that all cache lines accessed through read barriers were private to this thread for the duration of this transaction; no further validation is required. If some lines have been invalidated, the STM reverts to the software-based validation by processing the timestamps for the records in the read set. The software validation determines if the marked lines were evicted because of the limited capacity of the private caches or because of true conflicts between concurrent transactions. Additional mark bits can be used for write barriers to quickly eliminate redundant insertions into the undo log.

Unlike some of the hardware-based TM systems discussed later in this section, HASTM allows transactions to span seamlessly across context switches, page faults, and other interruptions. Before resuming from an interrupt, the STM runtime resets all mark bits. This does not abort the pending transaction – it merely causes a full software validation on commit. After resumption, the transaction benefits from marking for any subsequent accesses and hence gets accelerated. Nevertheless, the resumed transaction does not leverage the marking it performed before interruption. Similarly, HASTM allows a transaction to be suspended and its speculative state inspected and updated by a tool like a garbage collector (GC) without aborting the transaction. As long as the GC or tool does not change any of the transaction records, the suspended transaction will resume without aborting, but may lose some of its mark bits and perform a full software validation.

We can also accelerate STM conflict detection without modifying hardware caches. First-level caches are typically in the critical path of the processor core. They also interact with complex sub-systems such as the coherence protocol and pre-fetch engines. Even simple changes to caches can affect the processor clock frequency or increase its design and verification time. Signature-accelerated STM (SigTM) provides a similar filtering mechanism using hardware signatures maintained outside of the caches [9]. The signatures are hardware Bloom filters [6] that pessimistically encode sets of addresses provided by software. There are two signatures per hardware thread, one for read set and one for write set tracking. To implement conflict detection, the hardware monitors coherence requests for exclusive accesses that indicate data updates by concurrent transactions or requests for shared accesses that indicate data reads by concurrent transactions. The hardware tests if the address for such a request is potentially a member of the set encoded by the read set or write set signatures. If this is the case, a potential conflict is signaled. The STM can immediately abort the transaction or proceed with software validation.

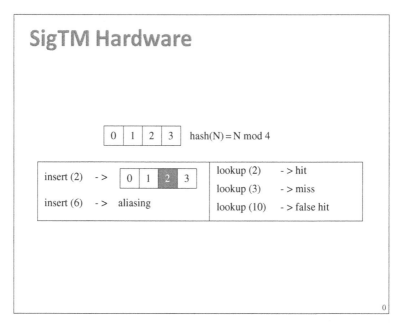

Fig. 5.6 Example illustrating the operation of SigTM hardware signatures for conflict detection

Figure 5.6 illustrates the operation of SigTM hardware signatures for read set tracking and conflict detection. For simplicity, we show a four-bit Bloom filter with a single hash function based on a modulo operation. On a read set barrier for address 2, the SigTM code inserts the address in the read set by first hashing it and setting the identified bit number 2 in the hardware filter. On a later read to address 6, the SigTM code will set the same filter bit again as the hash function leads to aliasing for addresses 2 and 6. As transactions on other processors commit their write set, the hardware signatures monitor the corresponding exclusive requests on the cache coherence protocol. For each request address, the hardware applies the same hash function and tests the identified filter bit. The filter lookup on an exclusive request to address 2 from a remote processor finds bit number 2 of the filter set and identifies a true conflict. A lookup for address 3 correctly identifies the lack of conflict as bit number 3 in the filter is not set. A lookup for address 10, however, identifies a false conflict as bit number 2 of the filter is set. There is no way for the hardware signature to determine if the filter bit was set on a read barrier to address 10 or any of the other addresses that alias to the same filter bit.

To minimize false conflicts, the hardware signatures must be 1024 and 64 bits wide for read set and write set tracking, respectively [9]. Read signatures need to be longer as most applications read significantly more shared data than they write [8]. Moreover, SigTM uses multiple hash functions in order to identify multiple filter bits to set or test per insertion or lookup in order to decrease the frequency of aliasing. Since the length of hardware signatures and the hash functions are not directly visible to user code, they can be changed with each processor generation to balance hardware cost and the probability of false conflicts given the common case behavior of representative workloads.

The filtering mechanisms of HASTM and SigTM can be used flexibly by the STM to track transaction records and reduce instrumentation overhead. Nevertheless, there are certain differences between the two systems. SigTM can encode read set and write set membership even if their size exceeds the capacity of private caches. Capacity and conflict misses do not cause software validation, as it is the case in HASTM. On the other hand, SigTM uses probabilistic signatures that will never miss a true conflict but may produce false conflicts due to address aliasing in the Bloom filter. Another implication of signature aliasing is that SigTM cannot filter redundant insertions into software structures such as the undo log. Since the hardware signatures are relatively short and easy to manipulate, they can be saved and restored as transactions are suspended and resumed for context switches and other interruptions. This is not the case in HASTM where the filter contents may be lost if the processor is used to run other tasks for a while. On the other hand, the SigTM signatures track physical addresses, which means that their content must be ignored after paging.

5.6.2 Hardware-Based Transactional Memory

TM can also be implemented fully in hardware, referred to as HTM (hardware transactional memory). An HTM system requires no read or write barriers within the transaction code. The hardware manages data versions and tracks conflicts transparently as software performs ordinary read and write accesses. Apart from reducing the overhead of instrumentation, HTM does not require two versions of the functions used in transactions and works with programs that call uninstrumented library routines.

HTM systems rely on the cache hierarchy and the cache coherence protocol to implement versioning and conflict detection [17, 27]. Caches observe all reads and writes issued by the cores, can buffer a significant amount of data, and are fast to search because of their associative organization. All HTMs modify the first-level caches, but the approach extends to lower-level caches, both private and shared. Figure 5.7 shows how a data cache tracks the transaction read set and write set by annotating each cache line with R and W bits that are set on the first read or write to the line, respectively. When a transaction commits or aborts, all tracking bits are cleared simultaneously using a gang or flash reset operation. Caches implement data versioning by storing the working set for the undo log or the data buffer for the transactions. Before a cache write under eager versioning, we check if this is the first update to the cache line within this transaction (W bit reset). In this case, the cache line and its address are added to the undo log using additional writes to the cache. If the transaction aborts, a hardware or software mechanism must traverse the log and restore the old data versions. In lazy versioning, a cache line written by the

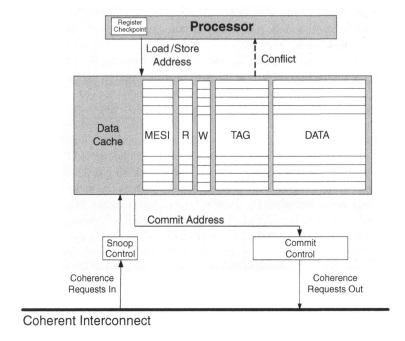

Fig. 5.7 The hardware support for transactional execution in an HTM system

transaction becomes part of the write buffer by setting its W bit. If the transaction aborts, the write buffer is instantaneously flushed by invalidating all cache lines with the W bit set. If the transaction commits, the data in the write buffer becomes instantaneously visible to the rest of the system by resetting the W bits in all cache lines.

To detect conflicts, the caches must communicate their read sets and write sets using the cache coherence protocol implemented in multicore chips. Pessimistic conflict detection uses the same coherence messages exchanged in existing systems. On a read or write access within a transaction, the processor will request shared or exclusive access to the corresponding cache line. The request is transmitted to all other processors that look up their caches for copies of this cache line. A conflict is signaled if a remote cache has a copy of the same line with the R bit set (for an exclusive access request) or the W bit set (for either request type). Optimistic conflict detection operates similarly but delays the requests for exclusive access to cache lines in the write set until the transaction is ready to commit. A single, bulk message is sufficient to communicate all requests.

The minimal software interface for hardware TM architectures is instructions that specify the beginning and end of transactions in user code [22]. Many HTM systems perform conflict detection and versioning for all load and store addresses used within a transaction. To allow software optimizations for thread-private or immutable data such as those discussed in the previous section, recent HTM architectures advocate the use of special load and store instructions to selectively use conflict detection and versioning mechanisms [25, 37]. Finally, HTM systems support software-based contention management and other runtime functions through transactional handlers [25]. Software can register handlers that are invoked automatically by hardware on important events such as transactional conflicts.

Figures 5.8 and 5.9 compare the performance of a full HTM with lazy versioning and optimistic conflict detection to that of the STM and SigTM systems for two

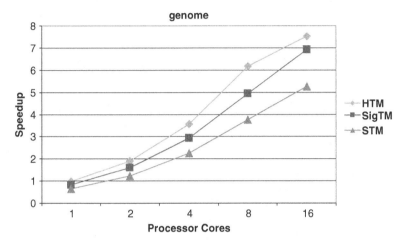

Fig. 5.8 The speedup for three TM systems on a simulated CMP architecture with up to 16 nodes for the genome application

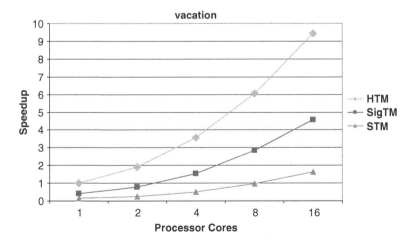

Fig. 5.9 The speedup for three TM systems on a simulated CMP architecture with up to 16 nodes for the vacation application

applications running on a simulated CMP [9]. Genome is a bioinformatics application and performs gene sequencing. Vacation implements a travel reservation system powered by an in-memory database. Both applications belong to the STAMP benchmark suite for transactional memory systems [8]. When using a single processor, the STM code is $1.5\times$ and $7\times$ slower than sequential code for genome and vacation, respectively. The hardware filtering in SigTM reduces the overhead of instrumentation to $1.2\times$ and $2.4\times$, respectively. HTM eliminates most sources of overhead for transactional execution and is typically within 2% of the sequential code when using a single processor. Running genome with 16 cores, STM leads to a speedup of $5.2\times$ over sequential execution without instrumentation, while SigTM provides a speedup of $6.9\times$. Running vacation with 16 cores, STM achieves a speedup of $1.6\times$, while SigTM accelerates transactional execution by more than a factor of two (speedup of $4.6\times$). With 16 processor cores, the HTM is $1.1\times$ to $2\times$ faster than SigTM, depending on the number of barriers per transaction needed for the SigTM code.

Even though HTM systems provide obvious performance advantages, they nevertheless introduce additional challenges. The caches used to track the read set, write set, and write buffer for transactions have finite capacity and may overflow on a long transaction. It is also challenging to handle transactional state in caches on interrupts, context switching, and paging events. Long transactions may be rare, but they still must be handled in a manner that preserves atomicity and isolation. Placing implementation-dependent limits on transaction sizes is unacceptable from the programmer's perspective. To address this shortcoming, most HTM systems have been extended with virtualization mechanisms [31]. Such mechanisms use hardware, firmware, or the operating system to map transactional state to virtual memory when hardware caches overflow [13, 11].

5.6.3 Hybrid Hardware–Software Techniques

An alternative virtualization technique is to use a hybrid HTM–STM implementation [24, 14]. Transactions start in the HTM mode using hardware mechanisms for conflict detection and data versioning. If hardware resources are exceeded, the transaction is rolled back and restarted in the STM mode using additional instrumentation. A challenge for hybrid TM systems is to perform conflict detection between concurrent HTM and STM transactions.

The hybrid approach is used by the upcoming Rock CMP by Sun Microsystem. Similar to the original HTM proposal [22], Rock provides a small transactional cache per core that acts as a separate write buffer for data versioning. The read set and write set are tracked by the regular data cache using additional metadata bits. These modest hardware resources are often sufficient to execute most transactions in a program without the need for instrumentation. The hardware is backed by an STM that is used when hardware resources are insufficient or on complex events such as context switches, interrupts, or paging. Hence, there is no need for a complex mechanism that virtualizes hardware TM resources.

The Rock design is also using the transactional cache to support other forms of speculative execution. For instance, the cache supports a double compare-and-swap instruction that accelerates non-blocking algorithms used for multithreaded data structures and fast thread scheduling. It also supports speculative lock-elision [30], a technique that speculatively executes lock-based regions as atomic transactions and falls back to conventional locking when there is no forward progress. Finally, Rock uses the hardware resources for TM for run-ahead execution in order to minimize the impact of low-latency cache misses [28]. These additional uses mitigate the risks of introducing hardware support for TM in a commercial chip. Other researchers are currently investigating the use of hardware TM mechanisms to accelerate or improve various debugging, security, availability, and memory management systems [12].

5.7 Conclusion

Multicore processors bring parallel programming and all its challenges to mainstream programmers. We will need new programming technologies that help programmers overcome the challenges of parallel programming. At the same time, multicore processors with their large transistor budgets and on-chip memories have an opportunity to provide support for such new programming technologies.

One of the challenges of multicore programming is synchronization using locks. Composing scalable parallel applications using locks is difficult and full of pitfalls. Transactional memory avoids many of these pitfalls and allows the programmer to compose applications safely and in a manner that scales. Transactions improve the programmer's productivity by shifting the hard concurrency control problems from the application developer to the system designer.

There has been significant research activity and progress on transactional memory recently, but many open challenges must be addressed before transactions can make it into the mainstream as first-class language constructs. This chapter drills down on one of the major challenges of implementing transactional memory: optimizing its performance using software and hardware techniques.

Deploying transactional memory will require pure software implementations that run on today's hardware. Implementing transactional memory efficiently requires support and new optimizations in the compiler and the language runtime. Although the performance of software transactional memory (STM) can scale with increasing number of cores, its overheads remain high. Compiler and runtime optimization techniques, such as the ones described in this chapter, can significantly reduce the overheads of STM. Even with these optimizations, overheads remain in a pure software implementation, providing an opportunity for hardware acceleration.

Ultimately, we predict that TM language constructs will be integrated into existing language specifications or replace locks as the main synchronization construct in new languages built for parallel programming. Although transactions are not a panacea for all parallel programming challenges, they do take a concrete step towards making parallel programming easier. Several vendors have already made prototype implementations of TM language extensions available to early adopters. These prototypes are based on production-quality compilers. Feedback from early adopters will help refine the TM language constructs and their semantics. Given the wide interest in TM, we expect more vendors to provide such prototypes.

Adoption of TM programming models will drive new hardware requirements in future processors as transactional memory performance can benefit greatly from architectural support. Pure hardware transactional memory (HTM) techniques can provide transactional semantics at the instruction level without any software over-heads, but hardware resource limitations constrain the size, semantics, and duration of hardware transactions. HTM has uses beyond acceleration of a TM programming model so it is a good first step. At this point in time, at least one hardware vendor has announced HTM support in an upcoming processor. Mainstream transactional memory implementations on future multicore processors will likely combine software and hardware techniques for performance using techniques such as Hybrid TM and HASTM.

References

1. M. Abadi, A. Birrell, T. Harris, and M. Isard. Semantics of transactional memory and automatic mutual exclusion. In *Proceedings of the Symposium on Principles of Programming Languages*, San Francisco, CA, Jan 2008.
2. A.-R. Adl-Tabatabai, C. Kozyrakis, and B. Saha. Unlocking Concurrency. *ACM Queue*, 4:10, Dec 2006/Jan 2007.
3. A.-R. Adl-Tabatabai, B. T. Lewis, V. S. Menon, B. M. Murphy, B. Saha, and T. Shpeisman. Compiler and runtime support for efficient software transactional memory. In *Proceedings of the Conference on Programming Language Design and Implementation*, Ottawa, Canada, June 2006.
4. C. S. Ananian, K. Asanovic, B. C. Kuszmaul, C. E. Leiserson, and S. Lie. Unbounded transactional memory. In *Proceedings of the 11th International Symposium on High-Performance Computer Architecture*, Feb 2005.
5. B. Blanchet. Escape analysis for object-oriented languages: Application to Java. In *Proceedings of the Conference on Object Oriented Programming Systems, Languages, and Architectures*, 1999.
6. B. Bloom. Space/Time Trade-Offs in Hash Coding with Allowable Errors. *Communications of ACM*, 13(7), July 1970.
7. C. Blundell, E. C. Lewis, and M. M. K. Martin. Deconstructing transactions: The subtleties of atomicity. In *Fourth Annual Workshop on Duplicating, Deconstructing, and Debunking*, 2005.
8. C. Cao Minh, J. Chung, C. Kozyrakis, and K. Olukotun. STAMP: Stanford Transactional Applications for Multiprocessing. In *Proceedings of the IEEE International Symposium on Workload Characterization*, Seattle, WA, October 2008.
9. C. Cao Minh, M. Trautmann, J. Chung, A. McDonald, N. Bronson, J. Casper, C. Kozyrakis, and K. Olukotun. An effective hybrid transactional memory system with strong isolation guarantees. In *Proceedings of the International Symposium on Computer Architecture*, San Diego, CA, June 2007.
10. B.D. Carlstrom, A. McDonald, H. Chafi, J.W. Chung, C.C Minh, C. Kozyrakis, and K. Olukotun. The ATOMO transactional programming language. In *Proceedings of the Conference on Programming Language Design and Implementation*, Ottawa, Canada, June 2006.

11. W. Chuang, S. Narayanasamy, G. Venkatesh, J. Sampson, M. Van Biesbrouck, M. Pokam, O. Colavin, and B. Calder. Unbounded page-based transactional memory. In *Proceedings of the 12th International Conference on Architecture Support for Programming Languages and Operating Systems*, San Jose, CA, Oct 2006.

12. J. Chung. *System Challenges and Opportunities for Transactional Memory*. Ph.D. thesis, Stanford University, June 2008.

13. J. Chung, C. Cao Minh, A. McDonald, T. Skare, H. Chafi, B. Carlstrom, C. Kozyrakis, and K. Olukotun. Tradeoffs in transactional memory virtualization. In *Proceedings of the 12th International Conference on Architecture Support for Programming Languages and Operating Systems*, San Jose, CA, October 2006.

14. P. Damron, A. Fedorova, Y. Lev, V. Luchangco, M. Moir, and D. Nussbaum. Hybrid transactional memory. In *Proceedings of the 12th International Conference on Architecture Support for Programming Languages and Operating Systems*, San Jose, CA, October 2006.

15. D. Dice, O. Shalev, and N. Shavit. Transactional locking II. In *Proceedings of the 20th International Symposium on Distributed Computing*, Stockholm, Sweden, September 2006.

16. L. Hammond, B. Carlstrom, V. Wong, B. Hertzberg, M. Chen, C. Kozyrakis, and K. Olukotun. Programming with transactional coherence and consistency (TCC). In *Proceedings of the Symposium on Architectural Support for Programming Languages and Operating Systems*, Boston, MA, October 2004.

17. L. Hammond, V. Wong, M. Chen, B. Hertzberg, B. Carlstrom, M. Prabhu, H. Wijaya, C. Kozyrakis, and K. Olukotun. Transactional memory coherence and consistency. In *Proceedings of the 31st Annual International Symposium on Computer Architecture*, Munich, Germany, June 2004.

18. T. Harris and K. Fraser. Language support for lightweight transactions. In *Proceedings of the Conference on Object Oriented Programming Systems, Languages, and Architectures*, Anaheim, CA, October 2003.

19. T. Harris, S. Marlow, S. P. Jones, and M. Herlihy. Composable memory transactions. In *Proceedings of the Symposium on Principles and Practice of Parallel Programming*, Chicago, IL, June 2005.

20. T. Harris, M. Plesko, A. Shinnar, and D. Tarditi. Optimizing memory transactions. In *Proceedings of the Conference on Programming Languages Design and Implementation*, Ottawa, Canada, June 2006.

21. M. Herlihy, V. Luchangco, M. Moir, and W. Scherer. Software transactional memory for dynamic-sized data structures. In *Proceedings of the Twenty-Second Annual ACM SIGACT-SIGOPS Symposium on Principles of Distributed Computing (PODC)*, July 2003.

22. M. Herlihy and E. Moss. Transactional memory: Architectural support for lock-free data structures. In *Proceedings of the 20th Annual International Symposium on Computer Architecture*, San Diego, CA, May 2003.

23. R. L. Hudson, B. Saha, A.-R. Adl-Tabatabai, B. C. Hertzberg. McRT-Malloc: A scalable transactional memory allocator. In *Proceedings of the International Symposium on Memory Management*, Ottawa, Canada, June 2006.

24. S. Kumar, M. Chu, C. J. Hughes, P. Kundu, and A. Nguyen. Hybrid transactional memory. In *Proceedings of the Symposium on Principles and Practices of Parallel Processing, Manhattan*, New York, March, 2006.

25. A. McDonald, J. Chung, B. Carlstrom, C. Cao Minh, H. Chafi, C. Kozyrakis, and K. Olukotun. Architectural semantics for practical transactional memory. In *Proceedings of the 33rd International Symposium on Computer Architecture*, Boston, MA, June 2006.

26. V. Menon, S. Balensiefer, T. Shpeisman, A.-R. Adl-Tabatabai, R. L. Hudson, B. Saha, and A. Welc. Practical weak atomicity semantics for Java STM. In *Proceedings of the Symposium on Parallelism in Algorithms and Architecture*, Munich, Germany, June 2008.

27. K. Moore, J. Bobba, M. Moravan, M. Hill, and D. Wood. LogTM: Log-based transactional memory. In *Proceedings of the 12th International Conference on High Performance Computer Architecture*, Austin, TX, Feb 2006.

28. O. Mutlu, J. Stark, C. Wilkerson, and Y. N. Patt. Runahead execution: An alternative to very large instruction windows for out-of-order processors. In *Proceedings of the 9th International Symposium on High-Performance Computer Architecture*, Anaheim, CA, Feb 2003.

29. Y. Ni, A. Welc, A.-R. Adl-Tabatabai, M. Bach, S. Berkowits, J. Cownie, R. Geva, S. Kozhukow, R. Narayanaswamy, J. Olivier, S. Preis, B. Saha A. Tal, X. Tian. Design and implementation of transactional constructs for C/C++. In *Proceedings of the Conference on Object Oriented Programming Systems, Languages, and Architectures*, Nashville, TN, Oct 2008.

30. R. Rajwar and J. Goodman. Speculative lock elision: Enabling highly concurrent multi-threaded execution. In *Proceedings of the 34th International Symposium on Microarchitecture*, Istanbul, Turkey, Nov 2002.

31. R. Rajwar, M. Herlihy, and K. Lai. Virtualizing transactional memory. In *Proceedings of the 32nd International Symposium on Computer Architecture*, Madison, WI, June 2005.

32. B. Saha, A.-R. Adl-Tabatabai, R. Hudson, C. C. Minh, B. Hertzberg. McRT-STM: A high performance software transactional memory system for a multi-core runtime. In *Proceedings of the Symposium on Principles and Practice of Parallel Programming*, New York, Mar 2006.

33. B. Saha, A.-R. Adl-Tabatabai, and Q. Jacobson. Architectural support for software transactional memory. In *Proceedings of the 39th International Symposium on Microarchitecture*, Orlando, FL, Dec 2006.

34. W. N. Scherer III and M. L. Scott. Advanced contention management for dynamic software transactional memory. In *Proceedings of the Symposium on Principles of Distributed Computing*, Las Vegas, USA, July 2005.

35. F. Schneider, V. Menon, T. Shpeisman, and A.-R Adl-Tabatabai. Dynamic optimization for efficient strong atomicity. In *Proceedings of the Conference on Object Oriented Programming Systems, Languages, and Architectures*, Nashville, TN, Oct 2008.

36. T. Shpeisman, V. S. Menon, A.-R. Adl-Tabatabai, S. Balensiefer, D. Grossman, R. L. Hudson, K. Moore, and B. Saha. Enforcing isolation and ordering in STM. In *Proceedings of the Conference on Programming Language Design and Implementation*, San Diego, USA, June 2007.

37. A. Shriraman, M. Spear, H. Hossain, V. Marathe, S. Dwarkadas, and M. Scott. An integrated hardware–software approach to flexible transactional memory. In *Proceedings of the 34th International Symposium on Computer Architecture*, San Diego, CA, June 2007.

38. M. Spear, M. Michael, and M. Scott. *Inevitability Mechanisms for Software Transactional Memory*. Presented at the Workshop on Transactional Computing, Salt Lake City, UT, Feb 2008.

39. C. von Praun, L. Ceze, and C. Cascaval. Implicit parallelism with ordered transactions. In *Proceedings of the Symposium on Principles and Practice of Parallel Programming*, San Jose, CA, Mar 2007.

40. C. Wang, W. Chen, Y. Wu, B. Saha, and A.-R. Adl-Tabatabai. Code generation and optimization for transactional memory constructs in an unmanaged language. In *Proceedings of the International Symposium on Code Generation and Optimization*, San Jose, USA, Mar 2007.

41. A. Welc, B. Saha, and A.-R. Adl-Tabatabai. Irrevocable transactions and their applications. In *Proceedings of the Symposium on Parallelism in Algorithms and Architecture*, Munich, Germany, June 2008.

Chapter 6
General-Purpose Multi-core Processors

Chuck Moore and Pat Conway

Abstract During the past several decades, the general-purpose microprocessor industry has effectively leveraged Moore's Law to offer continually increasing single-thread microprocessor performance and compelling new features. However, the amazing increase in performance was not free: many practical design constraints, especially power consumption, were pushed to their limits. The integration of multiple microprocessor cores into CPU chips has improved the capability of the single-CPU chip systems and extended the capability of the multiple-CPU chip systems in a very natural way. General-purpose multi-core processors have brought parallel computing into the mainstream and penetrated markets from laptops to supercomputers. This chapter discusses the history and trends behind this exciting development and future challenges and opportunities in hardware and software. This chapter presents the AMD Opteron^{TM} microarchitecture as a case study in how one company addressed the power, software stack, and infrastructure challenges posed by general-purpose CMP architectures across several product generations.

6.1 Motivating Trends and History

In 1965, Gordon Moore retrospectively observed that each new technology generation allowed designers to put twice as many transistor devices on the same-sized piece of silicon as the preceding generation. Perhaps more importantly, he also predicted that this trend would likely continue into the foreseeable future [1]. This compelling vision, commonly referred to as *Moore's Law*, has inspired generations of computer architects, technology developers, and companies to invest their time, intellect, and money to continue driving this exponential growth.

During the past several decades, the general-purpose microprocessor industry has effectively leveraged Moore's Law to offer continually increasing single-thread

C. Moore (✉)
Advanced Micro Devices Inc., Sunnyvale, CA, USA
e-mail: chuck.moore@amd.com

microprocessor performance and compelling new features. For example, in the 90 nm CMOS technology node, several companies were able to produce multi-instruction issue superscalar microprocessors with 64-bit addressing and up to a megabyte of cache memory running in excess of 2 GHz [2–4]. However, this amazing increase in performance was not free, pushing many practical design constraints to their limits. In particular, very significant increases in power consumption accompanied the dramatic increases in operating frequency and microarchitectural complexity. At the system level, these increases in power consumption resulted in extra costs for the power supplies and thermal mitigation solutions. In the end, these considerations set practical limits on how much power a given CPU socket can effectively handle in an economically feasible way. As the industry ran up to these limits, power efficiency and management emerged as a first-order design constraint in all types of systems.

At the same time, although Moore's Law continued to offer more integrated transistors with each new technology node, computer architects faced several more fundamental limitations. In particular, the prospects of deeper pipelining to increase frequency, more advanced microarchitectural structures to increase instruction-level parallelism, and larger caches to increase locality characteristics were all at the point of diminishing returns with respect to their costs. Figure 6.1 shows many of these trends qualitatively, and is supported by a large body of published work that more fully documents these issues [5, 6, 7, 8]. This led computer architects and designers to look beyond the microprocessor core for ways to improve performance and value.

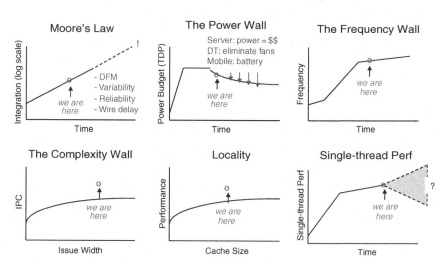

Fig. 6.1 Key trends

One important area on which computer architects immediately focused their attention was the memory system. There has always been an important balance between data availability from the memory system and computation. While this aspect of computer architecture had certainly not been completely ignored, the gap between cycle time of microprocessors and those of commodity DRAM chips had widened significantly (as illustrated in Fig. 6.2). This gap created very long latencies for accesses from the microprocessor core to the memory system. Clearly, it does not matter how powerful the execution capability of a processor is if the data cannot be supplied at an appropriate rate and/or time. More generally, for some larger workloads, the frequent waits for memory data return significantly degraded the overall performance of the system.

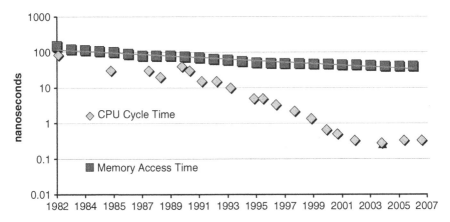

Fig. 6.2 Historical cycle times of CPUs and DRAM memories

In response to these issues, computer architects have focused a great deal of attention on improving the performance of the overall memory system. The use of on-chip multi-level cache hierarchies allowed designers to deploy larger overall cache footprints without disrupting the sensitive power and size constraints of the innermost L1 caches. The deployment of various hardware- and software-based prefetching mechanisms enabled the speculative access of memory ahead of the demand usage to minimize wait time and latency. The optimization of the memory controller to support more outstanding requests and to schedule requests that take maximum advantage of the DRAM banked structures enabled designers to shave additional latency from these critical memory references. The AMD Opteron processor, introduced in 2003, directly integrated the memory controller on the CPU die, eliminating an entire chip crossing in the path to memory.

Although these memory system optimizations offered some important performance improvements, they also hit the inevitable limits of diminishing returns. As each new technology node continued to offer more transistors, computer architects were compelled to find new ways to make the most effective use of them. Considering that many higher-end computer systems had made good use of multiple CPU chips to form symmetric multiprocessor (SMP) systems for decades, the integration of multiple microprocessor cores on the chip was the most natural next step. This signaled the start of the chip multiprocessing (CMP) era in computer architecture. Interestingly, the integration of multiple microprocessor cores in the CPU chips not only improved the capability of the single-CPU chip systems, it also extended the capability of the multiple-CPU chip systems in a very natural way, allowing them to build even larger SMP systems.

The integration of multiple microprocessor cores on the die raised some very interesting questions. First, given that multi-chip SMP systems had been available for decades before this point, were there aspects of these computing platforms that prevented the mainstream computer markets from broadly accepting them? Second, what sort of new optimizations become possible, and what sort of new challenges become exposed, when integrating multiple microprocessor cores on the same die?

The answer to the first question is that, although these multi-core SMP systems were gainfully accepted for use on certain higher-end technical and business workloads, a combination of market-driven economics and technical challenges slowed down the broad acceptance of these systems for mainstream deployment. In particular, the increased complexity of these systems tended to increase their time-to-market profile to the point at which it was difficult to compete with the ongoing increases in single-processor performance (bear in mind that this was during a time when single-thread processor performance was doubling every 18 months or so). In addition, the added costs at the system level due to multiple sockets, more exotic interconnects, and larger form factors put these systems into much higher priced and lower volume market segments. Finally, the availability of compelling workloads that made effective use of the multiple microprocessor cores was quite limited. As one might expect, the combination of the declining advancement of single-core performance and the integration of multiple microprocessor cores on a single die have helped to overcome many of the economic and market segment barriers. However, it is important to note that these systems still share many of the workload and scalability issues with their higher-end multi-chip cousins.

The second question is the topic addressed by the remainder of this chapter. In Sect. 6.2, we will describe some of the critical software and hardware challenges, as well as some of the interesting opportunities, in these emerging multiprocessor systems. In Sect. 6.3, we will walk through a case study of the AMD Opteron family of chips to illustrate some of the real-world tradeoffs that have gone into this design. Finally, in Sect. 6.4, we will offer some thoughts on the next round of challenges for microprocessor architects, along with some possible directions that these may take us.

6.2 CMP Design – Challenges and Opportunities

6.2.1 Power

As described earlier in this chapter, power has emerged as a first-order design constraint for nearly all types of computer systems. This is particularly true for the case of CMPs since these are often deployed in systems with fixed power budgets. When multiple cores are integrated on a single chip, the available power supply (as well as the heat dissipation structures) must be shared among all cores and other resources on the chip. This can have a dramatic impact on the power budget allocated to each core, especially when one considers that the cores tend to consume the most power of any of the on-chip components. In effect, power budgeting is a zero-sum game.

Power consumption, of course, is a function of capacitance, switching frequency, and the voltage squared. As a result, a relatively small reduction in voltage can dramatically reduce power. At the same time, a reduction in voltage also causes a linear reduction in operating frequency and, therefore, a reduction in single-core performance.[1] Chip designers quickly realized that they could trade off a couple of speed grades (about 10% of operating frequency) in each core to enable dual-core chips in the same chip-level power budget. As it turns out, this is a reasonable tradeoff for most end users. For many casual users, the throughput advantage of two cores offers a noticeable responsiveness improvement since the OS can spread the time multi-plexed processes between two cores instead of just one. For more advanced users with reasonably mature parallel applications, the performance advantage of the second core typically outweighs the performance loss due to the reduced frequency.

As CMPs grow in core count, the power constraint becomes more and more significant. For example, quad-core designs must share the available power budget among four cores, and six-core designs must share among six cores. In each of these cases, the operating frequency, and resulting performance, of each core in the CMP continues to drop. But it is perhaps even more important to recognize that CMOS technology has a practical limit on minimum operating voltage (V_{ddmin}). As CMP designs continue to drop voltage to make room for more cores, they will eventually hit this limit. Beyond this point, power reduction becomes a linear function of reducing the operating frequency. In other words, once V_{ddmin} has been reached, the doubling of core count will result in approximately halving the operating frequency.

The deployment of multi-core chips clearly escalates the importance of power optimization in microprocessor design to new levels. The design tradeoffs must take into account a smaller per-core budget and recognize that any power inefficiencies in the design are effectively multiplied by the core count. As a result, designers are forced to revisit previously reasonable tradeoffs. For example, the incremental performance benefits resulting from aggressive speculation must be rebalanced against the power consumed by the cases in which the speculation was incorrect. Similarly, the latency-hiding benefits of prefetching from memory must be carefully balanced from an accuracy and timeliness standpoint, since a considerable amount of system-level power can be expended on bus transactions, DRAM accesses, and cache activity. More generally, design tradeoffs involving the use of even small amounts of power must be carefully examined because these effects are additive in each core, and then multiplied by the number of cores.

The increased sensitivity to power as a chip-level resource also calls for more advanced types of power management. While some degree of fine-grain and

---

[1] It is interesting to note that the drive for higher and higher frequencies during the previous decade was augmented by *increasing* the voltage, which contributed significantly to the dramatic increases in single-chip power consumption.

coarse-grain clock gating[2] is common in most microprocessor designs today, the effectiveness of these techniques must continuously improve going forward [9]. In addition, it will become increasingly important for designs to make use of coarse-grain voltage islands and voltage-gating techniques. These will enable the chip-level power management controllers to shut off completely the power supply to unused sections of the chip at any given time. Of course, there are some interesting tradeoffs in the design of these controllers because the on/off cycles can involve significant latencies that ultimately decrease system-level responsiveness. Similarly, switching off the voltage distribution systems can have significant implications on the overall chip-level electrical stability, so the reliable design of these systems becomes very complex.

Some designs have taken the power management control systems to the next level by actually provisioning power as a resource. In these systems, the power control system monitors power consumption of the on-chip components and dynamically adjusts individual component clock frequencies and voltages to partition the power according to predefined profiles [10]. For example, early forms of these control systems enable one core in a dual-core system to operate at a higher frequency when the second core is idle or otherwise using very little power. We believe that these techniques, while complex, will grow in popularity as CMPs integrate more cores.

6.2.2 System-Level Software Stack

From a software perspective, chip multiprocessors simply represent an expansion of the existing multi-chip SMP systems that have existed for some time. However, the proliferation of these systems in mainstream markets, as well as the extended scalability of the multi-chip solutions, put some significant new demands on the software stack. Moreover, CMP designs can offer better user-observed responsiveness, increased task-level throughput, and higher performance on multi-threaded applications – but only if the system software stack and tools are in place to enable these improvements. In this section, we examine the implications for several key elements in that system-level software stack.

6.2.2.1 Operating System

The kernel of the operating system (OS) is responsible for resource management, scheduling, networking, device integration, and managing many of the low-level interfaces to the hardware. While most modern OSs make heavy use of time-sliced multi-tasking, the addition of multiple cores adds a new dimension to the system-level scope that the OS must manage. Although a detailed treatment of

---

[2]Fine-grain clock gating refers to the power-optimizing design practice of inhibiting the clock connected to one or more storage cells when it is known that their value is not going to change on any given cycle. Coarse-grain clock gating refers to the technique of inhibiting the clock on a larger, more autonomous chunk of logic, such as an execution unit.

the implications for the OS is beyond the intended scope of this chapter, the overall effectiveness of the resulting system is highly dependent on making the OS multiprocessor-aware.

Internal to the OS, several key data structures help the OS keep track of the dynamic state of the computer system. Data structures manage process scheduling, process migration, virtual memory translations, protection boundaries, I/O requests, and other forms of resource allocations. In a symmetric multiprocessor system, the OS can run on any processor in the system, and – in some cases – the OS can actually be running on multiple processors at the same time. To protect against data races in these situations, the OS places lock constructs on many of the key internal data structures. As the number of processors in the system increases, so does the potential for lock contention on these key data structures, which can seriously inhibit the overall scalability and throughput of the system.

As a result, modern OSs must tend towards the use of finer grain locks for these data structures. History and experience show that the appropriate placement and optimization of these finer grain locks is a non-trivial matter. High-end OSs have demonstrated significant gains in scalability, but this has taken years of optimization and tuning by teams of experts [11]. The gains in scalability often come with additional overhead even when operating without lots of CPU resources.

In addition, the OS *scheduler* must also be improved. Instead of time-slicing the active processes onto one CPU, the scheduler must now comprehend the dispatch to, and management of processes on, multiple CPUs. For a relatively small number of CPUs, this actually offers a significant improvement in responsiveness at the system level. However, when the available CPU count gets very high, the scheduler risks becoming a serialized bottleneck in the juggling of the active processes.

Moreover, the availability of multiple CPUs in the system also opens the door for parallel applications that put new demands on the scheduler. Ideally, these applications presume multiple threads running in different CPU cores at the same time. They also present a higher demand for the dynamic allocation and deallocation of threads as they implement fork and join-type primitives to develop and identify the parallel execution opportunities. Over time, it is likely that the optimal execution of many of these parallel applications will demand gang scheduling and location-aware assignment of resources across the system. In light of these emerging demands, it appears that the OS must pay special attention to the architecture and structure of the scheduler in future CMP systems.

6.2.2.2 Hypervisor

As systems grow to offer more compute capacity, so does the interest in using virtualization techniques to subdivide the system effectively. Virtualization on a multiprocessor system enables the logical partitioning of the resources of a single physical system, including cores, across multiple virtual systems that are isolated from one another. A *hypervisor* is a virtualization platform that allows multiple OSs to run on a host computer at the same time. The hypervisor plays a key role in managing the low-level system resources and presenting independent virtual system images to each OS that runs above it in the stack. As the number of CPUs and associated resources in the system increase, so do the challenges for the hypervisor.

Considering that most existing hypervisors are already structured to support multi-chip SMPs for larger scale servers, the introduction of CMPs really just calls for an extension of this existing support. However, as CMPs make their way into the commodity personal computer and notebook markets, the flexibility, ease of use, robustness, and utility model for virtualization is likely to expand.

6.2.2.3 Managed Runtime Environments

Today, managed runtime environments (MREs) such as the Java virtual machine (JVM) and common language runtime (CLR) are an increasingly popular development and execution paradigm for new applications. The available class libraries, sensible abstractions, automatic memory management, and device independence combine to offer a significant improvement in overall programmer productivity. While these runtime environments also come with a performance overhead, the gains in productivity are arguably a very good tradeoff for many classes of applications.

As CMP designs proliferate in the mainstream computer markets, there will be increasing demand for applications that make full and beneficial use of the additional compute capability. Considering that there are very few parallel applications available for mainstream use today, the only benefit that most consumers will get from CMP designs is the improvement in system-level responsiveness when they are running multiple applications concurrently. In the future, it is likely that the availability of compelling parallel applications will govern the growth in demand for increasingly capable CMP designs.

The problem is, of course, that developing and qualifying parallel applications is very hard. While there are ongoing and lasting debates in the computer science community about the fundamental nature of these challenges and their solutions, most people agree that the current methods impose too much complexity on the application programmer. It seems reasonably clear that a partial solution to this problem is to expand the role of the runtime environment to take some of this load off the application programmer.

Future runtime environments are expected to offer methods for lighter weight thread creation and management, increased determinism between the parallel threads, automated services for working with system resources with non-uniform characteristics, and optimized domain-specific libraries for common parallel constructs.

6.2.3 Application Software and Workloads

CMPs have the simultaneous effect of bringing multiprocessing into previously single-processor commodity systems, and extending the number of processors in high-end multi-chip SMP systems. The traditional application suites, or workloads, in these two environments can actually be quite different. In commodity systems, truly parallel workloads are very uncommon, but these systems are very adept at time multiplexing multiple applications onto available hardware to offer the appearance of concurrency. As additional processors are added to these systems, this will be first exploited by extending this capability by scheduling applications on the increased execution slots that come with more cores. Over time, more advanced applications will take advantage of the parallel and/or throughput opportunities that come with more cores, but this will require overcoming the challenges that have inhibited this sort of application development on their high-end SMP cousins.

In the higher end SMP systems, more applications can take advantage of true parallel processing. The development and quality validation of parallel applications is notoriously difficult due to several factors. First, although there has a been considerable amount of research into parallel programming techniques, the predominant starting point for most programmers still has a significant serial component to them. It is very difficult, maybe even unnatural, to convert a program originally conceived as serial into one that is parallel. Amdahl's Law, formulated in 1965, says that the maximum speedup of a program using multiple cores is bounded by the fraction of the program that is sequential. For example, if 80% of the program can be parallelized, the maximum speedup using multiple cores would be $5\times$, no matter how many cores are used [12]. Second, due to the inherent non-determinism in thread interaction, it is very difficult to prove that programs are correct under all circumstances. Third, the optimization of parallel programs is very difficult due to the sensitivity of lock granularity and inter-thread communication and synchronization. The underlying primitives that constitute these operations tend to be some of the slowest and, ironically, most serialized in modern computer systems. Although there has been some good progress in development of tools and best practices for parallel application development in recent years, the industry continues to be ill prepared for broad deployment of truly parallel applications [13].

As already mentioned, one of the more common uses of multi-core systems is the deployment of applications that maximize the *throughput* of independent subtasks. The simplest form of this involves only the optimized scheduling by the OS of multiple independent applications onto the available execution slots (time intervals on either a single processor or a collection of processors). More advanced applications are building on this technique to leverage the throughput opportunity in the context of a single application. These throughput-oriented workloads (for example, web servers, data indexing, video processing, and transactional databases) are increasingly important in today's data-rich and network-centric computing environments.

In addition to the program development challenges, there are also some important system-level scalability challenges with multiple cores. In addition to the scalability issues in the system-level software stack discussed in the previous section (hypervisor, OS, managed runtime environments), there are many hardware-level scalability challenges to consider. The addition of more cores puts increased stress on any shared resources, including caches, buses, switches, and I/O ports. As a result, it is important to characterize and understand workloads to understand their impacts on the scalability limits of these shared hardware structures. Some of the key metrics in this sort of analysis include working set size, memory latency sensitivity, memory bandwidth sensitivity, synchronization overhead, and the degree of available memory-level parallelism. Different workloads stress different aspects of the design and, in the case of CMPs, magnifies the scope and the importance of the system-level design details. The next section, on infrastructure scalability, will discuss these topics in more depth.

6.2.4 Infrastructure Scalability

This section describes several system-level parameters that have significant effects on the overall performance of general-purpose CMPs. In particular, the implications of shared caches, larger scale coherency management, rebalancing the system-level bandwidth, as well as the increased transactional load on the interconnect fabric and the memory system are examined and discussed. Most generally, the increased sharing of many of these system-level resources in multi-core designs puts new pressures on their capacity, bandwidth, and scalability. Optimized CMP designs have responded, and will continue to respond, with changes in these areas.

6.2.4.1 Caches

As the number of CPU cores on each chip increases, several important considerations come into play with caches. First, for a given chip's die size budget, there is a fundamental tradeoff between the number of cores and the total cache size appropriate to support those cores. Second, there are also interesting tradeoffs in distribution of the available cache size in the cache hierarchy and the circumstances in which the use of shared caches makes the most sense.

In a traditional cache hierarchy supporting a single core, the optimization has been primarily to make the upper-level caches (closest to the core) as large as possible without adding extra cycles (latency) or reducing the CPU clock frequency. The reasonably high hit rates of the L1 caches offer optimized latency for most accesses in many common workloads while also allowing the caches in subsequent levels of the hierarchy to make different tradeoffs with respect to latency, capacity, and bandwidth. As a result, the design of CMPs naturally leads to a re-evaluation of the traditional design approach of a uniprocessor cache hierarchy, which is basically making the upper-level caches as large as possible without adding additional pipeline stages to the processor core or reducing the CPU clock frequency (Fig. 6.3(a)).

First, of course, the system needs to contend with the increased memory access traffic that now comes from multiple CPUs, versus just one in the past. Subsequent sections on memory-level parallelism and system balance describe in more detail the issue of memory contention in CMPs.

Second, although most CPU designs will still benefit from private L1 caches, the presence of multi-level caches opens up the opportunity for cache sharing. For parallel workloads, it is clear how shared caching can help for instruction fetch, as one thread effectively prefetches code into the shared cache that is later used by a second

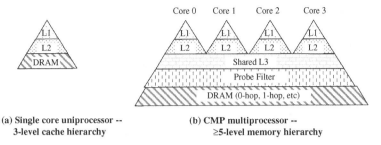

Fig. 6.3 Cache and memory hierarchy

thread. When multiple threads are reading a common data structure, cache sharing can offer similar benefits. In addition, shared caches can respond more flexibly to the varying demands of different workloads. For example, if one CPU is running a thread with a relatively light cache footprint, that leaves the thread on the other CPU to use up the extra capacity in the shared cache as needed. However, there are also some downsides to cache sharing. In particular, running multiple workloads that require large cache footprints can result in cache thrashing between the threads. This can cause unnatural increases in system-level activity, lower the overall performance of the system, and cause applications to experience annoying performance anomalies and system instability that vary depending on the conditions at runtime. Identifying how, when, and where the constructive effects of sharing outweigh the destructive effects of thrashing is a fundamentally important aspect of CMP cache design.

Third, because each CPU operates on a separate execution thread, the overall combined working set of the system is increased. In systems that support virtualization, this effect is magnified by multiple OSs and a hypervisor using the same hardware resources. In some sense, the additional compute capability of the CMP opens the door to these sorts of system configurations, which in turn creates the increase in overall working set size. In response to these situations, CMP designers are forced to look at more sophisticated replacement algorithms, dynamic cache partitioning algorithms, more advanced prefetch algorithms, alternative technologies for building more dense caches, and the management of non-uniform cache access latencies.

In addition to traditional instruction and data caches, modern CPUs also contain translation lookaside buffers (TLBs) that hold virtual address translation and protection information from the page tables. CMPs with increasing capability invite workloads with very large footprints that can stress the effectiveness of traditionally sized TLBs. An increase in TLB misses not only hurts the performance of the application that experiences the miss, it also causes a page table walk that results in numerous additional memory accesses at the system level. Although TLB coherency is typically maintained by the virtual memory manager in the OS, any changes to the page tables must be synchronized with all the system's TLBs. These tend to be highly serializing actions and, as a result, hurt the performance of CMPs more than their earlier single-core predecessors. In response to such challenges, CMP designers are forced to consider hierarchical TLB structures, support for larger page sizes, hardware broadcast of TLB-invalidate operations, and optimized table walk hardware.

6.2.4.2 Cache Coherency

Small-scale, high-volume one- to eight-socket commodity servers have traditionally relied on snoopy- or broadcast-based cache coherence protocols. Until recently, this brute-force approach to cache coherency has been effective and the associated traffic overhead has been modest and easily handled by over-provisioning system bandwidth.

There are two disadvantages to broadcast coherence protocols that effectively increase the latency under load and penalize performance: first, the latency to local memory can be delayed by the time it takes to probe all the caches in the system; and, second, the demand for greater bandwidth overhead from broadcasting probes and probe-response collection from all nodes in the system. These effects are more pronounced in eight-socket systems because these topologies are distributed and have larger network diameters and higher latencies. The overhead of broadcasting probes also grows quadratically with the number of sockets in the system.

To address these issues, next-generation CMPs will incorporate probe filters, also known as sparse directory caches, to track the state of cached lines in a system [23]. Probe filters reduce average memory latency and, by conserving system bandwidth, help provide more bandwidth *headroom* in the system, which allows more cores to be added to each CMP processor in the system.

On-chip interconnects and queuing subsystems must be improved to support the additional request-and-response throughput required by the second core. While some early implementations attempted simple bus-sharing techniques, it soon became obvious that more advanced switching and queuing were required to form a reasonably balanced system. In addition, the desire to scale up with multiple chips connected together for some high-end systems leads to additional demand on these switching and queuing structures. Moreover, as the effective diameter of the coherency domain continues to grow, the system performance can become constrained by coherency transaction latencies and bandwidths. As a result, a big challenge for future CMP-based systems involves managing the coherence traffic with snoop filters and, perhaps more generally, through the development of new and improved coherence schemes.

6.2.4.3 Memory-Level Parallelism

Although the rate of cycle time improvement in CPU cores has slowed down significantly, the emergence of multi-core CPUs continues to pressure the memory system. In particular, the increased rate of requests from the collection of on-chip CPU cores actually increases the effective gap between the CPU and the memory system (illustrated in Fig. 6.2). We fully expect this trend to continue and remain a significant challenge for future chip multiprocessors.

Managing the effect of high memory latency, the so-called *memory wall* [15], is an important issue in performance optimization. As more CPUs are added, the memory system must be prepared to handle increased throughput of memory-based transactions or risk becoming a significant bottleneck in overall performance. As a result, CMPs must recognize and optimize for memory-level parallelism (MLP).

Figure 6.4(a) illustrates the effects of a system with a low degree of MLP. In this hypothetical example, the processor issues ten cache-fill requests to memory. The access latency to memory for each request is 100 clock cycles (clk). The requests are fully serialized and the total execution time is 10×100, or 1000, clock cycles.

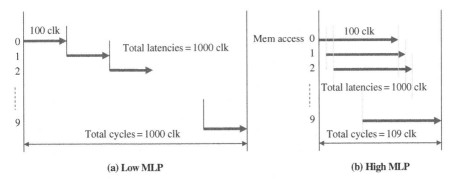

(a) Low MLP   (b) High MLP

Fig. 6.4 Memory-level parallelism

In contrast, Fig. 6.4(b) illustrates the effects of a system with a high degree of MLP. Here the accesses to memory are overlapped and pipelined. The total number of cycles to execute the ten operations is 109 clock cycles. Note that this sort of memory pipelining has the effect of increasing throughput and overall bandwidth, so it is critically important for CMPs that naturally stress the memory subsystem.

MLP is a characteristic of both the workload and the microarchitecture of the core it runs on. The degree of MLP exhibited by a workload can be defined as the rate at which the workload presents independent memory requests to the system.

Core architectures and applications that exhibit high MLP may not scale well as the number of cores grows. Relatively few cores with high MLP can soak up all the available bandwidth in the system, imposing a practical upper limit of how many cores can be effectively integrated on a single chip. Conversely, workloads that are inherently low in MLP are more sparing in their use of bandwidth and may scale better with an increasing number of cores in large-scale CMPs.

An application can increase the MLP potential by avoiding false dependencies between memory operations, organizing data structures that maximize the use of available cache-line capacities, or with prefetch instructions, among other techniques. In some cases, it is possible to structure multi-threaded applications such that the individual memory requests from each thread are nearly guaranteed to remain independent from one another. In these cases, as the application is capable of engaging increasing numbers of threads, the amount of aggregate traffic to memory can become dramatically higher than traditional single-threaded applications. These *throughput-oriented* applications are becoming increasingly common in the server domain, but there is also increasing evidence that many data-parallel applications can be structured in this manner.

To achieve this MLP potential, of course, the CPU cores and the overall memory system must be tuned for multiple concurrent requests in flight. A computer system can increase MLP throughput by using techniques such as speculative execution, memory address disambiguation, hardware prefetch, and deep buffering of outstanding requests. For truly throughput-oriented applications that require many concurrent threads to expose the MLP, hardware support for multi-threading has proven very effective [16]. These techniques may become increasingly important as the memory wall effects continue to grow.

6.2.4.4 System Bandwidth

An important use of CMPs is in multiprocessor configurations running commercial server workloads. These systems combine the CMP benefit of increased core count with increased I/O connectivity, high memory capacities, and a single-system image.

In many high-capacity and high-performance workloads, the available system bandwidth becomes an important aspect of optimization. System bandwidth, in this context, refers to overall throughput of the memory interface, the scalability ports, the I/O interface, and the on-chip buses, switches, and queues.

By virtue of the simple fact that chip multiprocessors support the concurrent execution of multiple programs, they increase the overall system bandwidth pressure. Moreover, since CMPs often run completely independent programs concurrently, the resulting composite transaction patterns can be very different from those of a single application that happened to stress the available system bandwidth. In addition, since most CMP designs do not have the luxury of increasing all aspects of the system bandwidth in proportion to the increased number of CPU cores, many resources need to be cooperatively shared.

The good news is that some things work in favor of increased system bandwidth efficiency for CMPs. In general, the system structures for many optimized single-core CPU designs were designed to work reasonably well during high-demand phases of program execution. In a sense, this can mean there is some reserve capacity already built into these systems for use during the less demanding phases. Also, since many of the buses and queues that support these structures are located on-chip, it is reasonable to increase their sizes. For example, by increasing an internal bus width from 64 bits to 256 bits, a transaction involving the movement of 64 bytes would only require two beats on the bus (versus eight in the narrower case). As a result, the trailing-edge effects of this bus resource, such as error checking before using the data, can be mitigated relatively easily.

6.2.4.5 Memory Bandwidth Demands

As expected, different workloads stress the system bandwidth in different ways. Figure 6.5 classifies several server and HPC workloads on two dimensions – cache miss rates and MLP. The cache miss rates assume a typical-sized shared L3 of 4–8 MB. Typically, commercial databases exhibit a low degree of MLP because they do a lot of pointer chasing in memory and because threads access data structures

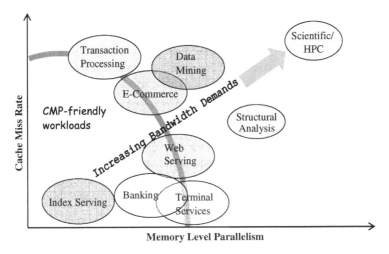

Fig. 6.5 Server workload classification by cache miss rate and MLP

in irregular access patterns that are difficult to prefetch; furthermore, the data-dependent nature of program control makes branch prediction difficult [14]. On the other hand, highly tuned scientific workloads that operate on large regular arrays in memory can exhibit near-perfect memory-level parallelism and, as a result, they are often constrained by the system bandwidth. Clearly, applications with high cache miss rates and high memory-level parallelism are the workloads that most stress interconnect and memory bandwidth and, therefore, system bandwidth. Workloads with low memory bandwidth demands can be expected to continue to scale well on large-scale CMPs as the number of cores grows. Workloads with high memory bandwidth will benefit more from scaling caches, memory capacity, I/O connectivity, and system bandwidth, making it unlikely that single-chip CMP architectures will ever displace scalable shared-memory multiprocessor architectures. A robust system design is one that can perform reasonably well across a broad set of application characteristics.

6.2.4.6 Processor Interface

The increasing number of cores on a CMP leads to a linear increase in off-chip processor interface and memory bandwidth. High-speed point-to-point links and high-bandwidth memory interfaces have helped scale up processor interface and memory bandwidth to track the growth in core count.

Memory bandwidth and capacity have also grown steadily, by a factor of 2× with each generation of DRAM. In addition, advances in packaging technology have allowed for increases in pin budget, which allows building more fully connected topologies and have a dramatic effect on increasing available coherent memory bandwidth and I/O bandwidth in the switch fabric. Balanced system design requires that increases in compute capacity resulting from growing core count be offset with corresponding increases in I/O connectivity, memory capacity, and system bandwidth.

6.3 Case Study: Multi-core AMD Opteron™ Processors

This section presents the AMD Opteron™ microarchitecture as a case study in how the power, software stack, and infrastructure challenges posed by general-purpose CMP architectures have been addressed in a contemporary design. The first-generation AMD Opteron microarchitecture was a superscalar uniprocessor designed to scale at the system level up to eight processors. The second- and third-generation AMD Opteron microarchitectures are best described as *CMP-based multiprocessors* that retrofit two- and four-core CMP processors into the original scalable platform infrastructure to provide significant performance uplift across the usage spectrum, from laptop to desktop to server. The software support for all of the challenges discussed earlier are critically important but beyond the scope of this case study.

6.3.1 First-Generation Single-Core AMD Opteron Chips

In 2003, AMD released the first generation of AMD Opteron processors. Although these chips were single-core solutions manufactured in 90 nm technology, they defined a platform with significant optimizations targeted specifically at commodity multiprocessing solutions. In particular, the design included an integrated memory controller, a set of three high-performance scalability links, an integrated high-performance crossbar switch, and a scalable coherence protocol called HyperTransport™ Technology (HT). AMD Opteron was also the first chip to introduce 64-bit extensions to the x86 architecture, offering the extended addressability needed for a growing number of high-performance applications [4, 17]. These features enabled the broad development and deployment of glueless two-, four-, and eight-socket symmetric multiprocessor systems. Some of the many possible system topologies are illustrated in Fig. 6.6.

The topology of these systems offered several important attributes for balanced multiprocessing solutions. First, the underlying platform and framework supported the flexibility to populate a variable number of sockets at any given time. This allowed customers to purchase lower-end systems initially, retaining the opportunity to upgrade the capacity of the system gracefully as their future needs grew. Second, as additional processor chips were added to the system, the overall capacity and bandwidth of the system grew. In particular, since each processor chip contained an integrated memory controller, the addition of another processor chip was accompanied by an increase in the available system-level memory bandwidth and memory capacity. Similarly, since each chip contained three dedicated scalability links, the addition of more processor chips also increased the overall throughput of the interconnect fabric. Third, and perhaps most importantly, the architecture defined a platform and coherence framework that set the stage for future chip multiprocessors. In particular, the design anticipated the possible addition of more CPU cores on-chip, the need to support a wide range of memory DIMM types, and the ability to offer increasingly capable cache structures.

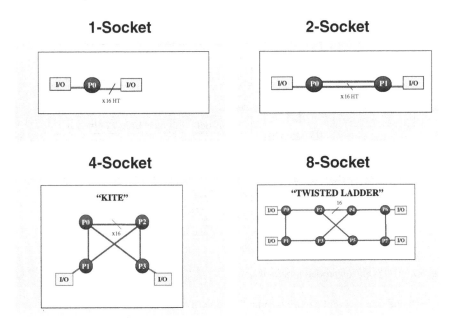

Fig. 6.6 Example of first-generation AMD Opteron system topologies

6.3.1.1 AMD Opteron Processor Core Microarchitecture

Figure 6.7 is a block diagram of the AMD Opteron core [4]. It is an aggressive out-of-order, three-way superscalar processor. It can fetch and decode up to three x86-64 instructions each cycle from the instruction cache. Like its predecessor, the AMD Athlon[TM] processor, it turns variable-length x86-64 instructions into fixed-length micro-operations (μops) and dispatches them to two independent schedulers – one for integer and one for floating-point and multimedia (MMX, SSE) operations. These schedulers can dispatch up to nine μops to the following execution resources: (a) three integer pipelines, each containing an integer execution unit and an address generation unit and (b) three floating-point and multimedia pipelines. In addition, load-and-store operations are dispatched to the load/store unit, which can perform two loads or stores each cycle. The AMD Opteron processor can reorder as many as 72 μops. AMD Opteron has separate instruction and data caches, each 64 KB in size and backed by a large on-chip L2 cache, which can be up to 1 MB in size. The caches and memory interface minimize latency and provide high bandwidth. The AMD Opteron processor's pipeline, shown in Fig. 6.8, is well balanced to allow high operating frequency. The following sections detail the key features.

Pipeline

The AMD Opteron processor's pipeline, shown in Fig. 6.8 [4], is long enough for high frequency and short enough for good inter-process communication (IPC). It is fully integrated from instruction fetch through DRAM access. It takes seven cycles

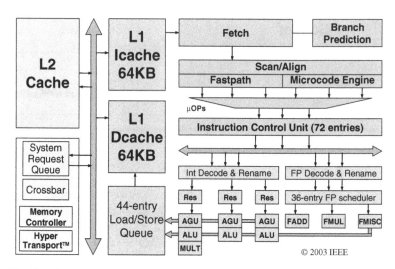

Fig. 6.7 AMD Opteron processor block diagram

to fetch and decode, and excellent branch prediction covers this latency. The execute pipeline is 12 stages for integer and 17 stages (or, longer or shorter depending on latency) for floating point. The data cache is accessed in pipestage 11 and the result is returned to the execution units in pipestage 12.

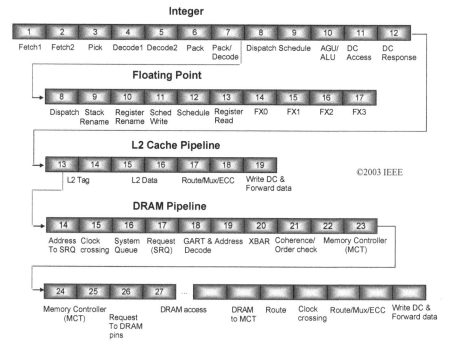

Fig. 6.8 Pipelines

In case of an L1 cache miss, the L2 cache is accessed. Picking the critical quad-word, ECC checking, and routing to the L1 cache takes two pipestages. A waiting load does not wait for the response to be written into the data cache; instead, the data cache forwards the data to the result bus at the same time the L1 data cache is updated.

On an L1 miss, a request is sent to the system request queue in parallel with the L2 cache access to get a head start in case of a L2 cache miss, but the request is canceled if the access hits in the L2 cache. It takes one cycle to cross the clock boundary in each direction. After the memory controller checks the page status, it schedules the command to DRAM. Note that all these pipestages are at the *same* frequency as the core. DRAM access in terms of CPU cycles is dependent on the frequency of the core, HT link speed, and the frequency of the DRAM chips used. Once data is returned, we need some pipestages for routing data across the chip to the L1 cache and ECC checking. As in the L2 cache hit case, system data can be forwarded to a waiting load while updating the cache.

Caches

The AMD Opteron processor has separate L1 data and instruction caches, each 64 KB in size. They are two-way set-associative, linearly indexed, and physically tagged with a cache-line size of 64 bytes. They are banked for speed with each way consisting of eight, 4 KB banks. They are backed by a large 1 MB L2 cache, which is mutually exclusive with the data in the L1 caches. It is 16-way set-associative and uses a pseudo-LRU replacement policy in which two ways are grouped into a sector and the sectors are managed by LRU. The way within each sector is managed efficiently to maintain performance, but only need half the number of LRU bits compared to a traditional LRU scheme. The standard MOESI protocol (M = modified, O = Owner, E = Exclusive, S = Shared, I = Invalid) is used for cache coherence.

The instruction and data caches each have independent L1 and L2 TLBs. The L1 TLB is fully associative and stores 32, 4 K plus eight, 2 M/4 M page translations. The L2 TLB is four-way set-associative with 512, 4 K entries. In a classic x86 TLB scheme, a change to the page table base pointer (CR3) flushes the TLB. AMD Opteron has a hardware flush filter that blocks unnecessary TLB flushes and flushes the TLBs only after it detects a modification of a paging data structure. Filtering TLB flushes can be a significant advantage in multiprocessor workloads by allowing multiple processes to share the TLB.

Instruction Fetch and Decode

The instruction fetch unit, with the help of branch prediction, attempts to supply 16 instruction bytes each cycle to the scan/align unit, which scans this data and aligns it on instruction boundaries. This is complicated because x86 instructions can be anywhere from 1 to 15 bytes long. To help speed up the scan process, some predecoding is done as a cache line is allocated into the instruction cache to identify the end byte of an instruction.

192 C. Moore and P. Conway

Variable-length instructions are decoded into internal fixed-length μops using either hardware decoders (also called fastpath decoders) or the microcode engine. Common instructions that decode to one or two μops are handled by the fastpath. As the name suggests, fastpath decode is indeed faster than microcode. Compared to the AMD Athlon, the AMD Opteron processor includes more fastpath decoding resources. For example, packed SSE instructions were microcoded in AMD Athlon processors, but they are fastpath in AMD Opteron processors. Overall, 8% fewer instructions in SpecInt and 28% fewer in SpecFP are microcoded compared to AMD Athlon processors.

Branch Prediction

We use a combination of branch prediction schemes [18] as shown in Fig. 6.9 [4].

The AMD Opteron branch-prediction logic has been significantly improved over the AMD Athlon processor. Figure 6.9 shows that AMD Opteron employs a branch selector array, which selects between static prediction and the history table. This array is big enough to cover the entire instruction cache. The global history table uses 2-bit saturating up/down counters. The branch target array holds addresses of the branch target and is backed by a two-cycle branch target address calculator to accelerate address calculations. The return address stack optimizes CALL/RETURN pairs by storing the return address of each CALL and supplying that as the predicted target address of the corresponding RETURN instruction. The penalty for a mispredicted branch is 11 cycles. When a line is evicted from

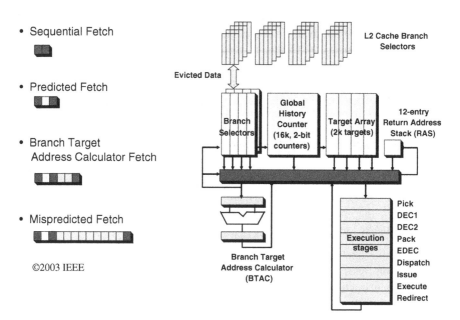

Fig. 6.9 Branch predictors

the instruction cache, the branch prediction information and the end bits are saved in the L2 cache in a compressed format. Since parity protection is sufficient for read-only instruction data, the ECC bits are used for this purpose. Thus, when a line is brought back into the instruction cache from the L2 cache, the branch history information is not lost. This unique and simple trick improved the AMD Opteron processor's branch prediction accuracy on various benchmarks by 5–10% compared to AMD Athlon.

Integer and Floating-Point Execution Units

The decoders dispatch up to three μops each cycle to the instruction control unit, which has a 72-entry re-order buffer (ROB) and controls the retirement of all instructions. Meanwhile, in parallel, the μops are issued to the integer reservation stations and the floating-point scheduler. These issue μops to the respective execution units once their operands are available. The AMD Opteron core has three integer execution units, three address generation units, and three floating-point units. The integer units have a full 64-bit data path and majority of the ops (32- and 64-bit add, subtract, rotate, shift, logical, etc.) are single-cycle. A hardware multiplier takes three cycles to do a 32-bit multiply and has a throughput of one per cycle. 64-bit multiplies take five cycles and can be issued every other cycle.

The floating-point data path is 64 or 80 bits (full extended) precision. Simple μops (e.g., compare and absolute, as well as most MMX ops) take two cycles; MMX multiply and sum of average take three. Single- and double-precision FP add and multiply take four cycles, while latency of divide and square root depends on precision. The units are fully pipelined with a throughput of one per cycle.

Load Store Unit

In addition to the integer and floating-point units, the decoders issue load/store μops to the load store unit, which manages load-and-store accesses to the data cache and system memory. It ensures that the architectural load-and-store ordering rules are preserved. Each cycle, two 64-bit loads or stores can access the data cache as long as they are to different banks. Load-to-use latency is three cycles when the segment base is zero. Uncommon cases, in which a load has a non-zero segment base, a misaligned load, or a bank conflict, take an extra cycle. While loads are allowed to return results to the execution units out of order, stores cannot commit data until they retire. Stores that have not yet committed forward their data to dependent loads. A hardware prefetcher recognizes a pattern of cache misses to adjacent cache lines (both ascending and descending) and prefetches the next cache line into the L2 cache.

The L2 cache can handle ten outstanding requests for cache misses, state changes, and TLB misses, including eight from the data cache and two from the instruction cache. The system interface consists of a victim buffer, which holds victim data, as well as snoop response data, a snoop buffer that holds snoop requests, and a write buffer that holds write-combining, uncacheable, and write-through data going out to the system.

Reliability Features

Reliability is very important in enterprise-class machines and AMD Opteron has robust reliability features. All large arrays are protected by either ECC or parity. The L1 data cache, L2 cache and tags, as well as DRAM, are ECC protected. We store eight ECC check bits for 64 bits of data, allowing us to correct single-bit errors and detect multi-bit errors (SECDED). DRAM also has support for Chipkill ECC, which is ECC checking on 128 bits of data using 16 ECC check bits. This provides the ability to correct 4-bit errors as long as the bits are adjacent, enabling correction of an error affecting an entire x4 DRAM chip. In addition, L1 data cache, L2 cache, and DRAM all have their own hardware scrubbers that steal idle cycles and clean up any single-bit ECC errors. The instruction cache data and tags, and level 1 and level 2 TLBs, are protected by parity. Hardware failures are reported via a machine check architecture that reports failures with sufficient information to determine the cause of the error.

Northbridge

Internally, the AMD Opteron integrates a set of functions collectively referred to as the *Northbridge* (NB), which includes all the logic outside of the processor core. Figure 6.10 is a simplified view of the NB microarchitecture that shows several of the key subsystems including the system request interface (SRI), the crossbar (XBAR), the memory controller (MCT), the DRAM controller (DCT), and the HT scalability ports [19].

The crossbar (XBAR) supports five ports: the SRI, MCT, and three HT scalability ports. Internally, transactions are logically separated into *command packet headers* and *data packets*, so there are actually separate XBAR structures for each. The command packet XBAR is dedicated to routing command packets (4–8 bits in size) and the data XBAR is dedicated for routing the data payload associated with commands (4–64 bits in size).

Memory Controller

The on-chip DDR memory controller provides a 128-bit DDR interface and supports both registered and unbuffered DIMMs. The controller is on a separate clock grid, but runs at the same frequency as the core. In low-power modes, the core clock can be turned off while keeping the memory controller's clock running to permit access to memory by graphics and other devices.

Other Northbridge (NB) functionality handled on-chip includes processing of CPU to I/O and DRAM to I/O requests as well as transaction ordering and cache

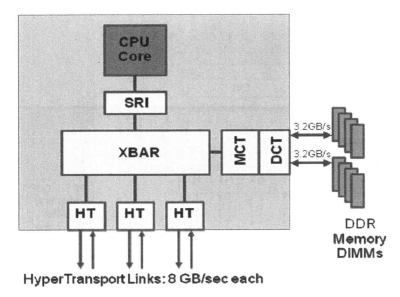

HyperTransport Links: 8 GB/sec each

Fig. 6.10 First-generation AMD Opteron Northbridge

coherence. Our first-generation design supports up to three point-to-point 16-bit HT links. Both memory latency and snoop throughput scale with CPU and HT link speed.

6.3.2 Second- and Third-Generation AMD Opteron Microarchitecture

In early 2005, AMD released the second generation of AMD Opteron processors, which featured two integrated CPU cores in each chip, manufactured with 65 nm CMOS silicon-on-insulator (SOI) technology. The CPU cores were connected through the integrated system request interface and crossbar switch, which allowed socket-level compatibility with the earlier single-core solutions. The chips were engineered to operate in the same power envelope and, because the power budget per core is reduced, each core in the dual-core chip runs at a slightly lower operating frequency than the high-end single-core solution. As a result, the platforms developed for the first-generation AMD Opteron were easily upgradable to this second-generation solution with twice the number of CPU cores. In late 2007, AMD announced their third-generation AMD Opteron CMP chips [20, 21]. This design doubled the number of cores in the CMP to four, and incorporated several additional features to improve the overall performance of the resulting single-chip CMP and multi-chip SMP topologies. A high-level diagram of this chip is illustrated in Fig. 6.11.

Each core in this new design was enhanced to include out-of-order load execution to expose subsequent L1 cache misses as early as possible. In many cases,

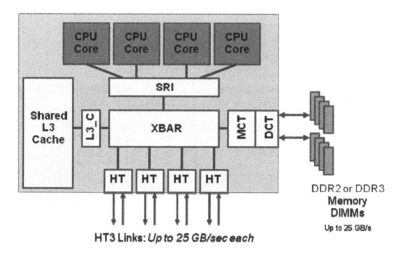

Fig. 6.11 Third-generation quad-core AMD Opteron CMP

the latency of these cache misses can be largely hidden by the out-of-order execution capability. The TLBs were enhanced to support 48 bits of physical address, translation support for 1 GB-sized pages, and an overall increase in capacity to 512 4-KB page entries + 128 2-MB page entries, to better support virtualized workloads, large footprint databases, and transaction processing. The major units of this design are labeled on the die photo in Fig. 6.12. Each core on this die has a dedicated 512 KB L2 and the four cores share a 6 MB L3. The die size is 17.4 × 16.5 mm (287.10 mm^2). It contains about 758 million transistors and is manufactured using AMD's 45 nm SOI CMOS process.

The design provides a second independent DRAM controller to provide more concurrency, additional open DRAM banks to reduce page conflicts, and longer burst length to improve efficiency. The DRAM controller tracks the status of open pages, which increases the frequency of page hits, decreases page conflicts, and enables history-based pattern prediction. Write bursting is supported to minimize read/write turnaround time. The DRAM prefetcher tracks positive, negative, and non-unit strides and has a dedicated buffer for prefetched data.

6.3.2.1 Virtualization

The second-generation AMD Opteron introduced virtualization extensions to the x86 architecture, allowing a virtual machine hypervisor to run an unmodified guest OS without incurring significant emulation performance penalties. These extensions include rapid virtualization indexing (RVI, formerly called nested paging) and tagged TLBs [22]. RVI is designed to reduce the overhead penalty associated with virtualization technologies by allowing the management of virtual memory in hardware instead of software. Other extensions include configuration of interrupt

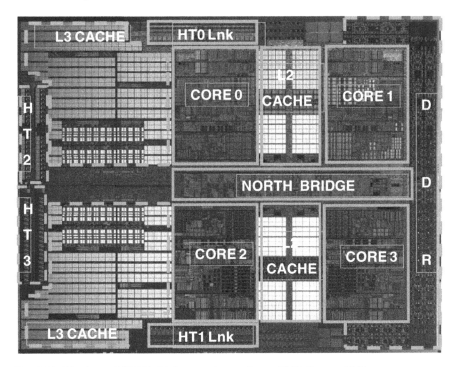

Fig. 6.12 Die photo of AMD's third-generation quad-core AMD Opteron CMP

delivery to individual virtual machines and an I/O memory translation unit for preventing a virtual machine from using DMA to break isolation.

6.3.2.2 Power Management

Internally, at the chip-level, the design includes seven clock domains and two separate power planes, one for the NB and one for the collection of CPU cores, as shown
in Fig. 6.13 [20]. Separate NB and CPU core power planes allow the processors to
reduce core voltage for power savings, while still allowing the NB to support the
system-level bandwidth and latency characteristics at full speed.

   Fine-grain power management provides the capability to adjust core frequencies dynamically and individually to improve power efficiency. For example, one
deployment of an AMD Opteron processor has a top speed of 2.6 GHz (state P0,
the highest power consuming state. If the CPU frequency is lowered to 1 GHz (state
P5, the lowest power consuming (idle) state), the voltage can be reduced from 1.4 V
(state P0) to 1.1 V (state P5). As noted earlier, power consumption is a function of
capacitance, switching frequency, and voltage squared. Therefore, a 79% reduction
in frequency and a 38% reduction in voltage results in a dynamic power savings of
75%. Furthermore, additional power savings result from reductions in short circuit

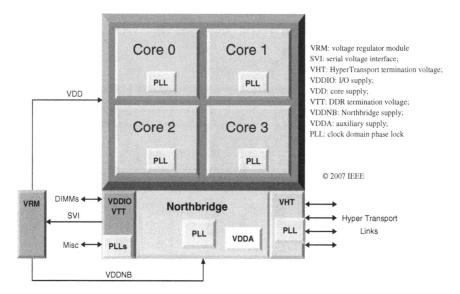

Fig. 6.13 Power planes and clock domains

and leakage power. Short-circuit power is reduced by decreasing frequency or voltage; leakage power is reduced by decreasing voltage.

6.3.2.3 Cache Hierarchy

The design added a shared L3 cache on the CMP as shown in Fig. 6.14 [20]. This cache is managed as a *victim cache*, whereby cache lines are installed in the L3 when they are cast out from any of the L2 caches dedicated to the four processor cores. The L3 cache is non-inclusive, which allows a line to be present in an upper-level L1

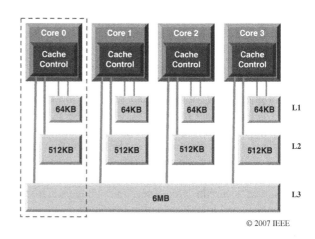

Fig. 6.14 Three-level cache
hierarchy

or L2 cache and not to be present in the L3. This serves to increase the overall cache effectiveness, since the total number of unique cache lines in the chip can be sum of the individual L1, L2, and L3 cache capacities (as opposed to the maximum number of distinct cache lines in an inclusive L3 scheme, which is simply the L3 capacity). The L3 cache has a sharing-aware replacement policy to optimize the movement, placement, and replication of data for multiple cores.

6.3.2.4 Infrastructure and System Topologies

This third-generation design defined a new socket infrastructure with significantly enhanced capabilities, but it also offered a compatibility mode that allowed the chip to drop into existing platforms. The new platform extended the system-level scalability of the design in several important ways. First, the design included a second independent DRAM controller to provide more memory concurrency. The improved DRAM controller supported additional open DRAM banks to reduce page conflicts and longer burst length to improve efficiency. In addition, a new DRAM prefetcher was included that tracks positive, negative, and non-unit strides. The prefetcher contains a dedicated buffer for prefetched data. Second, it introduced the HyperTransport 3.0 (HT3) interface, with the capability of running each bit lane with data rates up to 6.4 GT/s. It added a fourth HT3 link to the package, and enabled each of the four links to be configured as one 16-bit link (each direction), or as two 8-bit links (each direction).

The addition of this fourth HT3 link enabled a fully connected four-socket system, and the ability to split the links enabled a fully connected eight-socket system. These fully connected topologies provided several important benefits. In particular, network diameter (and therefore memory latency) was reduced to a minimum, links become more evenly utilized, packets traverse across fewer links to their destination, and there are more links for distributing the traffic. Reduced link utilization lowers queuing delay, in turn reducing latency under load. The performance benefit of fully connected SMP topologies can be demonstrated with two simple metrics for memory latency and coherent memory bandwidth. These metrics can be statically computed for any topology given the routing tables, the message visit count, and packet size:

> *Average diameter*: the average number of hops between any pair of nodes in the network.
> *XFIRE memory bandwidth*: the link-limited all-to-all communication bandwidth (data only). All processors read data from all nodes in an interleaved manner.

In the four-node system shown in Fig. 6.15, adding one extra HT port provides a 2× improvement in link-limited XFIRE bandwidth. This bandwidth scales linearly with link frequency; with HT3 running at 6.4 GT/s, the XFIRE bandwidth is increased 5× overall. In addition, the average diameter (memory latency) decreases from 1 hop to 0.75 hops.

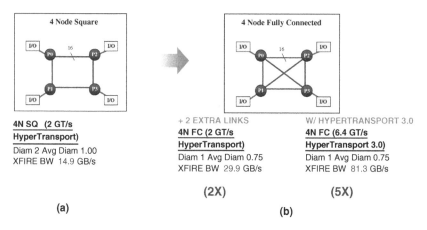

Fig. 6.15 Enhanced four-socket topology

The benefit of fully connected topologies is even more dramatic for eight-socket systems, as shown in Fig. 6.16. Each node in the eight-node fully connected systems has eight, 8-bit wide HT3 links. In this topology, each of the four, 16-bit wide HT3 links on a processor node is unganged into two independent 8-bit wide links. The XFIRE bandwidth increases 7× overall. In addition, the average diameter (memory latency) decreases significantly from 1.6 hops to 0.875 hops. Furthermore, the links are evenly utilized with this sort of access pattern, which is typical of many multi-threaded commercial workloads.

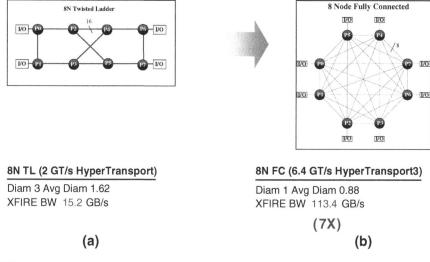

Fig. 6.16 Enhanced eight-socket topologies

6.3.2.5 Lessons Learned

One of the unexpected lessons from the first-generation AMD Opteron processors was that performance is very sensitive to virtual channel buffer allocation policy. The coherent HT protocol employs four virtual channels: request, response, probe, and posted write. The original design had a very flexible allocation scheme in which the buffer pool could be allocated in a completely arbitrary way, with the only requirement that each VC be allocated at least one buffer. Subsequent studies discovered that the optimum allocation is strongly related to the number of nodes in a system, the system topology, the coherence protocol, and the routing tables. Generally, the number of buffers allocated to the different VCs should be in the same proportion as the traffic on these VCs. As a result of these observations, the buffer allocation policies are now set as part of the topology configuration process.

Another challenge was tuning workloads to run well on the AMD Opteron processor's distributed shared-memory architecture. AMD Opteron processor-based systems can be configured at boot time to run with either a linear address map, which maps equal-sized contiguous regions of memory to each node, or with memory interleaved across nodes on 4 KB granularity. The choice of linear address mapping enables performance benefits from non-uniform memory access (NUMA) optimizations in the OS process scheduler and memory manager to create CPU-memory affinity between a thread and the memory it uses. On the other hand, if NUMA optimizations prove difficult or impossible, the fallback choice of node-interleaved mapping guarantees that no single memory controller will be a hotspot in the system and that memory accesses will be uniformly distributed across the system. We observed that some workloads benefit from NUMA optimizations with most accesses going to local memory; however, we also observed that the majority of the remaining accesses went to the low ~2 GB of memory, which is usually mapped to node 0, creating a memory hotspot on node 0 that can limit performance. This hotspot corresponds to page tables and other shared data structures in the OS/hypervisor, which are mapped early following machine boot. In response to this observation, we developed an effective hybrid-mapping scheme that interleaves the low 2 GB of memory to "smear" the hot data across all nodes while preserving the benefits of NUMA optimizations to the rest of memory.

6.4 Conclusions

General-purpose CMPs are the natural product of the convergence of Moore's Law, exploding demand for compute capacity in our web-based economy, and the availability of high-performance processor cores as IP blocks which can be used as building blocks in system-on-chip (SOC) designs. Today's general-purpose CMPs are high-volume commodity products that strive to address several conflicting goals. The first goal is to run the overwhelming majority of today's applications, which are single threaded, at ever-increasing performance levels; the second goal is to achieve high throughput on highly threaded and multi-programmed workloads; the third is to minimize power consumption and maximize hardware utilization in data centers.

The AMD Opteron™ processor addresses this challenge through a balanced combination of moderate-scale CMP, which integrates four superscalar x86-64 cores to serve the high-volume desktop market, coupled with a scalable system architecture that provides competitive performance in the high-volume 2P, 4P, and even 8P server space. Dynamic power management and hardware virtualization support help raise performance per watt and server utilization. The underlying AMD Opteron architecture is inherently customizable for different market segments, a fact that has helped drive high volumes and achieve competitive pricing, an economic imperative given the economics of building and operating deep submicron semiconductor fabrication facilities.

General-purpose CMPs will remain a good solution up to a point, but economic and technical constraints will govern how far (and how fast) the number of cores will ramp up. The next natural step is heterogeneous CMP, MP, and customized accelerators optimized to run specific workloads. Practical and productive parallel programming will need more hardware support and software layers to enable the majority of programmers, who are not experts in parallel programming, to easily exploit the potential of CMP architectures to solve problems beyond our grasp today.

References

1. G. Moore. Cramming More Components onto Integrated Circuits. *Electronics Magazine*, (38)8, April 19, 1965.
2. B. Sinharoy, R. Kalla, J. Tendler, R. Eickemeyer, and J. B. Joyner. POWER5 System Microarchitecture. *IBM Journal of Research & Development*, 49(4/5):505–521, July/Sep 2005.
3. R. Kessler. The Alpha 21264 Microprocessor. *IEEE Micro*, 19(2):24–36, Mar 1999.
4. C. Keltcher, K. McGrath, A. Ahmed, and P. Conway. The AMD Opteron Processor for Shared Memory Multiprocessor systems.*IEEE Micro*, 23(2):66–76, Mar/April 2003.
5. M. Flynn, P. Hung, and K. Rudd. Deep Submicron Microprocessor Design Issues. *IEEE Micro*, (19)4:11–22, July/Aug, 1999.
6. S. Borkar. Design Challenges of Technology Scaling, *IEEE Micro*, 19(4):23–29, July/Aug, 1999.
7. V. Agarwal, M. Hrishikesh, S. Keckler, and D. Burger. Clock Rate Versus IPC: The End of the Road for Conventional Microarchitectures. *International Symposium on Computer Architecture*, pp. 248–259, June 2000.
8. M. Flynn and P. Hung. Microprocessor Design Issues: Thoughts on the Road Ahead. *IEEE Micro*, 25(3):16–31, May/June 2005.
9. V. Tiwari, D. Singh, S. Rajgopal, G. Mehta, R. Patel, and F. Baez. Reducing Power in High-Performance Microprocessors. *Design Automation Conference*, pp. 732–737, June 1998.
10. C. Isci, A. Buyuktosunoglu, C. Cher, P. Bose, and M. Martonosi. An Analysis of Efficient Multi-core Global Power Management Policies: Maximizing Performance for a Given Power Budget, *International Symposium on Microarchitecture*, pp. 347–358, Dec 2006.
11. R. McDougall and J. Mauro. Solaris Internals: Solaris 10 and OpenSolaris Kernel Architecture. *Prentice Hall*, 2007.
12. G. Amdahl. Validity of the Single Processor Approach to Achieving Large-Scale Computing Capabilities. *AFIPS Conference Proceedings*, 30:483–485, 1967.
13. K. Asanovic, R. Bodik, B. Catanzaro, J. Gebis, P. Husbands, K. Keutzer, D. Patterson, W. Plishker, J. Shalf, S. Williams, and K. Yelick. The Landscape of Parallel Computing Research: A View from Berkeley. *Technical Report UCB/EECS-2006-183*, EECS Department University of California, Berkeley, December 2006.

14. L. Barroso, K. Gharacharloo, and E. Bugnion. Memory System Characterization of Commercial Workloads. *International Symposium on Computer Architecture*, pp. 3–14, 1998.

15. W. Wulf and S. Mckee. Hitting the Memory Wall: Implications of the Obvious. *Computer Architecture News*, 23(1):20–24, 1995.

16. J. Laudon, R. Golla, and G. Grohoski. Throughput-oriented Multicore Processors. In S. Keckler, K. Olukotun, and P. Hofstee (Eds.), *Multicore Processors and Systems*, Springer.

17. *AMD x86-64 Architecture Manuals*, http://www.amd.com

18. S. McFarling. Combining Branch Predictors. *WRL Technical Note TN-36*, June 1993.

19. *HyperTransport I/O Link Specification*, http://www.hypertransport.org/

20. P. Conway and W. Hughes. The AMD Opteron Northbridge Architecture. *IEEE Micro*, 27(2):10–21, Mar–Apr 2007.

21. J. Hennessy and D. Patterson. *Computer Architecture: a Quantitative Approach*. Fourth Edition, Morgan Kaufmann, 2007.

22. "AMD-V Nested Paging v1.0," white paper, July 2008, http://developer.amd.com/assets/NPT-WP-1%201-final-TM.pdf

23. D. Culler, J. Pal Singh, and A. Gupta, Parallel Computer Architecture, a Hardware/Software Approach. *Morgan Kaufmann*, 1999.

Chapter 7
Throughput-Oriented Multicore Processors

James Laudon, Robert Golla, and Greg Grohoski

Abstract Many important commercial server applications are throughput-oriented. Chip multiprocessors (CMPs) are ideally suited to handle these workloads, as the multiple processors on the chip can independently service incoming requests. To date, most CMPs have been built using a small number of high-performance super-scalar processor cores. However, the majority of commercial applications exhibit high cache miss rates, larger memory footprints, and low instruction-level parallelism, which leads to poor utilization on these CMPs. An alternative approach is to build a throughput-oriented, multithreaded CMP from a much larger number of simpler processor cores. This chapter explores the tradeoffs involved in building such a simple-core CMP. Two case studies, the Niagara and Niagara 2 CMPs from Sun Microsystems, are used to illustrate how simple-core CMPs are built in practice and how they compare to CMPs built from more traditional high-performance superscalar processor cores. The case studies show that simple-core CMPs can have a significant performance/watt advantage over complex-core CMPs.

7.1 Introduction

Many important commercial server applications are throughput-oriented, focused on servicing large numbers of requests within a given time interval. The rapid growth of the Internet has dramatically increased the demand for servers capable of providing good throughput for a multitude of independent requests arriving rapidly over the network. Since individual network requests are typically completely independent tasks, whether those requests are for web pages, database access, or file service, they can be spread across many separate processing threads. Chip multiprocessors (CMPs) are ideally suited to handle these workloads, as the multiple processors on

J. Laudon (✉)
Google, Madison, WI USA
e-mail: jlaudon@google.com

S.W. Keckler et al. (eds.), *Multicore Processors and Systems*, Integrated Circuits and Systems, DOI 10.1007/978-1-4419-0263-4_7, © Springer Science+Business Media, LLC 2009

the chip can each be servicing an incoming request. While symmetric multiprocessors (SMPs) also have multiple processors to service the incoming requests, CMPs have an advantage over SMPs that they do not need to provide expensive on-chip and off-chip resources such as secondary caches, main memory, and network interfaces for each processor. Instead, the caches, memory, and network bandwidth are all dynamically shared by on-chip processors while servicing the incoming requests.

The approach taken to date towards building CMPs has largely revolved around placing two or more conventional superscalar processors together on a single die [7, 16–18]. This approach to building CMPs has been driven by economic forces. There are huge developmental, marketing, and logistical advantages to being able to develop a single-processor core that is capable of being used in both uniprocessors and CMPs of differing sizes, supporting systems ranging from laptops to supercomputers. Since the majority of the computer volume is driven by the laptop and desktop market, where single-threaded application latency is important, choosing a superscalar core as the base unit to be leveraged from uniprocessors to many-core CMPs makes good economic sense.

However, these evolutionary CMPs are not necessarily ideal for the server space. A CMP designed specifically for throughput-oriented workloads can take advantage of the fact that while these workloads require high throughput, the *latency* of each request is generally not as critical [3]. Most users will not be bothered if their web pages take a fraction of a second longer to load, but they will complain if the website drops page requests because it does not have enough throughput capacity. Therefore, it becomes possible to think about building a throughput-oriented CMP that employs both a lower clock-rate and a simpler microarchitecture. This will allow many more processing cores to fit in the same area or power budget. While these simpler cores will take slightly longer on each individual task, the greater increase in the number of cores supported will more than compensate for the longer per-core latency and the result will be an improvement in overall performance. The rest of this chapter investigates how this simple-core performance advantage works in practice.

7.2 Building a CMP to Maximize Server Throughput

There are three main rules for building a throughput-oriented CMP. First, the processing cores should be made as simple as possible while still being capable of meeting any individual request latency requirements. This enables the CMP to maximize the number of cores that can be fit within the desired area or power envelope. Second, each core should be designed so that all the core resources are as highly utilized as possible. This is generally done by sharing underutilized resources, such as multiple hardware threads sharing a pipeline or multiple cores sharing outer-level caches and infrequently used resources such as floating-point units. Finally, sufficient bandwidth must be provided to ensure that the requests generated by the large number of hardware threads can all be satisfied without significant queuing delay.

7.2.1 The Simple Core Advantage

Cores can be made simpler both by reducing the pipeline depth and by simplifying the underlying microarchitecture. Until recently, many processors were extracting much of their performance from very deep pipelines [6]. There are multiple disadvantages to this. First, the dynamic power used by a processor has a linear dependence on frequency and a quadratic dependence on supply voltage. To run a deep pipeline at high frequencies, a high-supply voltage is often required to quickly switch the individual transistors within a pipestage. In a more modest pipeline, the supply voltage can be reduced in conjunction with the frequency. Assuming a voltage reduction linear with the frequency reduction, this results in a cubic reduction in dynamic power in response to what is at most a linear reduction in performance. In practice, the performance drop resulting from a lower frequency is less than linear, since request processing time is often limited by memory or disk latency, which remains unchanged. The second disadvantage of deep pipelines is that they incur a lot of overhead. This is because deeper and more complex pipelines require more transistors, switching at a higher frequency, to complete the same function as a shallower, simple pipeline. In addition to the simple linear increase in the number of the pipeline registers associated with adding pipeline stages, the cycle time overhead from issues such as pipeline stage work imbalance, additional register setup/hold times for each stage of registers, and the potential for higher clock skew at high frequencies causes a superlinear increase in the number of registers required in any real system simply due to overhead. These additional overheads are fixed regardless of cycle time, so as the processor becomes more deeply pipelined and the cycle time decreases, these overheads become a larger and larger portion of the clock cycle and make additional pipelining even more expensive in terms of the number of transistors (and therefore power) required [8]. A deeper and more complex pipeline also requires that more of the instructions be executed speculatively, and when that speculation does not pay off, the extra power consumed is wasted. Furthermore, additional power is required to get the processor back on the right execution path. As a consequence, more modest pipelines have an inherent advantage over deep pipelines, and indeed several of the more recent processors have backed off from deep pipelines [22].

In addition to a more modest pipeline depth and processor frequency, the processor microarchitecture can be made much more simple. These simple cores will not have the same peak performance as the more complex cores, but the lowered peak performance is not as critical for most throughput-oriented applications. Commercial server applications exhibit high cache miss rates, large memory footprints, and low instruction-level parallelism (ILP), which leads to poor utilization on traditionally ILP-focused processors [10]. Processor performance increases in the recent past have come predominately from optimizations that burn large amounts of power for relatively little performance gain in the presence of low-ILP applications. The current 3- to 6-way out-of-order superscalar server processors perform massive amounts of speculative execution, with this speculation translating into fairly modest performance gains on server workloads [14]. Simpler cores running these same

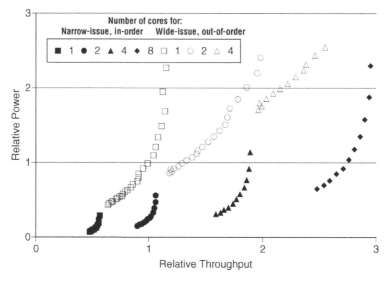

Fig. 7.1 Comparison of power usage by equivalent narrow-issue, in-order and wide-issue, out-of-order processors on throughput-oriented software, from [2]. Each type of processor is tested with different numbers of cores and across wide ranges of supply voltage and clock frequency

applications will undoubtedly take longer to perform the same task, but the relatively small reduction in latency performance is rewarded not only with higher overall throughput (due to the fact that more cores can be implemented on each processor chip) but also with better performance/watt (due to the elimination of power-wasting speculative execution).

Figure 7.1 shows a comparison between simple, in-order and complex, out-of-order processors with a single-thread per core across a wide range of voltages and clock rates, as described in [2]. For any given desired level of throughput (X axis), CMPs made from simple, in-order cores are able to achieve the same performance using only a fraction of the power of an equivalent machine made using complex, out-of-order processors, due to their significantly better performance/watt figures. For example, at the reference "throughput = 1" point, the in-order processor only requires about 20% of the power needed by the out-of-order processor to achieve the same performance, at the expense of requiring two parallel software threads instead of just one. Alternatively, looking at the same data from a position of fixed power limits, the eight-thread in-order and two-thread out-of-order processors require similar amounts of power across the voltage range studied, but the eight-thread system is 1.5–2.25 times faster than the two-thread system across the range.

7.2.2 Keeping the Processors Utilized

With most server applications, there is an abundant amount of application thread-level parallelism [10], but low instruction-level parallelism and high cache miss rates. Due to the high cache miss rates and low instruction-level parallelism, even a simple-core processor will be idle a significant portion of the time while running these applications. Clock gating can be used to reduce the active power of the idle blocks; however, the static and clock distribution power remains. Therefore, using *multithreading* to keep otherwise idle on-chip resources busy results in a significant performance/watt improvement by boosting performance by a large amount while increasing active power by a smaller amount. Multithreading is simply the process of adding hardware to each processor to allow it to execute instructions from multiple threads, either one at a time or simultaneously, without requiring OS or other software intervention to perform each thread switch—the conventional method of handling threading. Adding multithreading to a processor is not free; processor die area is increased by the replication of register files and other architecturally visible state (such as trap stack registers, MMU registers) and the addition of logic to the pipeline to switch between threads. However, this additional area is fairly modest. Adding an additional thread to a single-threaded processor costs 4–7% in terms of area [4, 5, 15], and each additional thread adds a similar percentage increase in area.

One question facing architects is what style of multithreading to employ. There are three major techniques:

1. Coarse-grained [1], where a processor runs from a single thread until a long-latency stall such as a cache miss, triggers a thread switch.
2. Fine-grained (or interleaved [12]), where a processor switches between several "active" (not stalled) threads every cycle.
3. Simultaneous multithreading [21], where instructions may be issued from multiple threads during the same cycle within a superscalar processor core.

Note that for a single-issue processor, fine-grained and simultaneous multithreading are equivalent. Coarse-grained multithreading has the drawback that short-latency events (such as pipeline hazards or shared execution resource conflicts) cannot be hidden simply by switching threads due to the multicycle cost of switching between threads [12], and these short-latency events are a significant contributor to CPI in real commercial systems, where applications are plagued by many brief stalls caused by difficult-to-predict branches and loads followed soon thereafter by dependent instructions. Multicycle primary caches, used in some modern processors, can cause these load-use combinations to almost always stall for a cycle or two, and even secondary cache hits can be handled quickly enough on many systems (10–20 cycles) to make hiding them with coarse-grained multithreading impractical. As a result, most multithreaded processors to date have employed fine-grained [8] or simultaneous multithreading [15, 19], and it is likely that future processors will as well.

Even after multithreading, there may be resources with low utilization, such as floating-point units, specialized accelerators, or possibly even the primary caches, and the sharing of these low-utilization resources between multiple cores can bring their utilization to the desired level [9].

7.2.3 Providing Sufficient Cache and Memory Bandwidth

The final component that is required for improved throughput and performance/watt from a server CMP is sufficient cache and memory bandwidth. When multithreading is added to a processor without providing sufficient bandwidth for the memory demands created by the increased number of threads, only modest gains in performance (and sometimes even slowdowns) will be seen [15].

7.2.4 Case studies of Throughput-Oriented CMPs

To show how these rules work in reality, two case studies of CMPs designed from the ground up for executing throughput-based workloads are examined: Sun's Niagara (UltraSPARC T1) and Niagara 2 (UltraSPARC T2) processors.

7.3 The Niagara Server CMP

The Niagara processor from Sun Microsystems [8], illustrated in Fig. 7.2, is a good example of a simple-core CMP that is designed specifically for high throughput and excellent performance/watt on server workloads. Niagara employs eight scalar,

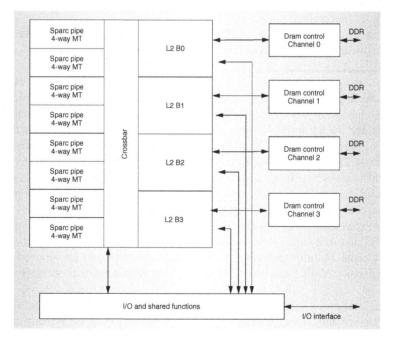

Fig. 7.2 Niagara block diagram. Reprinted with permission from [8]

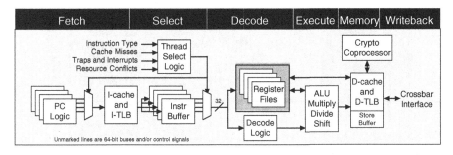

Fig. 7.3 Niagara core pipeline

shallow pipeline processors on a single die. The pipeline on Niagara is quite shallow, only six stages deep, and employs very little speculation, eschewing even simple static branch prediction. The Niagara pipeline is illustrated in Fig. 7.3.

Each Niagara processor supports four threads in hardware, resulting in a total of 32 threads on the CPU. The Niagara processor employs fine-grain multithreading, and the processor hides memory and pipeline stalls on a given thread by scheduling the other threads in the group onto the pipeline with the zero cycle switch penalty characteristic of fine-grain multithreading.

Niagara has a four-banked 3 MB L2 cache, with each bank of the level-two cache talking directly to a single memory controller. A crossbar interconnects the processor and the L2 cache on Niagara. The memory controllers on Niagara connect directly to DDR2 SDRAM. Niagara includes an on-chip IO controller, which provides a Sun-proprietary JBUS I/O interface. Each processor on the Niagara CMP contains a cryptography coprocessor to accelerate SSL processing.

Niagara is built in a 90 nm process from TI and runs at a frequency of 1.2 GHz. Niagara contains 279 million transistors in a 378 mm^2 die. A die photo of Niagara is shown in Fig. 7.4. A key challenge in building a CMP is minimizing the distance between the processor cores and the shared secondary cache. Niagara addresses this issue by placing four cores at the top of the die, four cores at the bottom of the die, with the crossbar and the four L2 tag arrays located in the center of the die. The four L2 data banks and memory I/O pads surround the core, crossbar, and L2 tags on both the left and right sides of the die. Filling in the space between the major blocks are the four memory (DRAM) controllers, the JBUS controller, internal memory-mapped I/O registers (I/O Bridge), a single-shared floating point (FPU), and clock, test, and data buffering blocks.

7.3.1 Multithreading on Niagara

Niagara employs fine-grained multithreading (which is equivalent to simultaneous multithreading for the Niagara scalar pipeline), switching between available threads each cycle, with priority given to the least-recently used thread. Threads on

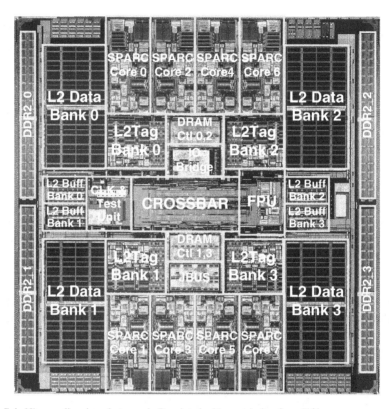

Fig. 7.4 Niagara die microphotograph. Reprinted with permission from [13]

Niagara can become unavailable because of long-latency instructions such as loads, branches, multiply, and divide. They also become unavailable because of pipeline "stalls" such as cache misses, traps, and resource conflicts. In Niagara, the thread scheduler assumes that loads are cache hits, and can therefore issue a dependent instruction from the same thread speculatively once the three-cycle load-to-use cost of Niagara has been satisfied. However, such a speculative thread is assigned a lower priority for instruction issue than a thread that can issue a non-speculative instruction. Along with the fetch of the next instruction pair into an instruction buffer, this speculative issuing of an instruction following a load is the only speculation performed by the Niagara processor.

   Figure 7.5 indicates the operation of a Niagara processor when all threads are available. In the figure, we can track the progress of an instruction through the pipeline by reading left-to-right along a row in the diagram. Each row represents a new instruction fetched into the pipe from the instruction cache, sequenced from top to bottom in the figure. The notation $S_{t0\text{-}sub}$ refers to a subtract instruction from thread 0 in the S stage of the pipe. In the example, the t0-sub is issued down the pipe. As the other three threads become available, the thread state machine selects thread1 and deselects thread0. In the second cycle, similarly, the pipeline executes

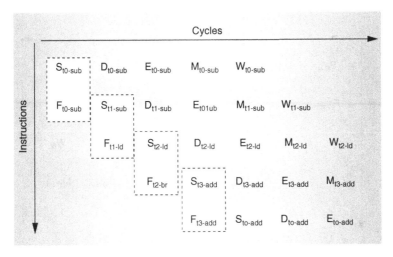

Fig. 7.5 Thread selection when all threads are available. Reprinted with permission from [8]

the t1-sub and selects t2-ld (load instruction from thread 2) for issue in the follow-ing cycle. When t3-add is in the S stage, all threads have been executed, and for the next cycle the pipeline selects the least-recently used thread, thread0. When the thread-select stage chooses a thread for execution, the fetch stage chooses the same thread for instruction cache access.

Figure 7.6 indicates the operation when only two threads are available. Here thread0 and thread1 are available, while thread2 and thread3 are not. The t0-ld in the thread-select stage in the example is a long-latency operation. Therefore, it causes the deselection of thread0. The t0-ld itself, however, issues down the pipe. In the second cycle, since thread1 is available, the thread scheduler switches it in. At this time, there are no other threads available and the t1-sub is a single-cycle operation, so thread1 continues to be selected for the next cycle. The subsequent instruction is a t1-ld and causes the deselection of thread1 for the fourth cycle. At this time, only thread0 is speculatively available and therefore can be selected. If the first t0-ld was a hit, data can bypass to the dependent t0-add in the execute stage. If the load missed, the pipeline flushes the subsequent t0-add to the thread-select stage instruction buffer and the instruction reissues when the load returns from the L2 cache.

Multithreading is a powerful architectural tool. On Niagara, adding four threads to the core increased the core area by 19–20% to a total of 16.35 mm^2 in a 90 nm process, when the area for the per-core cryptography unit is excluded. As will be shown later in this chapter, the addition of these four threads may result in a two and half- or even three-fold speedup for many large-scale commercial applications. These large performance gains translate into an increase in performance/watt for the multithreaded processor, as the power does not increase three-fold. Instead the core power increase is roughly 1.2•(static and clocking power) + 3.0•(dynamic power),

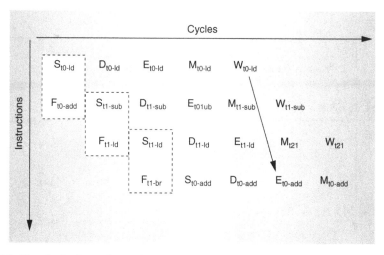

Fig. 7.6 Thread selection: only two threads available. The ADD instruction from thread 0 is speculatively switched into the pipeline before the hit/miss for the load instruction has been determined. Reprinted with permission from [8]

where the 1.2 increase in static and clocking power results from the 20% increase in processor core area for the addition of four threads. Note that the three-fold increase in core dynamic power is an upper bound that assumes perfect clock gating and no speculation being performed when a thread is idle. In addition to the core power increase, the multithreaded processor provides an increased load on the level-two (L2) cache and memory. This load can be even larger than the three-fold increase in performance, as destructive interference between threads can result in higher L1 and L2 miss rates. For an application experiencing a three-fold speedup due to multithreading, the L2 static power remains the same, while the active power could increase by a factor of greater than 3 due to the increase in the L2 load. In [11], the L2 dynamic power for a database workload was found to increase by 3.6. Likewise, for memory, the static power (including refresh power) remains the same, while active power could increase by a factor greater than 3. In [11], memory power for the database workload increased by a factor of 5.1. However, for both the L2 and memory, the static power is a large component of the total power, and thus even within the cache and memory subsystems, a three-fold increase in performance from multithreading will likely exceed the increase in power, resulting in an increase in performance/watt.

7.3.2 Memory Resources on Niagara

A block diagram of the Niagara memory subsystem is shown in Fig. 7.7 (left). On Niagara, the L1 instruction cache is 16 Kbyte, 4-way set-associative with a block size of 32 bytes. Niagara implements a random replacement scheme for area sav-

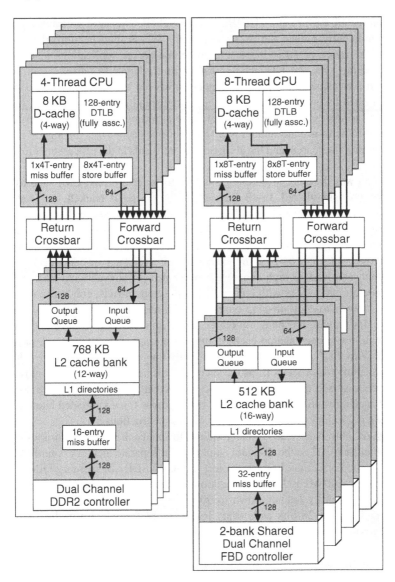

Fig. 7.7 (*Left*) Niagara memory hierarchy overview. (*Right*) Niagara 2 memory hierarchy overview

ings without incurring significant performance cost. The instruction cache fetches two instructions each cycle. If the second instruction is useful, the instruction cache has a free slot, which it can use to handle a line fill without stalling the pipeline. The L1 data cache is 8 Kbytes, 4-way set-associative with a line size of 16 bytes, and implements a write-through policy. To allow threads to continue execution past stores while maintaining the total store ordering (TSO) memory model, each thread has a dedicated, 8-entry store buffer. Even though the Niagara L1 caches are small, they significantly reduce the average memory access time per thread with miss rates in the range of 10%. Because commercial server applications tend to have large working sets, the L1 caches must be much larger to achieve significantly lower miss rates, so the Niagara designers observed that the incremental performance gained by larger caches did not merit the area increase. In Niagara, the four threads in a processor core are very effective at hiding the latencies from L1 and L2 misses. Therefore, the smaller Niagara level-one cache sizes are a good tradeoff between miss rates, area, and the ability of other threads in the processor core to hide latency.

Niagara was designed from the ground up to be a 32-thread CMP, and as such, employs a single, shared 3 MB L2 cache. This cache is banked four ways and pipelined to provide 76.8 GB/s of bandwidth to the 32 threads accessing it. In addition, the cache is 12-way associative to allow the working sets of many threads to fit into the cache without excessive conflict misses. The L2 cache also interleaves data across banks at a 64-byte granularity. Commercial server code has data sharing, which can lead to high-coherence miss rates. In conventional SMP systems using discrete processors with coherent system interconnects, coherence misses go out over low-frequency off-chip buses or links, and can have high latencies. The Niagara design with its shared on-chip cache eliminates these misses and replaces them with low-latency shared-cache communication. On the other hand, providing a single shared L2 cache implies that a slightly longer access time to the L2 cache will be seen by the processors, as the shared L2 cache cannot be located close to all of the processors in the chip. Niagara uses a crossbar to connect the processor cores and L2 cache, resulting in a uniform L2 cache access time. Unloaded latency to the L2 cache is 23 clocks for data and 22 clocks for instructions.

High off-chip bandwidth is also required to satisfy the L2 cache misses created by the multibank L2 cache. Niagara employs four separate memory controllers (one per L2 cache bank) that directly connect to DDR2 SDRAM memory DIMMs running at up to 200 MHz. Direct connection to the memory DIMMs allows Niagara to keep the memory latency down to 90 ns unloaded at 200 MHz. The datapath to the DIMMs is 128 bits wide (plus 16 bits of ECC), which translates to a raw memory bandwidth of 25.6 GB/s. Requests can be reordered in the Niagara memory controllers, which allow the controllers to favor reads over writes, optimize the accesses to the DRAM banks, and to minimize the dead cycles on the bi-directional data bus.

Niagara's crossbar interconnect provides the communication link between processor cores, L2 cache banks, and other shared resources on the CPU; it provides more than 200 GB/s of bandwidth. A two-entry queue is available for each source–destination pair, allowing the crossbar to queue up to 96 transactions in each direction. The crossbar also provides a port for communication with the I/O subsystem. Arbitration for destination ports uses a simple age-based priority scheme that ensures fair scheduling across all requestors. The crossbar is also the point of memory ordering for the machine.

Niagara uses a simple cache coherence protocol. The L1 caches are write through, with allocate on load and no-allocate on stores. L1 lines are either in valid or invalid states. The L2 cache maintains a directory that shadows the L1 tags. A load that missed in an L1 cache (load miss) is delivered to the source bank of the L2 cache along with its replacement way from the L1 cache. There, the load miss address is entered in the corresponding L1 tag location of the directory, the L2 cache is accessed to get the missing line and data is then returned to the L1 cache. The directory thus maintains a sharers list at L1-line granularity. A subsequent store from a different or same L1 cache will look up the directory and queue up invalidates to the L1 caches that have the line. Stores do not update the local caches until they have updated the L2 cache. During this time, the store can pass data to the same thread but not to other threads; therefore, a store attains global visibility in the L2 cache. The crossbar establishes TSO memory order between transactions from the same and different L2 banks, and guarantees delivery of transactions to L1 caches in the same order.

7.3.3 Comparing Niagara with a CMP Using Conventional Cores

Niagara-based systems (SunFire T1000 and SunFire T2000) became available in November 2005, allowing comparison against competitive systems in early 2006. These benchmark results highlight both the performance and performance/watt advantages of simple-core CMPs. Figure 7.8 (left) shows a comparison of SPECjbb 2005 results between the Niagara-based SunFire T2000 and three IBM systems based on CMPs using more conventional superscalar POWER or x86 cores: the IBM p510Q, IBM x346, and IBM p550. The SunFire T2000 has nearly twice the performance/watt of the closest system, the IBM p510Q, built from two 1.5 GHz Power5+ processors. The SunFire T2000 handily outperforms all three systems as well, despite the fact that the other systems all have two processor chips per box.

Similar performance and performance/watt advantages of the Niagara can be seen for SPECweb 2005. Figure 7.8 (right) shows the comparison between the

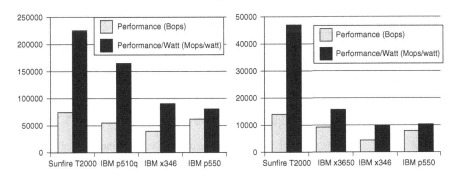

Fig. 7.8 (*Left*) SPECjbb 2005 performance. (*Right*) SPECweb 2005 performance

SunFire T2000, IBM x3650, IBM x346, and IBM p550 on SPECweb 2005. For SPECweb 2005, the performance/watt gap is even larger, with the SunFire T2000 having nearly three times the performance/watt of the nearest competitor, the IBM x3650, built from two dual-core 3.0 GHz Xeon 5160 processors. Again, the single-processor chip SunFire T2000 outperforms the competing dual-processor chip systems.

7.3.4 Niagara Performance Analysis and Limits

The large number of threads combined with limited speculation and a shallow pipeline allows Niagara to achieve excellent performance and performance/watt on throughput workloads. Figure 7.9 shows that processing an entire database work-load with only one thread running per core takes three times as long as processing the workload with four threads per core running, even though the individual threads execute more slowly in that case. Note that adding multithreading does *not* achieve the optimal four-times speedup. Instead, a comparison of the single-threaded and one-of-four thread time breakdowns allows us to see that destructive interference between the threads in the caches overcomes any inter-thread constructive cache interference and leads to increases in the portion of time that each of the multiple threads spends stalled on instruction cache, data cache, and L2 cache misses. There is also some time lost to contention for the shared pipeline, leading to an overall

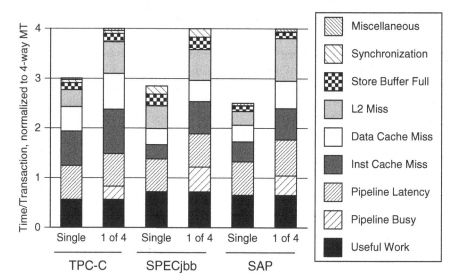

Fig. 7.9 TPC-C, SPECjbb2005, and SAP SD time breakdown within a thread, relative to the average execution time per task on a four-thread system. The overall single-threaded times show slowdown from the multithreaded case, while each thread in a four-thread system simply takes four times the average time per task, by definition

thread-versus-thread slowdown of 33%. However, the multithreading allows much of the increased cache latency and nearly *all* of the pipeline latency to be overlapped with the execution of other threads, with the result being a three-fold speedup from multithreading.

Similar results can be seen for SPECjbb 2005 and SAP SD. On both SPECjbb 2005 and SAP SD, the single thread is running at a higher efficiency than the database workload (CPIs of 4.0 and 3.8 are achieved for SPECjbb 2005 and SAP SD, respectively, compared to a CPI of 5.3 for the database workload), and as a result, the slowdown for each thread resulting from multithreading interference is larger, 40% for SPECjbb 2005 and 60% for SAP SD. As a result of the increased interference, the gains from multithreading visible on Fig. 7.9 are slightly lower for the two benchmarks, with SPECjbb 2005 showing a 2.85 times speedup and SAP SD a 2.5 times performance boost.

A closer look at the utilization of the caches and pipelines of Niagara for the database workload, SPECjbb 2005, and SAP SD shows where possible bottlenecks lie. As can be seen from Fig. 7.10, the bottlenecks for Niagara appear to be in the pipeline and memory. The instruction and data caches have sufficient bandwidth left for more threads. The L2 cache utilization can support a modestly higher load as well. While it is possible to use multiple-issue cores to generate more memory references per cycle to the primary caches, a technique measured in [5], a more effective method to balance the pipeline and cache utilization may be to have multiple single-issue pipelines share primary caches. Addressing the memory bottleneck is more difficult. Niagara devotes a very large number of pins to the four DDR2 SDRAM memory channels, so without a change in memory technology, attacking the memory bottleneck would be difficult. A move from DDR2 SDRAM to fully buffered DIMMs (FB-DIMM), a memory technology change already underway, is the key that will enable future Niagara chips to continue to increase the number

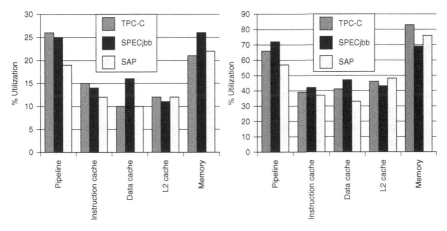

Fig. 7.10 (*Left*) Eight-thread (1 per processor) resource utilization. (*Right*) 32-thread (4 per processor) resource utilization

Fig. 7.11 Niagara 2 block diagram

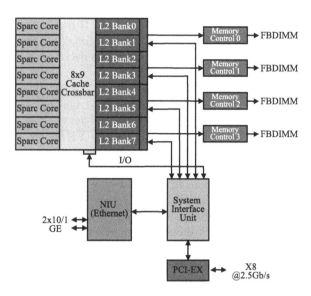

of on-chip threads without running out of pins. Niagara 2, the second-generation version of Niagara, uses both multiple single-issue pipelines per core to attack the execution bandwidth bottleneck and FB-DIMM memories to address the memory bottleneck. The next section looks at the CMP design produced following these changes.

7.4 The Niagara 2 Server CMP

Niagara 2 is the second-generation Niagara chip from Sun Microsystems. The design goals for Niagara 2 were to double the throughput of Niagara 1 and to include more system-on-a-chip (SOC) functionality. A block diagram of Niagara 2 is shown in Fig. 7.12. As with Niagara 1, Niagara 2 includes eight processor cores; however each core doubles the thread count to 8, for a total of 64 threads per Niagara 2 chip. In addition, each core on Niagara 2 has its own floating point and graphics unit (FGU), allowing Niagara 2 to address workloads with significant floating-point activity. The shared L2 cache in Niagara 2 increases to 4 MB and 16-way set-associativity. In addition, the L2 bank count doubles to eight to support the increased memory bandwidth requirements from the eight Niagara 2 processor cores. Niagara 2 retains four memory controllers, but each memory controller now connects to a dual-width FB-DIMM memory port, thereby greatly increasing the memory bandwidth while reducing the memory pin count. High-speed SERDES links, such as those employed in FB-DIMM interfaces, enable future Niagara chips to continue to increase the number of on-chip threads without running out of memory pins. A block diagram of the Niagara 2 memory subsystem is shown in Fig. 7.7 (right). The I/O interface on Niagara 2 changes from Niagara 1's JBUS to

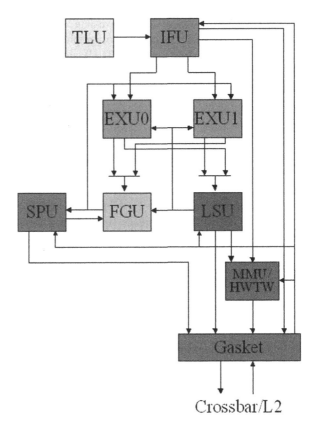

Fig. 7.12 Niagara 2 core block diagram. Reprinted with permission from [20]

an 8x PCI Express 1.0 port. Unlike Niagara 1, Niagara 2 includes a pair of on-chip 10/1 Gb Ethernet ports with on-chip classification and filtering.

7.4.1 Niagara 2 Core Details

While the Niagara 2 core is similar to Niagara 1, increasing the thread count necessitated re-engineering of most of the core components. As shown in Fig. 7.12, each core consists of several components. An instruction fetch unit (IFU) containing a 16 KB, 8-way set-associative instruction cache and a 64-entry, fully-associative ITLB supplies instructions to each of two independent integer units (EXU). The eight threads are partitioned into two groups of four threads. Each thread group has its own dedicated execution pipeline (integer unit), although the two thread groups share access to a single set of data and instruction caches, a load–store unit, and a floating-point unit. The integer units execute integer operations and generate addresses for load/store operations. The integer units also route floating-point instructions to the floating-point unit (FGU), and, for certain instructions like integer multiplies, provide data to the floating-point pipeline. The load/store unit (LSU) contains an 8 KB, 4-way set-associative data cache and a 128-entry, fully-associative DTLB. Each core also contains a cryptographic coprocessor, termed the stream processing unit (SPU), a trap logic unit (TLU), and a memory management unit (MMU) that performs hardware tablewalks to refill the ITLB and DTLB when a miss occurs. The Gasket arbitrates among the IFU, LSU, SPU, and MMU for the single crossbar port to the L2 and I/O subsystem.

7.4.2 Niagara 2 Instruction Fetch and Issue

Based on performance simulations, the size of the instruction cache on Niagara 2 is unchanged from Niagara 1's 16 KB instruction cache. Instruction cache miss rates are higher with eight threads versus four, but not enough to warrant a larger instruction cache. In order to supply enough instruction fetch bandwidth for eight threads, the instruction cache fetches up to four instructions each cycle. Associativity is increased to eight ways to help minimize potential set conflicts among the eight threads. To limit power consumption, the instruction cache is divided into quadrants, and only quadrants that contain instructions within the current cache line are enabled. The instruction cache feeds per-thread instruction buffers of eight entries each. In order to further improve performance, instruction fetching and instruction decoding are decoupled. The instruction fetch unit selects the least-recently fetched thread among ready-to-be-fetched threads, and seeks to keep each thread's instruction buffer full, independently of what thread is being picked for execution. A thread is ready to be fetched if it does not have a pending cache or TLB miss, and there is room in its instruction buffer. In order to limit speculation and L2 bandwidth, the instruction fetch unit prefetches only the next sequential line from the L2 cache. Unlike Niagara 1, branches are predicted not-taken, and instructions following the branch are allowed to flow down the pipeline until the branch is resolved by the integer unit.

7.4.3 Niagara 2 Instruction Execution

Based on performance simulations, the most area-efficient means of doubling Niagara 1's throughput is to increase the number of threads per core from 4 on Niagara 1 to 8 on Niagara 2. A second execution pipeline is added to each core in Niagara 2 in order to realize a doubling of throughput performance over Niagara 1. Each execution unit has its own corresponding thread state. Ideally, when picking instructions, the best two candidates from among all eight threads are examined. However, this is not feasible at the target frequency of Niagara 2. Instead, each pipeline is statically assigned a group of threads: pipeline 0 contains threads 0–3 and pipeline 1 contains threads 4–7. Picking within a given thread group is independent of the other thread group. On Niagara 2, independent picking achieves essentially the same performance as the ideal picking approach while satisfying frequency constraints.

Unlike Niagara 1, Niagara 2 allows independent instructions to be picked after a load instruction for a given thread. (Niagara 1 switched the thread out on a load regardless of whether the subsequent instructions depended on the load.) This can improve single-thread performance for the cases where the load actually hits into the data cache. A least-recently picked algorithm is used to select threads within each thread group to ensure fairness. Fairness is required in a multithreaded design such as Niagara 2 in order to avoid thread starvation issues. Independent picking between thread groups may lead to conflicts. Examples of this are when two threads are picked in the same cycle both requiring either the load/store unit or the FGU. In these cases, the conflict is resolved during the following pipeline cycle, the decode stage, using a favor bit. Each conflict point (e.g., load/store unit and FGU) has a favor bit. After picking, the favor bit is changed so that the thread group that lost arbitration has priority in the next conflict for that resource.

The FGU is completely redesigned on Niagara 2. It consists of a register file that contains 256 entries, divided into 32 registers for each thread, and several datapath components. The datapath consists of an add pipeline, which also performs certain graphics operations, a multiplier pipeline, which performs floating-point, integer, and certain cryptographic multiplies, and a divider. The divider can operate in parallel with the adder and multiplier pipelines, although the divider itself is not threaded. Unlike Niagara 1, Niagara 2 supports Sun's VIS 2.0 instructions directly in hardware. The FGU can produce one floating-point result each cycle. Furthermore, the latency of the FGU pipeline is less than the thread count, so even highly dependent floating-point code typically remains pipelined across eight threads. Coupled with the very high memory bandwidth of Niagara 2, this features leads to excellent performance on hard-to-optimize codes such as sparse matrix operations.

The cryptography coprocessor (a.k.a. streaming processing unit or SPU) is redesigned and enhanced on Niagara 2. The SPU supports two primary classes of operations: key generation and bulk operations. The key generation algorithms are enhanced to support elliptical curve cryptography (ECC), which provides a much higher degree of security for a given key size than an RSA key. Several improvements are added to the floating-point multiplier, such as a local bypass, XOR multiply capability for ECC, and 136-bit accumulator, so it more efficiently supports key generation computations. The SPU also incorporates support for common bulk cipher operations, such as RC4, DES/3DES, AES-128/192/256, SHA-1, SHA-256, and MD5. The SPU is designed to support wire speed cryptography across the two integrated 10 Gb Ethernet ports for all bulk cipher operations, assuming all cores' SPUs are employed at an operating frequency of 1.4 GHz.

7.4.4 Niagara 2 Memory Subsystem

Niagara 2's load–store unit has been enhanced to support eight threads. Replicating the load–store unit would have been too expensive in terms of area and power. Based on performance simulations, the size of the data cache on Niagara 2 is unchanged from Niagara 1's 8 KB data cache. Data cache miss rates are higher with eight threads versus four, but not enough to warrant a larger data cache. The DTLB is increased to 128 entries to cope with the increased miss rate of the additional threads. Store performance is enhanced in two ways. First, as in Niagara 1, the data cache is store-through and does not allocate on a store miss, and the L2 cache maintains global memory order and has a copy of the data cache tags. Stores thus have no need to interrogate the data cache tags enroute to the L2 cache from the store buffer. This frees up the data cache tags and data arrays to perform a fill operation whenever a load instruction is not coming down the pipeline. The second improvement is to pipeline store operations to the same L2 cache line. Allowing stores to pipeline to the same cache line improves single-thread performance. Stores crossing L2 banks cannot be pipelined, as independent L2 banks may process the stores out-of-order with respect to program order, violating the store ordering requirements of Sparc's TSO memory model.

The MMU is enhanced to support hardware TLB reloads. The MMU supports four page sizes (8 KB, 64 KB, 4 MB, and 256 MB), and four page tables. Typically, each page table will support pages of only one of the given page sizes, so the four tables can support all four page sizes. When a TLB miss occurs, the MMU therefore searches all page tables, as the missing page could be located in any table. It supports three modes of operation: burst, sequential, and prediction. Burst mode generally minimizes refill latency as all page tables are searched in parallel. However, if the TLB miss rate increases, burst mode can swamp the L2 cache, and impact the performance of other, unrelated threads. Sequential mode searches the page tables in a serial fashion. Sequential better utilizes L2 bandwidth but may result in longer TLB reload latencies. Prediction attempts to optimize both L2 bandwidth and TLB reload latency. In this mode, the MMU first hashes the virtual address to predict which page table to initially search and has been shown to improve throughput performance on high-TLB miss-rate applications.

7.4.5 Niagara 2 Physical Design

Niagara 2 is fabricated in TI's 65 nm process and is 342 mm^2. The use of FB-DIMM memory links and the high level of SOC integration allows Niagara 2 to keep the processor cores fed using only 711 signal pins (1831 total pins). A die of Niagara 2 is shown in Fig. 7.13. Niagara 2 retains the same basic layout as Niagara 1, with four cores on the top of the die, four at the bottom, with the L2 tags and crossbar located between the cores. Four L2 data banks and their corresponding pair of memory controllers lie on the left side of the cores, with another four L2 data banks and corresponding memory controller pair on the right side of the cores. FB-DIMM I/O pads line the left and right edges of the die. The PCI-Express port and pads are located in the lower left of the die, while the Ethernet controller and MAC lie in the right corner of the die, connecting to the Ethernet pads located

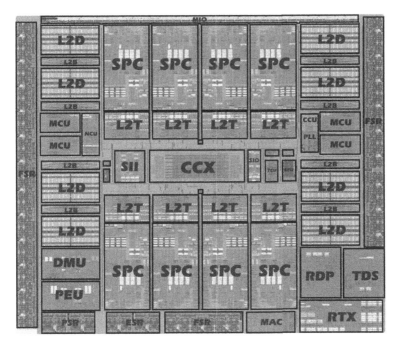

Fig. 7.13 Niagara 2 microphotograph. Reprinted with permission from [20]

next to the PCI-Express pads. Filling in the spaces between the major blocks are clocks, test circuits (Fuse), memory-mapped control registers, and the interconnect from the I/O blocks to memory.

Niagara 2 utilizes several different methods to minimize dynamic power. Wherever possible, clocks are extensively gated throughout the Niagara 2 design. In addition, Niagara 2 has the ability to throttle issue from any of the threads and keeps speculation to a minimum. Finally, speculation is kept to a minimum in the microarchitecture of Niagara 2.

Niagara 2 achieves twice the throughput performance of Niagara 1 within the same power envelope. It has improved single-thread performance, greatly improved floating-point and SIMD instruction performance, and more significant system-on-a-chip (SOC) features versus Niagara 1. Niagara 2's enhanced capabilities should service a broader part of the commercial server market versus Niagara 1.

7.4.6 Niagara 2 Performance Analysis

With the availability of real Niagara 2 systems (Sun SPARC Enterprise T5220), a variety of real results have become publicly available. These benchmark results show that Niagara 2 retains both the performance and performance/watt advantages of Niagara. Figure 7.14 shows a comparison of SPECjbb 2005 results between

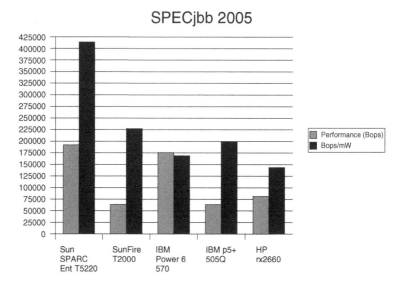

Fig. 7.14 SPECjbb 2005 performance

the Niagara 2-based Sun SPARC Enterprise T5220 and the Niagara-based SunFire T2000, an IBM system based on dual-core POWER 6 processors, an IBM system based on dual-core POWER 5+ processors, and a HP system built from dual-core Itanium 2 processors: the IBM Power 6 570, IBM p5+ 505Q, and HP rx2660. The Niagara 2-based Sun SPARC Enterprise T5220 achieved over three times the performance of the Niagara-based SunFire T2000 while improving performance/watt by over 80%. The Sun SPARC Enterprise T5220 also has over twice the performance/watt of the closest non-Niagara system, the IBM p5+ 505Q, built from two 1.65 GHz Power5+ processors. The Sun SPARC Enterprise T5220 outperforms all three non-Niagara systems as well, despite the fact that the other systems all have two processor chips per box.

Similar performance and performance/watt advantages of the Niagara 2 processor can be seen for SPECweb 2005. Figure 7.15 shows the comparison between the Sun SPARC Enterprise T5220, SunFire T2000, HP DL580 G5, HP DL380 G5, and HP DL585 G2 on SPECweb 2005. The Niagara 2-based Sun SPARC Enterprise T5220 achieved over 2.6 times the performance of the Niagara-based SunFire T2000 while improving performance/watt by over 25%. For SPECweb 2005, the Sun SPARC Enterprise T5220 has a 70% advantage in performance/watt over the nearest non-Niagara competitor, the HP DL380 G5, built from two quad-core 2.66 GHz Xeon 5355 processors. Again, the single-processor chip Sun SPARC Enterprise T5220 handily outperforms the competing non-Niagara dual- (HP DL380 G5) and quad-processor (HP DL580 G5 and DL585 G2) chip systems on the SPECweb 2005 composite metric.

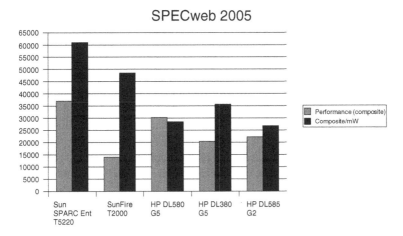

Fig. 7.15   SPECweb 2005 performance

7.5 Simple-Core Limitations

Of course, there is no such thing as a free lunch, and simple-core CMPs such as Niagara and Niagara 2 do have some limitations.

As has already been discussed, low single-thread performance could be a potential disadvantage of the simple-core approach. Most large-scale commercial applications, such as e-commerce, online transaction processing (OLTP), decision support systems (DSS), and enterprise resource planning (ERP) are heavily threaded, and even for non-threaded applications there is a trend towards aggregating those applications on a common server pool in a grid computing fashion. Multithreaded, simple-core CMPs like Niagara, can have a significant advantage in both throuput performance and performance/watt on these heavily threaded workloads. For workloads where the parallelism is low, however, the highest performance will be achieved by a CMP built from more complex cores capable of exploiting instruction-level parallelism from the small number of available software threads.

Another disadvantage of highly threaded designs are their requirements for large cache and memory bandwidths. Put another way, miss traffic typically scales with throughput. A large number of threads put a large demand on the throughput requirements of shared structures such as L2/L3 caches and memory buses. Additional area is required to satisfy the bandwidth requirements. The density of the Niagara/Niagara 2 L2 cache is reduced compared to a monolithic L2 due to the need to support multiple banks. Each L2 cache bank has separate input queues, fill and write-back buffers, and control state machines. Similarly, massive memory bandwidth requirements mean a good fraction of the chip is devoted to off-chip communication circuits.

An additional possible limitation results from the area efficiency of the simple core. Because it is possible to place many more simple cores on the same die, the total thread count of the CMP can become quite large. This large thread count is an advantage when the workload has sufficient parallelism, but when coupled with the lower single-thread performance can become a liability when insufficient parallelism exists in the workload. A single Niagara 2 chip has roughly the same number of threads as yesterday's large-scale symmetric multiprocessors. Future Niagara processors will likely include more threads per chip and support multiple chips in a single shared-memory system, leading to very large number of active threads implemented by hardware even in small, cost-effective systems. While many commercial applications have sufficient parallelism to be able to scale to several hundreds of threads, applications with more limited scalability will only be able to use a fraction of the threads in these future systems. Of course, these systems can run multiprogrammed workloads consisting of many of these more limited scalability applications. In addition, the Niagara processor line, along with many of the more recent processors, has hardware support for virtualization, and multiple operating systems (referred to as the guest OS) can be run on a single Niagara system, with each guest operating system running their own set of application workloads. As single systems become capable of running what used to require multiple dedicated mainframes, this ability to consolidate multiple workloads, each running under their own guest operating system fully protected from the effects of other guest operating systems, will become extremely important.

7.6 Conclusions

Throughput-oriented applications have become very important for commercial servers. Building throughput-oriented servers from simple-core CMPs can result in significant throughput and power advantages over systems employing complex-core CMPs, at the cost of lower per-thread performance. Commercial products such as Niagara and Niagara 2 have demonstrated the throughput advantages of simple-core CMPs in both performance and performance/watt. We saw that systems employing Niagara were able to achieve 2–3 times the performance/watt as systems employing complex-core CMPs from the same time frame. Niagara 2 systems achieve 1.7–2 times the performance/watt as current complex-core CMP systems.

Looking forward several years, it will be feasible to build CMPs with thousands of threads. Will software be able to use all these threads? While a number of key applications are capable of running with thousands of threads, many applications still run with one or a handful of threads. Hopefully a combination of new programming languages, more advanced compilers and run-time optimizers, and more sophisticated parallel programmers will result in the breakthrough that allows a significant fraction of application software to be able to take full advantage of the thousand-thread CMP, enabling us to solve problems that currently lie beyond our reach.

References

1. A. Agarwal, J. Kubiatowicz, D. Kranz, B.-H. Lim, D. Yeung, G. D'Souza, and M. Parkin. Sparcle: An evolutionary processor design for large-scale multiprocessors. *IEEE Micro*, pages 48–61, June 1993.
2. T. Agerwala and S. Chatterjee, Computer architecture: challenges and opportunities for the next decade. *IEEE Micro*, 25(3): 58–69, May/June 2005.
3. L. A. Barroso, K. Gharachorloo, R. McNamara, A. Nowatzyk, S. Qadeer, B. Sano, S. Smith, R. Stets, and B. Verghese. Piranha: A scalable architecture based on single-chip multiprocessing. In *Proceedings of the 27th International Symposium on Computer Architecture (ISCA-27)*, pages 282–293, June 2000.
4. J. Clabes, J. Friedrich, M. Sweet, J. DiLullo, S. Chu, D. Plass, J. Dawson, P. Muench, L. Powell, M. Floyd, B. Sinharoy, M. Lee, M. Goulet, J. Wagoner, N. Schwartz, S. Runyon, G. Gorman, P. Restle, R. Kalla, J. McGill, and S. Dodson. Design and implementation of the POWER5^{TM} microprocessor. *IEEE International Solid-State Circuits Conference (ISSCC)*, Feb 2004.
5. J. D. Davis, J. Laudon, and K. Olukotun. Maximizing CMT throughput with mediocre cores. In *Proceedings of the 14th International Conference on Parallel Architectures and Compilation Techniques*, pages 51–62, Sep 2005.
6. M. Hrishikesh, D. Burger, N. Jouppi, S. Keckler, K. Farkas, and P. Shivakumar. The optimal logic depth per pipeline stage is 6 to 8 FO4 inverter delays. In *Proceedings of the 29th Annual International Symposium on Computer Architecture*, pages 14–24, May 2002.
7. S. Kapil. UltraSPARC Gemini: Dual CPU processor. In *Hot Chips 15*, http://www.hotchips.org/archives/, Stanford, CA, Aug 2003.
8. P. Kongetira, K. Aingaran, and K. Olukotun. Niagara: A 32-way multithreaded SPARC processor. *IEEE Micro*, pages 21–29, Mar/Apr 2005.
9. R. Kumar, N. Jouppi, and D. Tullsen. Conjoined-core chip multiprocessing. In *Proceedings of the 37th International Symposium on Microarchitecture (MICRO-37)*, pages 195–206, Dec 2004.
10. S. Kunkel, R. Eickemeyer, M. Lip, and T. Mullins. A performance methodology for commercial servers. *IBM Journal of Research and Development*, 44(6), 2000.
11. J. Laudon. Performance/Watt: The New Server Focus. In *ACM SIGARCH Computer Architecture News*, 33(4), pages 5–13, Nov 2005.
12. J. Laudon, A. Gupta, and M. Horowitz. Interleaving: A multithreading technique targeting multiprocessors and workstations. In *Proceedings of the 6th International Symposium on Architectural Support for Parallel Languages and Operating Systems*, pages 308–318, Oct 1994.
13. A. Leon, J. Shin, K. Tam, W. Bryg, F. Schumacher, P. Kongetira, D. Weisner, and A. Strong. A power-efficient high-throughput 32-thread SPARC processor. *IEEE International Solid-State Circuits Conference (ISSCC)*, Feb 2006.
14. J. Lo, L. Barroso, S. Eggers, K. Gharachorloo, H. Levy, and S. Parekh. An analysis of database workload performance on simultaneous multithreaded processors. *Proceedings of the 25th Annual International Symposium on Computer Architecture*, pages 39–50, June 1998.
15. D. Marr. Hyper-Threading Technology in the Netburst® Microarchitecture. In *Hot Chips*, 14, http://www.hotchips.org/archives/, Stanford, CA, Aug 2002.
16. T. Maruyama. SPARC64 VI: Fujitsu's next generation processor. In *Microprocessor Forum*, San Jose, CA, Oct 2003.
17. C. McNairy and R. Bhatia. Montecito: The next product in the Itanium processor family. In *Hot Chips*, 16, http://www.hotchips.org/archives/, Stanford, CA, Aug 2004.
18. C. Moore. POWER4 system microarchitecture. In *Microprocessor Forum*, San Jose, CA, Oct 2000.
19. S. Naffziger, T. Grutkowski, and B. Stackhouse. The Implementation of a 2-core Multi-Threaded Itanium® Family Processor. *IEEE International Solid-State Circuits Conference (ISSCC)*, pages 182–183, Feb 2005.

20. U. Nawathe, M. Hassan, K. Yen, A. Kumar, A. Ramachandran, and D. Greenhill. Implementation of an 8-Core, 64-Thread, Power-Efficient SPARC Server on a Chip. *IEEE Journal of Solid-State Circuits*, 44(1), pages 6–20, Jan 2008.
21. D. Tullsen, S. Eggers, and H. Levy. Simultaneous multithreading: Maximizing on-chip parallelism. In *Proceedings of the 22nd Annual International Symposium on Computer Architecture*, pages 392–403, June 1995.
22. O. Wechsler. Inside Intel®Core™ Microarchitecture Setting New Standards for Energy-Efficient Performance. http://download.intel.com/technology/architecture/new_architecture_06.pdf

Chapter 8
Stream Processors

Mattan Erez and William J. Dally

Abstract Stream processors, like other multi core architectures partition their functional units and storage into multiple processing elements. In contrast to typical architectures, which contain symmetric general-purpose cores and a cache hierarchy, stream processors have a significantly leaner design. Stream processors are specifically designed for the stream execution model, in which applications have large amounts of explicit parallel computation, structured and predictable control, and memory accesses that can be performed at a coarse granularity. Applications in the streaming model are expressed in a *gather–compute–scatter* form, yielding programs with explicit control over transferring data *to* and *from* on-chip memory. Relying on these characteristics, which are common to many media processing and scientific computing applications, stream architectures redefine the boundary between software and hardware responsibilities with software bearing much of the complexity required to manage concurrency, locality, and latency tolerance. Thus, stream processors have minimal control consisting of fetching medium- and coarse-grained instructions and executing them directly on the many ALUs. Moreover, the on-chip storage hierarchy of stream processors is under explicit software control, as is all communication, eliminating the need for complex reactive hardware mechanisms.

8.1 Principles of Stream Processors

Stream processors (SPs) embrace the principles of explicit locality and parallelism found in multi core architectures to an extreme, achieving high performance with high efficiency for applications that use the stream programming model, rather than the traditional parallel extensions to the von Neumann execution model. SPs typically have a sustained performance that is a factor of 5–15 better than conventional architectures, while being 10–30 times more efficient in terms of performance per unit power. By allowing streaming software to take full advantage of the parallelism

M. Erez (✉)
The University of Texas at Austin, 1 University Station, Austin, TX 78712, USA
e-mail: mattan.erez@mail.utexas.edu

S.W. Keckler et al. (eds.), *Multicore Processors and Systems*, Integrated Circuits and Systems, DOI 10.1007/978-1-4419-0263-4_8, © Springer Science+Business Media, LLC 2009

and locality inherent to the problem being solved, SP hardware is freed from many of the responsibilities required of processors when executing execute traditional programs. Thus, stream processors are designed for the strengths of modern VLSI technology with minimum execution overheads, providing efficiency on par with application-specific solutions (such as ASICs and ASSPs) [36]. The stream architecture focuses on effective management of locality (state) and bandwidth (communication) and optimizing throughput rather than latency. The principles of the stream architecture are based on the key properties of many performance-demanding applications and the characteristics of modern VLSI technology.

Many of today's performance-demanding applications come from the signal processing, image processing, graphics, and scientific computing domains. Applications in all of these domains exhibit high degrees of data parallelism, typically have structured control, and can exploit locality. These properties are exposed by the stream programming style, in which programs are expressed as a sequence of kernels connected by streams of data. In its strict form, stream processing is equivalent to the synchronous dataflow model (SDF) [41, 10]. In its generalized form, the stream programming model is a *gather–compute–scatter* model, where the programmer has more explicit control over the movement of data *to* and *from* on-chip memory (*gather* and *scatter*, respectively). This model will be described in greater detail in Sect. 8.2. Stream applications are written in a high-level language and a stream programming system targets the architectural features of an SP.

Today's fabrication technology allows numerous execution resources, such as thousands of arithmetic-logical units (ALUs) or hundreds of floating-point units (FPUs), to be integrated into a single chip, such that execution resources are not the factor limiting the performance or efficiency of processors. For example, Table 8.1 lists the estimated area and energy requirements of a 64-bit floating-point unit spanning a number of technology generations,[1] as well as the maximum potential performance that can be achieved with a die area of $100\,mm^2$ and a power budget of $100\,W$, which can be economically mass produced. Thus, the challenge is not in designing an architecture with high peak performance, but rather in efficiently and effectively providing the ALUs with data and instructions to sustain high perfor-

Table 8.1 Estimated area of a 64-bit multiply–add floating-point unit, the energy to compute a single result (logic only), and the maximum performance possible on a $100\,mm^2/100\,W$ chip, across technology nodes

Node	FPU Area	Energy/FLOP	FPUs/chip	FPU Frequency	GFLOP/s/chip
90 nm	$0.50\,mm^2$	100 pJ	200	1.0 GHz	200 GFLOP/s
65 nm	$0.26\,mm^2$	72 pJ	383	1.4 GHz	530 GFLOP/s
45 nm	$0.13\,mm^2$	41 pJ	800	2.0 GHz	1,600 GFLOP/s
32 nm	$0.06\,mm^2$	24 pJ	1,562	2.8 GHz	4,200 GFLOP/s

---

[1] These illustrative estimates are based on the results presented in [36, 21] extrapolated using the parameters presented in [52].

mance. Broadly speaking, three potential bottlenecks must be overcome in order to meet this goal:

1. First, many independent instructions must be identified in order for the ALUs to operate concurrently as well as for tolerating latencies (latencies can be hidden with concurrent useful work).
2. Second, the resources devoted to controlling the ALUs must be restricted to a fraction of the resources of the ALUs themselves. In many modern general-purpose CPUs, in contrast, the area and power devoted to the arithmetic units is a small fraction of the overall chip budget. Estimates for an Intel® Core™ 2 and an AMD Phenom™ are that less than about 5% of the area and the peak power are consumed by arithmetic.
3. Third, the bandwidth required to feed the ALUs must be maintained, which is perhaps the most difficult challenge. The cost of data transfer grows at least linearly with distance in terms of both the energy required and available bandwidth [18, 29]. As a result, global bandwidth, both on and off-chip, is the factor limiting the performance and dominating the power of modern processors.

Stream processors directly address all three challenges and potential bottlenecks described above. The stream architecture is guided by the abundant parallelism, locality, and structured control of applications and the fabrication properties that make compute resources cheap and bandwidth expensive. There are three key principles that work together to enable the high performance and efficiency of the stream architecture:

1. **Storage organized into a *bandwidth hierarchy*** – This hierarchy ranges from high-bandwidth registers local to each ALU through on-chip SRAM to off-chip DRAM. Each level of the bandwidth can provide roughly 10 times as much bandwidth at 1/10 the power requirement as the level above it, but has a much smaller total capacity. This bandwidth hierarchy enables high-bandwidth data supply to the ALUs even with limited global and off-chip bandwidths.
2. **Exposed communication** – All communication and data movement in an SP is under explicit software control, including all allocation of state within the storage hierarchy as well as all transfers between levels. This architectural feature allows software to maximize locality and be fully aware of the cost and importance of data movement and bandwidth usage. Even if data is not reused multiple times, exposing communication facilitates exploiting producer–consumer locality. Additionally, exposing allocation and communication enables software to take on the task of hiding latency, relieving hardware of this responsibility, and allowing for efficient throughput-oriented hardware design.
3. **Hierarchical control optimized for throughput and parallelism** – SPs use a hierarchical control structure where a stream application is expressed using two levels of control. At the high level, *kernels* represent atomic bulk computations that are performed on an entire stream or block of data, and *stream loads*

and stores process asynchronous bulk data transfers that move entire streams between on-chip and off-chip storage. The system manages these coarse-grained bulk operations as atomic units with more efficient hardware mechanisms than if all the computations were to be managed at a word granularity. At the lower level, many parallel ALUs, in different *processing elements* (PEs), are controlled by a single kernel microcode sequencer, amortizing instruction delivery overheads. Additionally the stream memory system utilizes the asynchronous bulk semantics to optimize expensive and scarce off-chip bandwidth.

Like other multi core architectures, SPs partition the ALUs and storage into multiple PEs. In contrast to typical designs, which contain symmetric general-purpose cores and a cache hierarchy, SPs have a significantly leaner design. Relying on large amounts of explicit parallel computation, structured and predictable control, and memory accesses performed in coarse granularity, the software system is able to bear much of the complexity required to manage concurrency, locality, and latency tolerance. Thus, SPs have minimal control, which fetches medium- and coarse-grained instructions and executes them directly on the many ALUs. Moreover, the on-chip storage hierarchy is under explicit software control, as is all communication, eliminating the need for complex reactive hardware mechanisms. SPs share some of these characteristics with other recent processor designs, such as the Cell Broadband Engine^{TM} [30] and graphics processing units from NVIDIA® and AMD [46, 7].

Section 8.2 explains how programs written in the stream programming model take advantage of the SP hardware, and Sect. 8.3 provides greater detail on the SP architecture. Section 8.4 describes software tools developed specifically for SPs, including StreamC/KernelC, Brook, and Sequoia. Section 8.5 discusses the specific stream processor designs of the Stanford Imagine that targets media and signal processing, the Stanford Merrimac Streaming Supercomputer, and the Stream Processors Inc. Storm 1. Section 8.6 and 8.7 conclude with remarks on the scalability of SPs and a summary.

8.2 Stream Programming Model

The stream programming model provides an efficient representation for compute- and memory-intensive applications and enables the software system to effectively target efficient stream processing hardware. As explained below, stream programming is well-matched to the three stream architecture principles and expresses multiple levels of locality, structured communication, and a hierarchy of control.

Stream programs express computation as coarse-grained *kernels*, which process input and output data that is organized into *streams*. A *stream* is a sequence of similar data elements, such as primitive data types or user-defined records (structures). Streams are used both as high-level data containers and for expressing communication. Streams are used to express communication of values between two kernels as well as to represent memory transfers required for kernel inputs and outputs.

These streams are commonly large collections of hundreds or thousands of elements. A *kernel* is a coarse-grained operation that processes entire input streams of data and produces output streams. In the stream model, kernels can only access input and output defined by stream arguments and do not access arbitrary memory locations. This partitioning of communication enables a two-level control hierarchy. Because all off-chip references are grouped into stream transfers, stream loads and stores encapsulate a large number of individual word transfers in a single-compound memory operation. Similarly, once all kernel inputs have been brought on chip, kernel execution can proceed without stalls and can be regarded as a single-compound compute operation. This in stark contrast to the traditional reactive caching model, in which loads and stores process for single words and cache misses can lead to rapid and non-deterministic performance degradation.

Stream programs are represented as a *stream flow graph* (SFG) and are compiled at this higher granularity of complex kernel operations (nodes) and structured data transfers (edges). Figure 8.1 describes the SFG of an n-body particle method, which interacts each of the n particles in the system (a central particle) with particles that are close enough to affect its trajectory (the neighbor particles). The compound kernel operation typically loops through the sequence of input elements producing elements on its output streams. In this way, the model exposes *instruction-level parallelism* (ILP) within the processing of a stream element. In addition, kernels expose a level of locality (*kernel locality*) as all accesses made by a kernel refer to data that is local to the kernel or that is passed to the kernel in the form of a stream. The streams that connect kernels expose *producer–consumer locality*, as one kernel consumes the output of a kernel it is connected to, shown by the time-stepping kernel consuming the forces produced by the interaction kernel. Streams connecting kernels and memory encapsulate all memory transfers and communication. The structured style of stream programs also expresses *task-level parallelism* (TLP), as the dependencies between multiple kernels are clearly defined. Finally, *data-level parallelism* (DLP) can also be conveyed with state-less kernels.

The stream programming model has its roots in a strict form represented by a synchronous dataflow graph [41, 10]. An SDF is a restricted form of dataflow where all rates of production, consumption, and computation are known statically and where nodes represent complex kernel computations and edges represent sequences of data elements flowing between the kernels. A compiler can perform rigorous analysis on an SDF graph to automatically generate code that explicitly expresses parallelism and locality and statically manages concurrency, latency tolerance, and state allocation [14, 54]. The key properties of the strict stream code represented by an SDF are large amounts of parallelism, structured and predictable control, and data transfers that can be determined well in advance of actual data consumption.

The same properties can be expressed and exploited through a more flexible and general stream programming style that essentially treats a stream as a sequence of blocks as opposed to a sequence of elements and relies on the programmer to explicitly express parallelism [38]. This style directly matches the execution model of SPs.

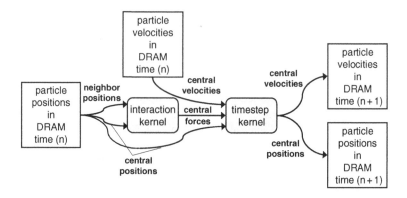

```
// declaration of input data in memory (assume initialized)
// molecule is a struct for holding oxygen and 2 hydrogen 3-vectors
memory<molecule> M_central_pos[N_molecules];
memory<molecule> M_velocities[N_molecules];
memory<molecule> M_neighbor_pos[N_neighbors];

// helper structures that contain blocking information
memory<int>      M_central_blocks[N_blocks+1];
memory<int>      M_neighbor_blocks[N_blocks+1];

// streams for holding kernel inputs and outputs
stream<molecule> S_central_pos;
stream<molecule> S_neighbor_pos;
stream<molecule> S_forces;
stream<molecule> S_velocities;
stream<molecule> S_new_pos;
stream<molecule> S_new_velocities;


// high-level stream control for n-body code
for (int i=0; i<N_blocks; i++) {
  streamGather(in M_central_pos[M_central_blocks[i]:
                                M_central_blocks[i+1]],
               out S_central_pos);
  streamGather(in M_neighbor_pos[M_neighbor_blocks[i]:
                                 M_neighbor_blocks[i+1]],
               out S_neighbor_pos);

  interaction_kernel(in S_central_pos, in S_neighbors_pos,
                     out S_forces);

  streamGather(M_velocities[M_central_blocks[i]:
                            M_central_blocks[i+1]],
               out S_velocities);

  timestep_kernel(in S_central_pos, in S_velocities,
                  out S_new_pos, out S_new_velocities);

  streamScatter(in S_new_pos,
                out M_central_pos[M_central_blocks[i]:
                                  M_central_blocks[i+1]]);
  streamScatter(in S_new_velocities,
                out M_velocities[M_central_blocks[i]:
                                 M_central_blocks[i+1]]);
}
```

Fig. 8.1 Stream flow graph and pseudo-code of an n-body particle method. Shaded nodes represent computational kernels, clear boxes represent locations in off-chip memory, and edges are streams representing stream load/store operations and producer–consumer locality between kernels

Figure 8.1 shows the stream flow graph and pseudo-code for an n-body method, which demonstrates how locality and parallelism are expressed. The interaction kernel processes an entire block of particles (central particles) and a block of the particles that interact with them (neighbor particles). Because the calculation of each particle is independent of the rest, DLP is exposed by the nature of the data and ILP can be extracted from the kernel body. The dependencies between the interaction and time-step kernel are explicit and express pipeline, or task, parallelism between the two kernels. Expressing this dependency through the explicit communication of values from one kernel to another also exposes producer–consumer locality that can be exploited, as well as the reuse of central-molecule position values. The interaction kernel has the flexibility to reuse position values if they appear in the neighbor lists of multiple central particles to further improve locality, which can be done by hardware within the memory system or by a software pre-processing step that renames off-chip memory locations to on-chip locations [23]. In addition, the read and write rates of the streams need not be fixed and each particle may have a different number of neighbors. This variable rate property is impossible to express in strict SDF, but can be supported by SP systems as explained in [33, 48, 47, 23]. The greater flexibility and expressiveness of this *gather, compute, scatter* style of generalized stream programming is not based on a solid theoretical foundation as SDF, but it has been successfully implemented in experimental compilation and hardware systems [43, 19, 34, 17, 21].

Whether using the strict or general form, the stream programming system extracts information about multiple types and levels of locality, explicit communication, and hierarchical control. Mapping an application to a SP involves two steps: *kernel scheduling*, in which the operations of each kernel are scheduled and compiled to a binary representation; and *stream scheduling*, in which kernel executions and bulk data transfers are scheduled as coarse-grained operations to use the on-chip bandwidth hierarchy efficiently and to maximize data locality. In the next section we describe an abstract stream processor and explain how streaming software and hardware interact for economic high-performance computation using the same n-body example shown above.

8.3 Stream Processor Architecture

Stream processors (SPs) are designed, and optimized, specifically for the stream execution model described in Sect. 8.2. The hardware architecture provides mechanisms for software to take on responsibilities for scheduling, allocation, and latency hiding, which gives the hardware greater efficiency without sacrificing performance. Furthermore, the explicit control given to streaming software enables more predictable performance, increasing the effectiveness of programmer and compiler optimizations. In this section, we will present how a stream application is executed on an SP and explain how the key principles of stream processors: hierarchical control, bandwidth hierarchy, exposed communication, and throughput-oriented design (see Sect. 8.1) are manifested in the architecture.

8.3.1 Execution Overview and Hierarchical Control

An SP has three major components as shown in Fig. 8.2:

Scalar unit - A general-purpose processing core that is responsible for the overall execution and control of a stream application (see Sect. 8.3.1).

Stream unit - Operates at a lower control hierarchy level and executes the computational kernels on the *compute clusters* and stream memory operations on the *stream load/store* units (SLSs). The *compute clusters* are the stream unit's processing elements (PEs) and contain the large number of ALUs required to attain high bandwidth, as well as an on-chip storage hierarchy (see Sections 8.2 and 8.3.3.1). The *address generators* handle asynchronous bulk data transfers between off-chip and on-chip storage.

Interconnect - Includes data and control paths.

Memory and I/O - DRAM, network, and I/O interfaces (see Sect. 8.3.3.2).

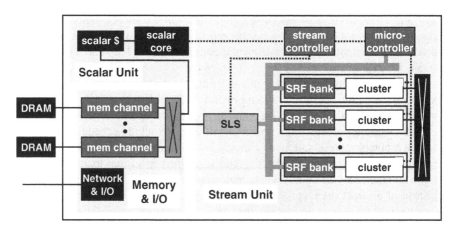

Fig. 8.2 Control hierarchy and structures of a stream processor

A stream application consists of a high-level stream control program and a collection of kernel objects. The stream control program is executed on the **scalar unit** and is expressed in a conventional ISA with word-granularity load, store, arithmetic, logic, and control instructions augmented with coarse-grained stream memory and kernel instructions. When the scalar unit encounters a stream instruction, it does not execute it directly, but rather issues the stream instruction to the stream unit. Thus, an SP has a hierarchical control structure with the scalar unit at the top and the stream unit's compute clusters and SLS at the bottom.

The bulk of a computation is contained within the coarse-grained stream instructions that are embedded as part of the scalar execution flow. The *stream controller* serves as the interface between the scalar and stream units, and coordinates the execution of stream memory operations, kernels, and their synchronization. The clusters in an SP are controlled in a SIMD fashion by a *microcontroller*, which is at the lower level in the control hierarchy and is responsible for sequencing the arithmetic instructions that were produced at the kernel scheduling step and represented as a binary (see Sect. 8.3.3.1 for details). Stream loads and stores, which typically transfer hundreds or thousands of words from memory to on-chip storage, are similarly off-loaded from the scalar unit (which executes the stream control program) and executed by the SLS units. The microcontroller and SLS operate at the granularity of individual word memory and arithmetic instructions, and through the stream controller, decouple this fine-grained execution from the control program. The SP control hierarchy and its asynchronous bulk stream operations allow specialization of the hardware tuning the different control structures to work most effectively. For example, sophisticated bulk memory transfers do not require a full hardware context or thread and are executed by dedicated and efficient SLS units.

While we defer the details of the storage bandwidth hierarchy to Sect. 8.3.2, the example depicted in Fig. 8.3 demonstrates how the storage and control hierarchy come together in stream execution. Coarse-grained stream loads and stores in the stream control program initiate data transfer between global memory and the on-chip storage. The memory system is optimized for throughput and can handle both strided loads and stores and indexed gathers and scatters in hardware. Kernel instructions in the stream control program cause the stream execution unit to start running a previously loaded kernel binary. Kernel execution is controlled by the microcontroller unit and proceeds in a SIMD manner with all clusters in the same node progressing in lockstep. Data transfers between various levels of the on-chip memory hierarchy are controlled from the microcode as explained in Sects. 8.3.2 and 8.3.3.1. The stream control program uses a coarse-grained software pipeline to load more data into on-chip storage and save output streams to main memory

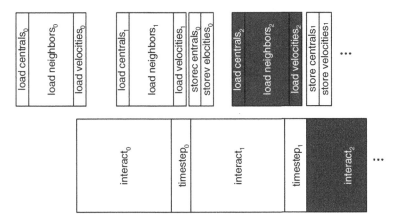

Fig. 8.3 Stream level software controlled latency hiding. The software pipeline of the stream instructions corresponding to the blocked n-body SFG of Fig. 8.1

concurrent with kernel execution, hiding memory access latencies with useful computation.

8.3.2 Locality and the Bandwidth Hierarchy

As VLSI technology scales, and the distance an operand traverses from storage to functional unit grows, the operand bandwidth decreases [18]. Therefore, to support high sustained operand throughput, stream architectures are strongly partitioned into PEs, which are *clusters* of functional units with a hierarchy of storage. The storage hierarchy directly forms the bandwidth hierarchy of an SP and provides the mechanisms required for the support of exposed-communication software.

The lowest level in the storage hierarchy is the *local register file* (LRF) within each cluster, which is directly connected to the functional units over short high bandwidth wires (see also Sect. 8.3.3.1). The LRF exploits *kernel locality*, which is short-term producer–consumer within kernels. As we will see in the case studies later in this chapter, a stream processor can economically support up to 16 clusters of five 32-bit ALUs (which can act as ten 16-bit ALUs) in the stream processors Inc. (SPI) Storm 1 processor for media applications or 16 clusters of four 64-bit FPUs in the Merrimac streaming supercomputer chip. The efficient LRF structure [51] can sustain the required operand bandwidth, and is able to provide over 90% of all operands in most applications by exploiting kernel locality between instructions within a function or inner-loop body [17, 4].

The second level in the storage hierarchy is the *stream register file* (SRF), which is a fast local memory structure. The SRF has a larger capacity and is partitioned across the clusters (PEs). We refer to the portion of the SRF local to a cluster as an *SRF lane*. Being local to a cluster enables the SRF to supply data with high throughput on the chip. The SRF has a local address space in the processor that is separate from global off-chip memory, and requires software to *localize* data into the SRF before kernel computation begins.

Explicitly managing this distinct space in software enables two critical functions in the stream architecture. First, the SRF can reduce pressure on off-chip memory by providing an on-chip workspace and explicitly capturing long-term producer–consumer locality between kernels. Second, explicit control allows software to stage large granularity asynchronous bulk transfers in to and out of the SRF. Off-chip memory is thus only accessed with bulk asynchronous transfers, allowing for a throughput optimized memory system and shifting the responsibility of tolerating the ever-growing memory latencies from hardware to software. SPs rely on software controlled bulk transfers to hide off-chip access latency and restrict cluster memory accesses to the local SRF address space. This result in predictable latencies within the execution pipeline, enabling static compiler scheduling optimizations.

Beyond the SRF is a global on-chip level of the storage hierarchy composed of an interconnection network that handles direct communication between the clusters and connects them with the memory system described towards the end of this section.

8.3.3 Throughput-Oriented Execution

Effectively utilizing the many ALUs of a stream processor requires extensive use of parallelism: ALUs must be run concurrently on distinct data; ALUs must be provided with instructions; small execution pipeline latencies must be tolerated; and long memory latencies must be overlapped with computation. Stream processors offer a hierarchy of control and decoupled execution to achieve these goals. The stream processor architecture focus on efficient throughput-oriented design is apparent in both the arithmetic clusters that process kernel operations and the memory system, which executes stream loads and stores.

8.3.3.1 Arithmetic Clusters

The microcontroller is responsible for issuing kernel instructions to the ALUs, and it relies on explicit software parallelism to maintain high-instruction bandwidth. SPs utilize DLP across the clusters and the microcontroller issues a single instruction to all clusters in a SIMD fashion. Each of these kernel instructions, however, is a VLIW instruction that controls all ALUs within the cluster as well as the operation of the register files, LRF, and SRF. This is depicted in Fig. 8.4, which shows how control of the ALUs is structured along the DLP and ILP dimensions of parallelism.

Figure 8.5 shows a single-generic stream processor arithmetic cluster. The cluster has four major component types: a number of arithmetic functional units (ALUs or FPUs; shaded in light gray), supporting units (SRF-in/out and communication; shaded in dark gray), the distributed LRF, and the intra-cluster switch. The microcontroller transmits a single VLIW instruction to all the clusters, and the VLIW independently controls all the units across all four cluster components. This cluster organization is enabled by the stream execution model, which decouples the unpredictable service time of off-chip memory requests from the execution pipeline. Because the clusters can only access the SRF, all latencies are known statically, and the compiler can optimize for ILP and DLP across the functional units and clusters.

The static knowledge of latencies and abundant parallelism enables aggressive compilation and permits a spartan and efficient pipeline hardware design that does not require non-execution resources such as a branch predictor, a dynamic out-of-order scheduler, and even bypass networks (ample parallelism). Moreover, the cluster exposes all communication between functional units and does not rely on a central register file to provide connectivity. This partitioning of the register file into storage, within the LRFs that are directly attached to the functional units and communication provided by the intra-cluster switch saves significant area and energy [51]. In addition to the LRF, switch, and arithmetic units,

Fig. 8.4 ALU organization along the DLP and ILP axes with SIMD and VLIW (all ALUs with the same shading execute in SIMD). The figure shows four clusters of four ALUs each (*shaded boxes*), and the chip interconnect partitioned into inter- and intra-cluster switches

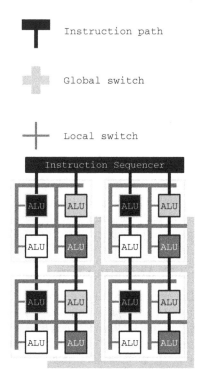

Fig. 8.5 Generic cluster organization with four ALUs (or FPUs), four in/out units for interfacing with a lane of the SRF local to the cluster, and a communication unit for connecting the cluster to the inter-cluster switch

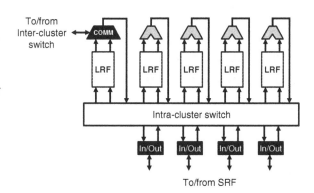

the cluster also contains supporting functional units. These include in/out units for accessing the SRF and a unit for connecting the cluster to the inter-cluster switch and providing explicit information exchange between all clusters without accessing the memory system. The case studies in Sect. 8.5 provide more details on the specific organization of the arithmetic cluster in various stream processors.

8.3.3.2 Stream Memory System

Stream memory systems are tuned for throughput in contrast to traditional memory systems that are often optimized for latency. Stream memory systems provide high throughput of multiple words per cycle using conventional DRAM technology by processing all requests in bulk asynchronous DMA-style transfers [6, 15]. One or more address generators process bulk stream requests from the processor and issue them as potentially reordered fine-granularity word accesses to memory. This allows the memory system to support complex addressing modes including gathers, scatters, and atomic data-parallel read-modify-writes (scatter-add) [5].

Because latency hiding is orchestrated by software in the coarse granularity of kernels and stream loads and stores, the memory system provides very deep buffers and performs *memory access scheduling*. *Memory access scheduling* (MAS) is used to reorder the DRAM commands associated with pending memory accesses to enhance locality and optimize throughput [50]. This technique maximizes memory bandwidth utilization over the expensive and power-hungry off-chip I/O interfaces and pins.

8.4 Software Tools for Stream Processors

Stream processors are fully programmable and require a set of software tools to compile a program expressed in high-level language to the stream control program and kernels. In this section, we focus on the tools developed for the generalized stream programming and execution model and not on tools restricted to synchronous dataflow [14, 54]. We first describe the software tools based on the StreamC/KernelC language developed specifically for Imagine (Sect. 8.5.1) and then the more general Brook and Sequoia that were developed as part of the Merrimac Streaming Supercomputer project (see Sect. 8.5.3).

8.4.1 StreamC/KernelC

The Imagine project established the potential of the stream processor architecture, part of which included developing software tools for the generalized stream programming and execution models [43, 19]. The Imagine software system automates both the kernel and stream scheduling tasks so that an SP can be programmed entirely in C/C++ without sacrificing efficiency. Imagine program programs are written in *KernelC* and *StreamC*, which are languages based on C/C++ and used to express kernels and the high-level stream control, respectively, as explained below.

Kernels are expressed using *KernelC*, which has C-like syntax that is tuned for compiling to the arithmetic clusters of an SP that are operated in SIMD, and can only access the SRF, and require static scheduling of the VLIW functional units and switches. A KernelC program describes SIMD execution implicitly, where the same code runs on all clusters. The Imagine version of KernelC restricts control flow to loops with predicated select operations. This limitation was removed with the SPI version of KernelC as part of the RapiDev^{TM} suite, which allows arbitrary C code and control flow within kernels. Both versions of KernelC disallow the use of pointers and global variables because kernels can only access their explicit input and output arguments, which are either streams stored in the SRF or scalars. The kernel scheduler takes a kernel and uses communication scheduling [44] to map each operation to a cycle number and arithmetic unit, and simultaneously schedule data movement over the intra-cluster switch to provide operands. The compiler software pipelines inner loops, converting data parallelism to instruction-level parallelism where it is required to keep all operation units busy. To handle conditional (if-then-else) structures across the SIMD clusters, the compiler uses predication and conditional streams [33].

While KernelC is used for kernels to enable low-level scheduling for the clusters, *StreamC* is an extension to C++ for describing the high-level control program with the definition of streams, data movement, and kernel invocations. StreamC adds class templates for declaring streams, which are equivalent to arrays of C structures (data records), syntax for deriving sub-streams based on strided or indexed access patterns, and a modifier to identify certain function calls as kernel invocations. Kernel functions accept stream and sub-stream arguments, and the stream flow is expressed through the dependencies on these kernel arguments. The streaming constructs are embedded within a standard C++ program with its full control flow. The stream scheduler schedules not only the transfers of blocks of streams between memory, I/O devices, and the SRF, but also the execution of kernels [19]. This task is comparable to scheduling DMA (direct memory access) transfers between off-chip memory and I/O that must be performed manually for most conventional processors, such as DSPs. The stream scheduler accepts the C++ stream program and outputs machine code for the scalar processor including stream load and store, I/O, and kernel execution instructions. The stream scheduler optimizes the block size so that the largest possible streams are transferred and operated on at a time, without overflowing the capacity of the SRF. This optimization is similar to the use of cache blocking on conventional processors and to strip-mining of loops on vector processors.

8.4.2 Brook

The Brook streaming language [11] is an extension of standard ANSI C and is designed to incorporate the ideas of data parallel computing and arithmetic intensity of streaming into a familiar, efficient language. While StreamC was developed for Imagine as a replacement to its low-level hardware API, the Brook language was developed as an abstract and portable stream language. We developed Brook specifically for scientific applications, but strived to maintain a level of abstraction that will not tie it down to a specific platform. Brook was developed from the start both as the language of choice for Merrimac benchmarks and as a software system targeting programmable GPUs [13, 12]. Brook was one of the first languages used for general-purpose GPUs, was downloaded nearly 20,000 times, and an extended version, Brook+, has been incorporated into the AMD stream computing SDK [9].

The Brook execution model and language [11] refer to computational kernels that are applied to streams, which are unordered collections of data elements that serve. A stream is used to express DLP, and a kernel applied to a stream can potentially process all the stream elements in parallel. Kernels can only use a subset of C and can only reference arguments passed to the kernel and locally declared kernel variables. Stream arguments to kernels are presented to the kernels as single elements, and kernels do not retain any state other than explicitly expressed reductions. These unordered streams and kernels, in conjunction with the ability to perform reductions, allow the programmer to effectively express DLP. Additionally, kernels operating on streams capture the ideas of kernel locality, producer–consumer locality, and a two-level locality hierarchy of off-chip memory and an on-chip SRF. To ensure locality, streams in Brook do not refer to a specific memory location.

Given the above restrictions on stream allocation and access to stream elements in kernels, Brook offers a variety of stream manipulation operators for scientific computing. First, Brook streams can be assigned multi dimensional rectangular shapes. Second, the streams can be modified with stream operators, including multi dimensional stencils, multi dimensional groups, gathers/scatters, and atomic stream memory operations such as scatter-add. These features of Brook are similar to SDF languages, but Brook breaks away from such restrictions with the availability of variable output streams and data-dependent stream operators.

8.4.3 Sequoia

Building on the success of Brook, the Sequoia programming system [25] was developed to improve upon Brook and address many of Brook's limitations. Sequoia extends the streaming model by generalizing kernels into *tasks* and allowing nesting of streams to support arbitrarily deep storage hierarchies. Sequoia is a programming language designed to facilitate the development of memory hierarchy aware parallel programs that remain portable across modern machines with different memory hierarchy configurations. Sequoia abstractly exposes hierarchical memory in the programming model and provides language mechanisms to describe communication vertically through the machine and to localize computation to particular memory locations within it, extending the two-level memory hierarchy of the generalized streaming model. While Sequoia was never used for Merrimac, the project implemented a complete programming system, including a compiler and runtime systems for Cell Broadband Engine^{TM} processor-based blade systems and distributed memory clusters, and demonstrated efficient performance running Sequoia programs on both of these platforms.

The principal idea of Sequoia is that the movement and placement of data at all levels of the machine memory hierarchy should be under explicit programmer control via first class language mechanisms. The Sequoia programming model focuses on assisting the programmer in structuring bandwidth-efficient parallel programs that remain easily portable to new machines. The design of Sequoia centers around the following key ideas:

- The notion of hierarchical memory introduced directly into the programming model to gain both portability and performance. Sequoia programs run on machines that are abstracted as trees of distinct memory modules and describe how data is moved and where it resides in a machine's memory hierarchy.
- *Tasks*, which are a generalization of kernels, are used as abstractions of self-contained units of computation that include descriptions of key information such as communication and working sets. Tasks isolate each computation in its own local address space and also express parallelism.
- To enable portability, a strict separation between generic algorithmic expression and machine-specific optimization is maintained. To minimize the impact of this separation on performance, details of the machine-specific mapping of an algorithm are exposed to programmer control.

Sequoia takes a pragmatic approach to portable parallel programming by providing a limited set of abstractions that can be implemented efficiently and controlled directly by the programmer. While the compiler implementation described in this paper does not make heavy use of automatic analysis, we have taken care to ensure that Sequoia programs are written within a framework that is amenable to the use of advanced compiler technology.

Sequoia effectively addresses three significant limitations of Brook with regards to flexibility and expressibility. First, Brook's programming model only allows the user to express a single dimension of data parallelism, because all elements of a stream can be processed in parallel, and because kernels and streams cannot be nested the use of non-nested, kernels and streams also limits the programmer to expressing only three levels of locality: streams which exist in the memory namespace, short-term kernel locality within the stateless Brook kernels, and producer–consumer locality expressed as streams connecting two kernels. While this abstraction is sufficient for GPUs and a single Merrimac node, the rich hierarchies of full supercomputers may require greater expressibility for optimal performance. Finally, Brook relied on sophisticated compiler analysis because the programmer does not express blocking and does not directly control on-chip state. These concepts must be garnered from the Brook code by understanding and analyzing the use of stream operators.

8.5 Case Studies

8.5.1 *The Imagine Stream Processor*

The Imagine processor research project at Stanford University focused on developing and prototyping the first stream processor system. The goals were to develop a programmable architecture that achieves the performance of special-purpose hardware on graphics and image/signal processing. This was accomplished by exploiting stream-based computation at the application, compiler, and architectural level. At the application level, several complex media applications such as polygon rendering, stereo depth extraction, and video encoding were cast into streams and kernels. At the compiler-level, programming languages for writing stream applications were developed and software tools that optimize their execution on stream hardware were implemented. Finally, at the architectural level, the Imagine stream processor was designed and tested. Imagine is a novel architecture that executes stream applications and is able to sustain over tens of GFLOP/s over a range of media applications with a power dissipation of less than 10 W. Design of Imagine at Stanford University started in November 1998 and the silicon prototype, fabricated jointly with Texas Instruments in a 0.15 μm standard cell technology, was functioning in the lab in June 2002.

8.5.1.1 Imagine Processor Architecture

Imagine [4, 34, 37, 49] shown in Fig. 8.6 was fabricated in a 0.15 μm CMOS process and follows the architecture and organization described in Sect. 8.3. Because of the 0.15 μm technology node, the scalar control core is not integrated within the Imagine processor itself, and Imagine acts as an external co-processor to a PowerPC-based host mounted on the Imagine development board. The Imagine SP and the PowerPC host have separate memory spaces and communicate using memory-mapped I/O. The host writes bulk stream instructions to the stream controller at a maximal rate of 2 million instructions per second and can also query and set status and control registers.

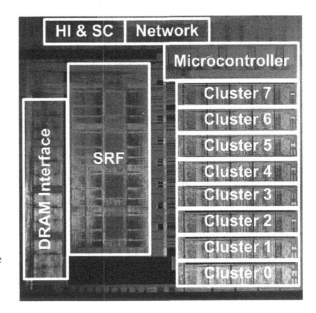

Fig. 8.6 Imagine prototype die photo (16×16 mm^2 in Texas Instruments. 15 μm ASIC CMOS). Reprinted with permission from U. J. Kapasi et al., "Programmable stream processors," IEEE Computer, August 2003 (© 2003 IEEE)

Imagine contains eight arithmetic clusters, each with six 32-bit floating-point arithmetic units: three adders, two multipliers, and one divide-square-root (DSQ) unit. With the exception of the DSQ unit, all units are fully pipelined and support 8-, 16-, and 32-bit integer operations, as well as 32-bit floating-point operations. The LRF in Imagine consists entirely of 1-read/1-write port register files. Each input of each arithmetic unit has a separate local register file of sixteen or thirty-two 32-bit words. The SRF has a capacity of 32 K 32-bit words (128 KB) and can read 16 words per cycle (two words per cluster). The clusters are controlled by a 576-bit microinstruction. The microcode store holds 2 K such instructions. The memory system interfaces to four 32-bit-wide SDRAM banks and reorders memory references to optimize bandwidth. Imagine also includes a network interface and router for connection to I/O devices and to combine multiple Imagines for larger signal-processing tasks. The network interface acts similarly to the memory system in that it accepts bulk stream transfer instructions from the scalar control code. These streams send and receive operations transfer streams of data from the SRF of one Imagine chip to the SRF of another Imagine chip using source routing.

Imagine's bandwidth hierarchy enables the architecture to provide the instruction and data bandwidth necessary to efficiently operate 48 ALUs in parallel. The hierarchy consists of a streaming memory system (2.1 GB/s), a 128 KB stream register file (25.6 GB/s), and direct forwarding of results among arithmetic units via local register files (435 GB/s). Using this hierarchy to exploit the parallelism and locality of streaming media applications, Imagine is able to sustain performance of up to 18.3 GOP/s on key applications. This performance is comparable to special-purpose processors; yet Imagine is still easily programmable for a wide range of applications.

8.5.1.2 Analysis

A detailed analysis of the Imagine prototype appears in prior publications [4, 36] and we summarize several highlights below.

Imagine Processor Power

The Imagine prototype runs at 200 MHz with a 1.8 V supply, sustaining a measured peak performance of 25.4 GOP/s of 16-bit arithmetic operations or 7.96 GFLOP/s for single-precision floating-point arithmetic and only dissipating a measured maximum of 8.53 W. Imagine achieves a peak performance efficiency of 862 pJ per floating-point operation (1.13 GFLPO/s/W). When normalized to a modern 0.13 μm 1.2 V process technology, Imagine would achieve 277 pJ/FLOP on this metric, between 3 and 13 times better than modern commercial programmable DSPs and microprocessors targeted for power efficiency in similar process technologies. For example, based on its published peak performance and power dissipation in a 0.13 μm 1.2 V technology, the 225 MHz TI C67x DSP dissipates 889 pJ/FLOP [53], while the 1.2 GHz Pentium®M dissipates 3.6 nJ/FLOP [31]. Furthermore, the improved design methodologies and circuit designs typically used in these commercial processors would provide additional improvement in the achieved power efficiency and performance of Imagine, demonstrating the potential of stream processors to provide over an order of magnitude improved power efficiency when compared to commercial programmable processors.

Application Performance

Table 8.2 lists the overall performance for four applications: DEPTH is a stereo depth extractor that processes images from two cameras [32]; MPEG encodes three frames of 360×288 24-bit video images according to the MPEG-2 standard; QRD converts a 192×96 complex matrix into an upper triangular and an orthogonal matrix, and is a core component of space–time adaptive processing; and RTSL renders the first frame of the SPECviewperf 6.1.1 advanced visualizer benchmark using the Stanford real-time shading language [47]. The first column in the table, which lists the number of arithmetic operations executed per second, shows that Imagine sustains up to 7.36 GOP/s and 4.81 GFLOP/s over entire applications. If we consider all operations, not just arithmetic ones, then Imagine is able to sustain over 40 instructions per cycle on QRD. In fact, for all three video applications, Imagine can easily meet real-time processing demands of 24 or 30 frames per second. These high absolute performance numbers are a result of carefully managing bandwidth, both in the programming model and in the architecture. Imagine dissipates between 5.9 W and 7.5 W while executing these applications at their highest performance levels.

Table 8.2 Measured Imagine application performance (peak IPC is 48). Reprinted with permission from J.H. Ahn et al., "Evaluating the Imagine Stream Architecture," Proceedings of the 31st Annual International Symposium on Computer Architecture (© 2004 IEEE)

Application	ALU	IPC	Performance summary	Power (W)
DEPTH	4.91 GOP/s	33.3	90 frames/s	7.49
MPEG	7.36 GOP/s	31.7	138 frames/s	6.80
QRD	4.81 GFLOP/s	40.1	326 QRD/s	7.42
RTSL	1.30 GOP/s	17.7	44 frames/s	5.91

Imagine's arithmetic to memory bandwidth ratio for floating-point computation is over 20:1. It can perform over 20 floating-point operations for each 32-bit word transferred over the memory interface. This is five times higher than the 4:1 ratio typical of conventional microprocessors and DSPs [53, 31]. Yet Imagine is still able to sustain half of its relatively high peak performance on a variety of applications.

Imagine is able to achieve good performance with relatively low memory bandwidth for two reasons. First, exposing a large register set with two levels of hierarchy to the compiler enables considerable locality (kernel locality and producer–consumer locality) to be captured that is not captured by a conventional cache. This locality is evidenced by the measured LRF to memory bandwidth ratio of over 350:1 across four applications shown in Fig. 8.7. The LRF to SRF to memory bandwidths

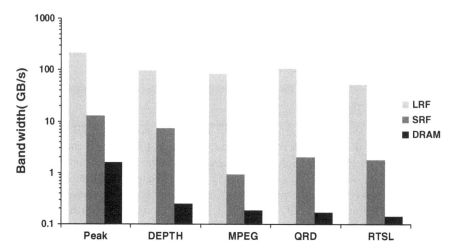

Fig. 8.7 Bandwidth demands of Imagine applications across the storage hierarchy. Reprinted with permission from J.H. Ahn et al., "Evaluating the Imagine Stream Architecture," Proceedings of the 31st Annual International Symposium on Computer Architecture (© 2004 IEEE)

demands show how a good match between the applications and the architecture can lead to efficient and effective performance.

Second, by scheduling stream loads and stores to hide latency, the Imagine memory system can be designed to provide the *average* bandwidth required by applications without loss of performance. In contrast, conventional processors are highly sensitive to memory latency and hence providing memory bandwidth well in excess of the average is required to avoid *serialization latency* on each memory access. Although the memory system of a conventional processor is idle much of the time, reducing its bandwidth would increase memory latency (by the additional cycles required to transfer a cache line across the lower bandwidth interface) and hence increase execution time.

8.5.1.3 Summary

The Imagine project demonstrated the potential of stream processors in providing a high-performance and highly efficient processor that can support a wide variety of applications (ranging from molecular dynamics to video compression). Moreover, the Imagine project developed tools to enable programming entirely in a high-level language, with no assembly performance tuning. The stream and kernel compilers are able to keep an array of 48 floating-point units busy using a combination of data-level parallelism (eight-way SIMD) and instruction-level parallelism (six-way VLIW) – achieving over 50% utilization at the kernel level (and over 80% on many inner loops) as reported in the IPC column of Table 8.2. The same compilation tools are able to efficiently exploit the two-level register hierarchy of Imagine keeping the register to memory ratio over 350:1 across all four applications that were studied in depth, allowing a modest off-chip memory bandwidth to keep the large number of functional units busy.

The success of Imagine included a working prototype developed within Stanford by a small group of graduate students. This, along with the performance and programming achievements, has influenced other designs in both academia and industry. Two such stream processors are the Storm 1 family of processors from stream processors Inc. and the Merrimac Streaming Supercomputer design described in Sect. 8.5.3.

8.5.2 SPI Storm 1

Stream Processors Inc. (SPI) is commercializing the stream processor architecture targeting the sharply increasing media and signal processing requirements. Standards, such as H.264 video compression, save disk space and network bandwidth, but at the cost of increased computational needs. At the same time, there is a growing need for market-specific tuning and competitive differentiation, especially with the continued evolution of standards. Those goals are more easily met in software than in fixed-function hardware. However, even with more efficient multi core designs, traditional processors and DSPs have not been successful at approaching the higher efficiency of fixed-function hardware. SPI's solutions leverage the well-matched hardware and software execution model of the stream architecture to move beyond conventional approaches to a new location on the flexibility/performance trade-off curve. The SPI stream processors are fully flexible and programmable at the high-level of a stream application, while the hardware is carefully designed, laid-out, and optimized for the VLSI process technology and streaming applications.

8.5.2.1 SPI Storm 1 Processor Architecture

The SPI stream processor architecture builds on top of the research results of Imagine and targets both the architecture and tools towards the embedded digital signal processing domain. The overall architecture of the SPI Storm 1 processor family is shown in Fig. 8.8 and is described in detail in [35]. Each Storm 1 processor is a *system-on-a-chip* (SoC) that integrates two general-purpose cores onto the same die with a streaming unit, memory system, and peripheral I/O. SPI includes a host CPU (System MIPS) for system-level tasks on top of the DSP coprocessor subsystem, where a second dedicated MIPS core runs the stream program that make kernel and stream load/store function calls to the stream unit, which SPI refers to as the data parallel unit (DPU).

The stream unit itself follows the overall stream architecture as exemplified by Imagine and is composed of 16 arithmetic clusters (*lanes* in SPI terminology). Each cluster, or lane, uses a 12-wide VLIW instruction to control five 32-bit MAC (multiply-accumulate) ALUs as well as the intra-cluster local switch, the LRFs (referred to as *operand register files* – ORFs), and supporting in/out and communication units. The bandwidth hierarchy comprises 304 32-bit registers aggregate

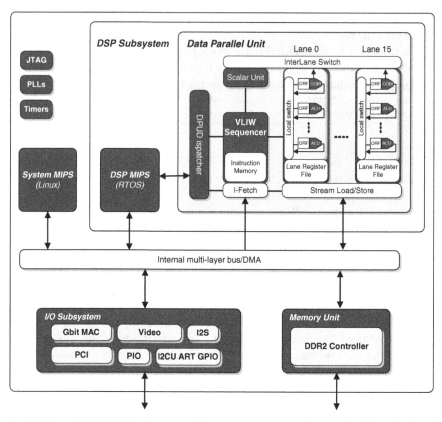

Fig. 8.8 Storm 1 SP16 block diagram, a 16-lane stream processor. Reprinted with permission of Stream Processors Inc. (© 2007 Stream Processors Inc.)

per ORF that are distributed across the ALU inputs and can supply 240 words/cycle of bandwidth across the chip (words being 32- bit wide); the SRF (referred to as a lane register file in SPI terminology) that can store 16 KB per cluster and supply 64 words/cycle across the chip; the inter-cluster switch with an aggregate bandwidth of 16 words/cycle; and the memory system with a bandwidth of 10.7 GB/s. The memory subsystem includes the stream load/store units (AGs and memory access scheduling) and interfaces with DDR2 external memory. Each of the 80 32-bit ALUs can perform sub-word arithmetic on four 8-bit values or two 16-bit values for a peak of 320 operations per cycle.

The energy efficiency of SPs is also evident in the Storm 1 processor, where a peak of 256 16-bit GOP/s (320 16-bit multiply-accumulate operations/cycle, 700 MHz) was measured in the lab with a maximum power dissipation of only 9.2 W, or 41 pJ per operation, which is a factor of 10–20 better than traditional architectures, such as a Texas Instruments C64x+ [1] or an AMD Quad-Core Opteron^{TM} [8] when normalized to the same process technology. In addition, Storm 1 does well when compared to ASIC-style solutions and is within a factor of 4 of the power efficiency of the XETAL-II low-power image processing engine [40] when normalized for process technology.

8.5.2.2 SPI Development Tools

SPI's RapiDev™ Tools Suite leverages the architecture's predictability to provide a fast path to optimized results. An eclipse-based IDE provides a convenient environment for project management, compiling, profiling, and debugging, including target simulation and hardware device execution.

The stream processor compiler (SPC) generates the VLIW executable and pre-processed C code that is compiled/linked via standard gcc for MIPS. SPC allocates streams and provides kernel dependency information. Performance features such as software pipelining and loop unrolling are supported. The cycle-accurate visual profiler shows utilization with data loads and execution and provides a link to the corresponding line of code. A fully-featured debugger provides programmable breakpoints and single-stepping capabilities in both control and kernel code.

8.5.2.3 Analysis

The Storm 1 processor is designed to provide high sustained performance within a low-power envelope. Figure 8.9 shows a die photograph of the chip, which is fabricated in a 130 nm TSMC process and operates at 700 MHz with a 1.2 V supply. The massive parallelism and high sustained ALU bandwidth provides the basis for excellent compute performance. At the instruction level, a 15-stage pipeline feeds the five 32-bit ALUs with VLIW instructions that extract instruction-level parallelism. Combined with the 16 clusters and 4×8 or 2×16 sub-word arithmetic at the data-parallel level, the Storm 1 SP16HP-G220 part can achieve a peak performance of 224 gops while requiring 41 mW per GOP/s, including the overhead of the scalar cores and I/O subsystems (10 times better than a modern Texas Instruments C64x DSP [1]).

As with Imagine, the stream architecture of Storm 1 allows a wide range of applications to execute effectively. The predictable behavior of the architecture and the deep storage hierarchy coupled with the stream programming model allow the stream compiler to schedule the application and sustain a high fraction of peak performance, often over 80% of peak on challenging applications. Table 8.3 summarizes the sustained performance of several DSP and video processing kernels targeting the Storm 1 SP.

8.5.3 The Merrimac Streaming Supercomputer

The Merrimac Streaming Supercomputer [17, 21] was a research project at Stanford that started in 2002 and set out with an ambitious goal of improving the performance

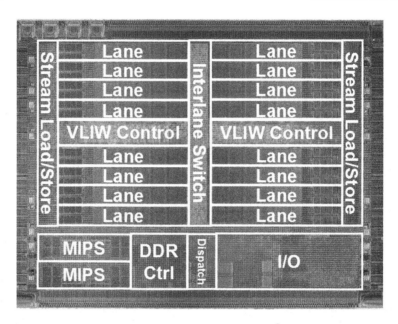

Fig. 8.9 Storm 1 SP16 SoC operating with an area of 155 mm² in a 130 nm CMOS process. Reprinted with permission from B. Khailany et al., "A Programmable 512 GOPS Stream Processor for Signal, Image, and Video Processing," Solid-State Circuits, IEEE Journal, 43(1):202–213, 2008 (© 2008 IEEE)

Table 8.3 Application performance on SPI Storm 1 SP16HP-G220 processor, running at 700 MHz with a peak of 224 GOP/s (8-bit MACs or 16-bit operations). Reprinted with permission from B. Khailany et al., "A Programmable 512 GOPS Stream Processor for Signal, Image, and Video Processing," Solid-State Circuits, IEEE Journal, 43(1):202–213, 2008 (© 2008 IEEE)

Description	Performance	Sustained ALU utilization (%)
8×8 Forward DCT	4.56 cycles/8×8 block	82.6
3×3 Convolution filter	0.085 cycles/3×3 block	71.6
Floyd-Steinberg error diffusion	0.89 cycles/pixel	42.5
4×4 Motion estimation full search (H.264)	0.22 cycles/search-pint	88.6
4×4 Forward integer transform and quantization (H.264)	1.8 cycles/4×4 block	84.1
Deblocking filter (H.264)	58.3 cycles/macroblock	64.4

per unit cost and performance per unit power by two orders of magnitude when compared to supercomputers built around commodity general-purpose CPUs. Scientific applications, just like media applications, are, for the most part, inherently parallel at both the algorithm and the code implementation levels. Additionally, scientific codes have been developed with locality in mind from the outset. These properties suggest that the basic ideas of stream processing developed in the Imagine project can be adapted for supercomputing.

Unlike media processing, however, scientific codes often have irregular control and require overall performance that is orders of magnitude higher. Merrimac extended the architecture developed in the Imagine project to handle the demands of supercomputing. By combining architecture research with development of software tools and applications, Merrimac demonstrated that more than an order of magnitude improvement over current supercomputer designs can actually be achieved. The Merrimac system is a flexible design that is scalable from a 128 GFLOP/s single chip (90 nm technology) through a 2 TFLOP/s workstation, to a 2 PFLOP/s supercomputer.

The rest of this section touches upon the main aspects of the Merrimac project as they relate to multi core architectures: processor architecture, software tools, and application analysis.

8.5.3.1 Merrimac Processor Architecture

Merrimac follows the same basic architecture described in Sect. 8.3 and implemented by Imagine and SPI. The Merrimac processor was never fully designed or fabricated and the specification called for a 128 GFLOP/s, 64 GB/s processor consuming under 65 W in 90 nm technology. Merrimac (see Fig. 8.10) has 16 arithmetic

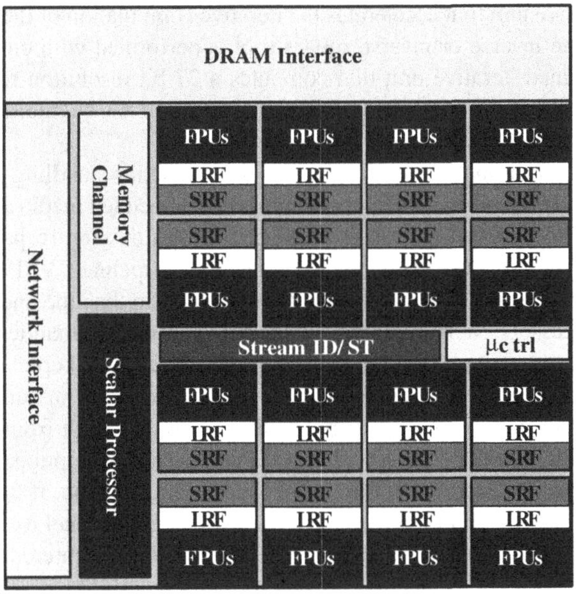

Fig. 8.10 Merrimac processor block diagram. Merrimac was designed to achieve 128 GFLOP/s (double precision) of peak performance using 16 arithmetic cluster while consuming under 65 W and 144 mm^2 in 90 nm CMOS.

clusters, each with four 64-bit fused multiply–add floating-point units (with denormalized number support), a specialized floating-point unit that accelerates divide and square-root operations, four in/out units for accessing the 1 MB SRF, and a communication unit for accessing the inter-cluster switch. The memory system is a stream memory system, which directly connects a Merrimac chip to eight XDR DRAM [20] chips and to the system-wide interconnection network to enable a single address space for the entire supercomputer. The general-purpose unit in Merrimac is a MIPS20kC [45] core with fast memory-mapped and co-processor interfaces to the stream unit.

Beyond the basic stream architecture Merrimac introduced several major enhancements to SPs that are particularly relevant to the scientific computing domain and are explained below: designing novel optimized divide and square-root operations for physical modeling; introducing a high-throughput exception model to emulate precise exceptions in the highly parallel SP; developing an indexable SRF to better support irregular data access and computation found in graph and mesh-based codes; incorporating a stream cache into the memory system to further improve the performance of irregular and unstructured codes; developing stream (bulk) atomic memory operations to enhance parallel reductions and scans; and expanding the architecture of SPs to a large-scale supercomputing platform that can support thousands of nodes.

Numerical codes often require very high throughput divide and square-root computation support. Such high performance is not economically achievable with pipelined or replicated hardware units. Merrimac took a different approach by combining a hardware unit that accelerates the iterative computation of these operations. For example, an inverse or inverse square-root is performed with one operation of the fully pipelined iterative unit that computes a 27-bit resolution result that then requires a single Newton-Raphson iteration using the 64-bit multiple–add units [21, 42].

Scientific programmers often rely on precise exception handling for arithmetic operations that is not of interest to programmers of embedded media and signal processing systems. Supporting precise arithmetic exceptions for 16 parallel clusters each with potentially over 30 instructions in flight (pipelined VLIW instructions in each cluster) requires a different exception mechanism than the one employed by traditional processors. Merrimac relies on the bulk execution semantics of the stream model to efficiently support arithmetic exceptions. Because a kernel cannot access off-chip memory directly and its inputs and outputs are entire stream blocks in the SRF, an entire kernel can be considered as an atomic operation from the programmers perspective. If an operation within a kernel raises an exception, the exception is simply logged and reported at the end of the kernel execution. If the exception is not masked, it will be reported to software, in which case kernel execution can be restarted and stepped until the precise exception point is encountered.

The third major enhancement is allowing arbitrary indexed accesses into the SRF. The applications driving the development of Imagine were mostly purely streaming applications and the SRF was optimized for streaming accesses of providing the next consecutive element in the stream. Scientific applications for Merrimac take better advantage of the bulk execution model and rely less on fully sequential streams – in essence, Merrimac often deals with streams of coarse-grained blocks rather than just streams of records. Utilizing this execution model fully, requires random access to data within each block, which implies allowing indexed accesses into the SRF. This indexed feature, which had very little area overhead in Merrimac, enables efficient implementation of several irregular applications.

The fourth important extension is incorporating a stream cache into the memory hierarchy. The stream cache is not intended to reduce latency, because latency is explicitly hidden, but rather to increase memory throughput for applications that perform irregular memory accesses with temporal locality. Along with the cache, an effective coherence mechanism was developed that, like the exception handling technique, utilizes the bulk semantics of the execution model to allow hardware and software to work together to ensure the most up to date value is returned for memory requests. Another difference between Merrimac and Imagine is the extension of the memory address space to single global address space by combining the memory and network subsystems (more in the next subsection).

Another memory system related feature of Merrimac is the addition of a *scatter-add* operation [5]. A *scatter-add* is a stream memory operation that performs a collection of atomic read–modify–write operations. This operation was inspired by the prevalent use of the superposition principle in scientific applications and proved invaluable in simplifying coding and improving the performance of molecular dynamics and finite-element method programs [22, 2].

Finally, when designing a supercomputer that contains up to 16 K nodes, reliability and correctness of the final result is a major concern. Research on Merrimac developed a novel approach to soft-error reliability [24]. Merrimac requires different soft-error fault tolerance techniques than control-intensive CPUs because of its throughput-oriented design. The main goal of the fault-tolerance schemes for Merrimac is to conserve the critical and costly off-chip bandwidth and on-chip storage resources, while maintaining high peak and sustained performance. Following the explicit and exposed architecture philosophy of stream processors, Merrimac allows for reconfigurability in its reliability techniques and relies on programmer input. The processor is either run at full peak performance employing software fault-tolerance methods, or reduced performance with hardware redundancy. We present several methods, their analysis, and detailed case studies.

8.5.3.2 Application Analysis

Most scientific application have abundant DLP and locality, which are the necessary characteristics for efficient stream processing. In order to cast an application into the streaming style, it must be converted into a gather–compute–scatter form. In the *gather* phase, all data required for the next compute phase is *localized* into the SRF. Once all the data has been loaded to on-chip state, the *compute* stage performs the arithmetic operations and produces all data within the SRF. Further, once all data from the localized data, and any intermediate results, have been processed, the results are written back off-chip in the *scatter* phase. For efficient execution, the steps are pipelined such that the computation phase can hide the latency of the communication phases.

This type of transformation is common in scientific computing, where many applications are designed to also run on distributed memory computers. For such machines, a *domain decomposition* is performed on the data and computation to assign a portion of each to every node on the system. This is similar to the localization step, but the granularity is typically much greater in traditional systems. The SRF in Merrimac is 128 KWords total, and only 8 KWords in each compute cluster, whereas compute nodes in distributed memory machines often contain several GB of memory. The smaller granularity, also results in more data movement operations in Merrimac. Other critical considerations due to Merrimac's unique architecture are mapping to the relaxed SIMD control of the clusters and taking advantage of the hardware acceleration for sequential SRF accesses and low-cost inter-cluster communication.

The Merrimac benchmark suite is composed of simple applications that were chosen to be representative of the Stanford ASCI center in particular and scientific applications in general. In addition, the programs were chosen to cover a large space of execution properties including regular and irregular control, structured and unstructured data access, and varying degrees of arithmetic intensity. A low arithmetic intensity stresses the memory system and off-chip bandwidth. Table 8.4 summarizes the programs and more information is presented in [21, 22]. This diverse set of characteristics highlights the enhancements of Merrimac over the original SP architecture of Imagine.

Table 8.5 summarizes the performance analysis using the cycle accurate Merrimac processor simulator. In the results, only floating-point operations and datawords transferred from memory that are required by the algorithm are counted towards the performance numbers. MATMUL is the most arithmetic-intensive application (see column 6 in Table 8.5) and also achieves the highest performance at 92% of peak. CONV2D and FEM also have high arithmetic intensity (24 floating-point operations for every word transferred to or from off-chip memory) and can sustain almost 60% of peak performance. While MATMUL and CONV2D worked well on Imagine, FEM relies heavily on several of Merrimac's features because it uses irregular data structures, reductions, and requires very high-throughput divide and square-root units. MD and FFT3D have lower arithmetic intensity and stress the memory system more, sustaining about a third of peak performance and two-thirds of peak memory bandwidth. As with FEM, MD differs from FFT3D in its heavy use of divides, square-roots, reductions, and irregular memory and SRF accesses. All of the five applications described above utilize the compute clusters for over 87% of execution and only perform useful computation with no speculation in either the pipeline or memory system. Note that FEM and MD require a large number of divide and square-root operations, which reduce their reported performance because both operations are counted as a single floating-point instruction, but require significantly more execution resources. CDP and SPAS can only sustain small fractions of Merrimac's peak performance due to their very low arithmetic intensity. These

Table 8.4 Merrimac evaluation benchmark suite.

Benchmark	Regular control	Structured access	Arithmetic intensity	Description
CONV2D	Y	Y	High	2D 5×5 convolution.
MATMUL	Y	Y	High	Blocked dense matrix–matrix multiply (DGEMM).
FFT3D	Y	Y	Medium	3D double-precision complex FFT.
FEM	Y	N	High	Streaming FEM code for fluid dynamics based on the Discrete Galerkin formulation for magnetohydrodynamics equations with linear interpolation (9664) elements.
MD	N	N	Medium	Molecular dynamics simulation of a 11,475-molecule water system using algorithms from GROMACS [55].
CDP	N	N	Low	Finite volume large eddy flow simulation for a 29,096-element mesh with adaptive mesh refinement.
SPAS	N	N	Low	Sparse matrix-vector multiplication.

Table 8.5 Merrimac performance evaluation summary

App.	Dataset	GFLOP/s	% Busy	BW (GB/s)	Arith. Intens.
MATMUL	2048^2	117.3	98	15.6	60.1
CONV2D	512^2	78.6	99	28.5	24.2
FEM	MHD	69.1	89	22.7	24.3
MD	11 475	46.5	87	42.3	12.0
FFT3D	128^3	37.3	89	43.5	6.8
CDP	AMR	8.6	39	34.3	2.1
SPAS	1 594	3.1	14	37.0	0.7

programs are entirely memory bandwidth bound and present very challenging and irregular access patterns that limit DRAM performance [6]. Merrimac's streaming memory system is able to utilize over 55% of the peak DRAM bandwidth with these nearly random traffic patterns, which would not be possible if the system had long cache lines.

It is also interesting to do a direct comparison between Merrimac and one of the fastest general-purpose CPUs that was fabricated in the same 90 nm technology as Merrimac, the Intel® Pentium®4. We use fully optimized versions of the Merrimac benchmarks on the Pentium®4 for MATMUL, MD, and FFT3D, and the results are that Merrimac outperforms the Pentium®4 by factors of 17, 14, and 5 times, respectively, all while consuming significantly less power achieving a factor of 25 improvement in sustained performance per unit power.

8.6 Scalability of Stream Architectures

The case studies described in this chapter show stream processors with 48–64 ALUs, but the stream architecture can scale to much larger functional unit counts. To do so efficiently, it is necessary to add the third, *task-level parallelism* (TLP), dimension to SPs, which so far have been described as exploiting ILP and DLP only with the SIMD-VLIW ALU organization (Sect. 8.3.3).

To address TLP, we provide hardware MIMD support by adding multiple instruction sequencers and partitioning the inter-cluster switch. This is shown in Fig. 8.11, which extends Fig. 8.4 with multiple sequencers, each controlling a *sequencer group* of clusters (in this example four clusters of two ALUs each). In some ways, this extra level of hierarchy is more similar to a other multi core design. The hierarchical control of a stream program with bulk operations and kernels with fine-grained control, however, is maintained and remains unique when compared to other multi core designs. Within each cluster the ALUs share an intra-cluster switch, and within each sequencer group the clusters can have an optional inter-cluster switch. The

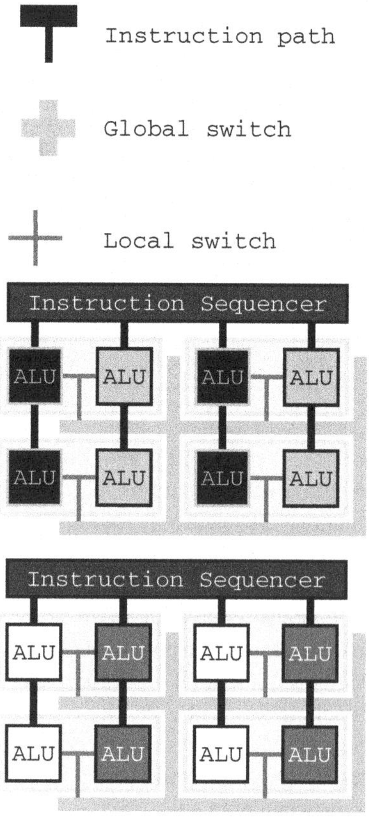

Fig. 8.11 ALU organization along the DLP, ILP, and TLP axes with SIMD, VLIW, and MIMD (all ALUs with the same shading execute in SIMD). The figure shows two sequencer groups with two clusters of four ALUs each (*shaded boxes*), and the chip interconnect partitioned into inter- and intra-cluster switches. Sequencer groups do not communicate directly and use the memory system (not shown). © 2007 ACM, Inc. Included here by permission

inter-cluster switch can be extended across multiple sequencers, however, doing so requires costly synchronization mechanisms to ensure that a transfer across the switch is possible and that the data is coherent.

In the following subsections, we address the hardware costs and implications on application performance of adding MIMD support and the trade-offs of scaling along the three axes of parallelism.

8.6.1 VLSI Scaling

We explore the scalability of stream processors by studying how the area is affected by the number and organization of ALUs. Our hardware cost model is based on an estimate of the area of each architectural component for the different organizations normalized to a single ALU. We focus on area rather than energy and delay for two reasons. First, as shown in [39], the energy of a stream processor is highly correlated to area. Second, a stream processor is designed to efficiently tolerate latencies by relying on software and the locality hierarchy, thereby reducing the sensitivity of performance to pipeline delays. Furthermore, as we will show below, near optimal configurations fall within a narrow range of parameters limiting the disparity in delay between desirable configurations. We only analyze the area of the structures relating to the ALUs and their control. The scalar core and memory system performance must scale with the number and throughput of the ALUs and do not depend on the ALU organization. Based on the implementation of Imagine and Storm I [39, 35] and the design of Merrimac, we estimate that the memory system and the scalar core account for roughly 40% of a stream processor's die area.

The full details of this study are reported in [3, 2], which also discuss the VLSI cost models in detail. An interesting point is that because stream processors are strongly partitioned, we need only focus on the cost of a given sequencer group to draw conclusions on scalability, as the total area is simply the product of the area of a sequencer group and the number of sequencers.

The main components of a sequencer group are the sequencer itself, the SRF, the clusters of functional units, LRF, and intra-cluster switch. A sequencer may also contain an inter-cluster switch to interconnect the clusters.

The sequencer is the main additional cost of adding MIMD capability to the stream architecture and includes the SRAM for storing the instructions and a datapath with a simple and narrow ALU (similar in area to a non-ALU numerical functional unit) and registers. The instruction distribution network utilizes high metal layers and does not contribute to the area. In this study, we assume dedicated per-sequencer instruction storage to overcome potential contention for a central shared instruction store.

Because the area scales linearly with the number of sequencer groups, we choose a specific number of threads to understand the trade-offs of ALU organization along the different parallelism axes, and then evaluate the DLP and ILP dimension using a heatmap that represents the per-ALU area of different *(C,N)* design space points relative to the area-optimal point (Fig. 8.12). The horizontal axis in the figure corresponds to the number of clusters in a sequencer group (C), the vertical axis to the number of ALUs within each cluster (N), and the shading represents the relative cost normalized to the area of a single ALU in the area-optimal configuration. The area given is per ALU and can be extrapolated for any number of ALUs on a chip by setting the degree of TLP. For the 64-bit case, the area optimal point is (8-DLP, 4-ILP) requiring 1.48×10^7 grids per ALU accounting for all stream execution elements. This corresponds to $94.1 \, \text{mm}^2$ in a 90 nm ASIC process with 64 ALUs in a (2-TLP, 8-DLP, 4-ILP) configuration. For a 32-bit datapath, the optimal point is at (16-DLP, 4-ILP).

Figure 8.12 indicates that the area dependence on the ILP dimension (number of ALUs per cluster) is much stronger than on DLP scaling. Configurations with an ILP of 2–4 are roughly equivalent in terms of hardware cost, but further scaling along the ILP axis is not competitive because of the N^2 term of the intra-cluster switch and the increased instruction store capacity required to support wider VLIW instructions. Increasing the number of ALUs utilizing DLP leads to better scaling. With a 64-bit datapath (Fig. 8.12a), configurations in the range of 2–32 and four ALUs are within about 5% of the optimal area. When scaling DLP beyond 32 clusters, the inter-cluster switch area significantly increases the area per ALU. A surprising observation is that even with no DLP, the area overhead of adding a sequencer for every cluster is only about 15% above optimal. We conclude that the scaling behavior is determined mostly by the scaling of the switching fabric and not the relative overheads of ILP and DLP control. Our results were obtained using full crossbar switches and it is possible to build more efficient switches. Analyzing scaling with

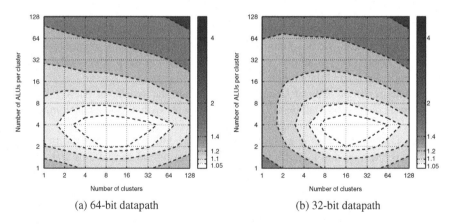

(a) 64-bit datapath   (b) 32-bit datapath

Fig. 8.12 Heatmap showing the relative area per ALU normalized to the optimal area per ALU organization including all communication and instruction sequencing overheads: (8-DLP, 4-ILP) and (16-DLP, 4-ILP) for 64-bit and 32-bit datapaths, respectively; the total number of ALUs desired is achieved with the TLP dimension. *Darker shades* represent less efficient design points, which have a larger area per ALU than the optimal organization. © 2007 ACM, Inc. Included here by permission

sparser connectivity will require extensions to the communication scheduling algorithm of the kernel compiler.

Both trends change when looking at a 32-bit datapath (Fig. 8.12b). The cost of a 32-bit ALU is significantly lower, increasing the relative cost of the switches and sequencer. As a result only configurations within a 2–4 ILP and 4–32 DLP are competitive. Providing a sequencer to each cluster in a 32-bit architecture requires 62% more area than the optimal configuration.

8.6.2 Performance Scaling

To understand the effects of different ALU configurations on application performance, we used a cycle-accurate simulator and measured execution time characteristics for six applications and six ALU configuration. We modified the Merrimac cycle-accurate simulator [17] to support multiple sequencer groups on a single chip. The memory system contains a single wide address generator to transfer data between the SRF and off-chip memory. We assume a 1 GHz core clock frequency and memory timing and parameters conforming to XDR DRAM [20]. The memory system also includes a 256 KB stream cache for bandwidth amplification, and scatter-add units, which are used for fast parallel reductions [2]. Our baseline configuration uses a 1 MB SRF and 64 multiply–add floating-point ALUs arranged in a (1-TLP,16-DLP,4-ILP) configuration along with 32 supporting iterative units, and allows for inter-cluster communication within a sequencer group at a rate of one word per cluster on each cycle.

Figure 8.13 shows the runtime results for our applications on six different ALU organizations. The runtime is broken down into four categories: all sequencers are busy executing a kernel corresponding to compute bound portion of the application; at least one sequencer is busy and the memory system is busy, indicating load imbalance between the threads, but performance bound by memory throughput; all sequencers are idle and the memory system is busy during memory bound portions

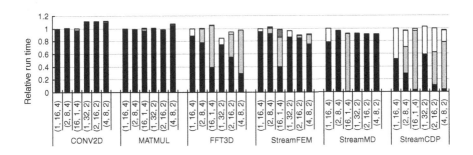

Fig. 8.13 Relative runtime normalized to the (1-TLP, 16-DLP, 4-ILP) configuration of the six applications on six ALU organizations. © 2007 ACM, Inc. Included here by permission

of the execution; at least one sequencer is busy and the memory system is idle due to load imbalance between the execution control threads.

Overall, we see that the performance results are not very sensitive to the ALU organization. The amount of ILP utilized induces up to a 15% performance difference (FFT3D), while the gain due to multiple threads peaks at 7.4% (FEM). We also note that in most cases, the differences in performance between the configurations are under 5%. After careful evaluation of the simulation results and timing we conclude that much of that difference is due to the sensitivity of the memory system to the presented access patterns [6].

Ahn et al. present a detailed analysis of the effects of alternate ALU configurations on application performance in [3]. They conclude that ILP, DLP, and TLP impact performance through six factors that are explained below: sensitivity of the compiler's VLIW scheduling algorithms; relation between the ALU organization and partitioning of the SRF banks; effects of MIMD support on applications with irregular control flow; use of MIMD for improved memory-latency hiding; detrimental impact of partitioning the inter-cluster switch; and concerns regarding synchronization and load balancing costs with MIMD.

The first factor impacting performance is that the VLIW kernel scheduler performs better when the number of ALUs it manages is smaller, because a smaller number of register files is easier to allocate and less parallelism needs to be extracted by the compiler. This is the reason for the relatively large performance difference in the case of FFT3D and MD.

An opposite trend with respect to ILP is that the SRF capacity per cluster grows with the degree of ILP since the SRF size required to support an ALU is independent of the ALU configuration. Therefore, a two-ILP configuration has twice the number of clusters of a four-ILP configuration or half the SRF capacity per cluster. The SRF size per cluster can influence the degree of locality that can be exploited and reduce performance. We see evidence of this effect in the runtime of CONV2D.

Third, some applications including MD and CDP require irregular control, which can benefit from TLP [23]. MD and CDP are examples of such applications from the Merrimac benchmark suite. MIMD support can significantly improve the amount of computation performed by the stream processor as explained in [23]. With respect to overall performance, however, we found little benefit to TLP support for the applications studied. The two main reasons for the reduced impact of MIMD are two-fold. In some cases, such as CDP, performance can be limited by the memory system rather than arithmetic capabilities. In addition, the effects of reduced communication flexibility resulting from partitioning the clusters into sequencer groups can diminish the benefits of more effective control as we explain below.

Fourth, allowing independent control of each cluster with TLP can mask transient memory stalls with useful work with applications that have irregularity at the coarse-grained stream operation level and hence cannot hide all latency with static scheduling. The structure of the FEM application limits the degree of double buffering that can be performed in order to tolerate memory latencies. As a result, computation and memory transfers are not perfectly overlapped, leading to performance degradation. We can see this most clearly in the (1-TLP, 32-DLP, 2-ILP) configuration of FEM in Fig. 8.13, where there is a significant fraction of time when all clusters are idle waiting on the memory system (white portion of bar). When multiple threads are available this idle time in one thread is masked by execution in another thread, reducing the time to solution and improving performance by 7.4% for the (2-TLP, 16-DLP, 2-ILP).

Fifth, one detrimental effect of using TLP is that the inter-cluster switch is partitioned and is only used within a sequencer group. This noticeably impacts the performance of MATMUL and MD. The performance of MATMUL is directly related to the size of the blocks processed by the algorithm. With an inter-cluster switch, a single block can be partitioned across clusters leading to an overall larger block size for a given SRF capacity and improved performance. The overall effect is quite small in our configuration because of the relatively high bandwidth of the Merrimac memory system. MD had irregular data accesses to a high-degree mesh structure and can utilize the SRF to remove duplicate memory accesses and improve performance. With the higher total SRF capacity available, the data is shared across SRF lanes via the inter-cluster switch [21]. This greater flexibility in managing on-chip memory balances the improvement in performance due to irregular control support of TLP. If the SRF size is increased to support the working set of MD within a single sequencer group, or if the inter-cluster switch is removed, TLP-related control benefits rise to 10% speedup.

Finally, a second detrimental effect of TLP relates to the synchronization costs resulting from load imbalance between the threads. This is clearly visible in the case of FEM, which contains several barrier synchronization points. The increased runtime due to load imbalance increases the synchronization time. This is most apparent with the extreme (16-TLP, 1-DLP, 4-ILP) configuration that has 16 sequencer groups.

8.6.3 Summary

Our evaluation can be summarized as two important concepts. First, continuous increases of transistor count force all the major parallelism types – DLP, ILP, and TLP – to be exploited in order to provide performance scalability as the number of ALUs per chip grows. We demonstrate that from a VLSI cost perspective, performance can be scaled linearly with area at the modest cost of adding additional sequencer groups. We estimate the area overhead of multiple sequencers to be as low as 15% of the compute cluster area for an optimal organization with 64-bit floating-point units.

The second important point is that the achieved performance of applications on SPs, can also scale with ALU number and is relatively insensitive to how parallelism is exploited across ILP, DLP, and TLP. For the applications we studied, all of which feature parallelism that scales with data set and code structure that is amenable to streaming, performance ranged from factors of 0.9 to 1.15 across all configurations and applications.

8.7 Conclusions

In conclusion, Stream Processors offer very high and very efficient scalable performance by embracing the properties of modern VLSI technology and matching them with the characteristics of important applications. The resulting architecture focuses on highly tuned and throughput-oriented computation and data transfer while exposing state management, latency hiding, and parallelism control mechanisms directly to software. This redefinition of the responsibilities of hardware and software results in very predictable performance from the software system perspective enabling a stream compiler to utilize the hardware mechanisms provided. The stream compiler targets high-level stream languages, which expose the importance of locality, parallelism, and communication to the programmer and algorithm designer. Therefore, stream processors typically have a sustained performance advantage of 5–15 times and a power/performance advantage of 10–30× compared to conventional designs.

Stream processors, such as Imagine, Storm 1, and Merrimac demonstrate that a wide variety of applications from the multimedia and scientific computing domains can benefit from stream processing and offer performance, efficiency, and flexibility that is an order of magnitude or more greater than traditional general-purpose style processors. Imagine established the potential of stream processors, including the ability to compile high-level code to the exposed communication of SPs. Storm 1 is a commercial SoC design that provides efficiency on par with application-specific processors while remaining fully programmable using a C-like language. Finally, Merrimac demonstrated the potential of SPs for applications beyond the media-processing domain and targeted large-scale physical modeling and other scientific computation.

Interesting avenues for future work started with the implementation of streaming irregular applications and the Sequoia programming model and system as part of the Merrimac project. Continuing work on irregular applications is focused on mutable and dynamic irregular data structures for which exploiting parallelism and locality is more challenging. Sequoia builds on the principles of stream programming and provides portable performance across a range of architectures, and future work includes expanding the types of architectures to include multi core CPUs and GPUs while simplifying the programming interface. In addition, other programming and architecture efforts use the stream model to target and improve conventional CPUs [28, 27, 26]. Finally, while the stream processor architecture and software tools were initially developed at Stanford University, several commercial processors utilize the technology and principal ideas, including the Cell Broadband Engine™ family of processors [30, 16] and recent GPUs such as the NVIDIA® GeForce® 8 and 9 series [46] and the AMD ATI Radeon™ HD 2000 and 3000 series [7].

Acknowledgments We would like to thank Steve Keckler for his insightful comments as well as the contributions of Jung Ho Ahn, Nuwan Jayasena, and Brucek Khailany. In addition, we are grateful to the entire Imagine and Merrimac teams and the projects' sponsors.

Imagine was supported by a Sony Stanford Graduate Fellowship, an Intel Foundation Fellowship, the Defense Advanced Research Projects Agency under ARPA order E254 and monitored by the Army Intelligence Center under contract DABT63-96-C0037 and by ARPA order L172 monitored by the Department of the Air Force under contract F29601-00-2-0085.

The Merrimac Project was supported by the Department of Energy ASCI Alliances Program, Contract LLNL-B523583, with Stanford University as well as the NVIDIA Graduate Fellowship program.

Portions of this chapter are reprinted with permission from the following sources:

- U. J. Kapasi, S. Rixner, W. J. Dally, B. Khailany, J. H. Ahn, P. Mattson, and J. D. Owens, "Programmable Stream Processors," *IEEE Computer*, August 2003 (©2003 IEEE).
- J. H. Ahn, W. J. Dally, B. K. Khailany, U. J. Kapasi, and A. Das, "Evaluating the imaginestream architecture," *In Proceedings of the 31st Annual International Symposium on Computer Architecture* (© 2004 IEEE).
- Stream Processors Inc., "Stream Processing: Enabling the New Generation of Easyto Use, High-Performance DSPs," *White Paper* (© 2007 Stream Processors Inc.).
- B. K. Khailany, T. Williams, J. Lin, E. P. Long, M. Rygh, D. W. Tovey, and W. J. Dally, "A Programmable 512 GOPS Stream Processor for Signal, Image, and Video Processing," Solid-State Circuits, *IEEE Journal*, 43(1):202–213, 2008 (© 2008 IEEE).
- J. H. Ahn, M. Erez, and W. J. Dally, "Tradeoff between Data-, Instruction-, and Thread-level Parallelism in Stream Processors," *In Proceedings of the 21st ACM International Conference on Supercomputing (ICS'07)*, June 2007 (DOI10.1145/1274971.1274991). © 2007 ACM, Inc. Included here by permission.

References

1. S. Agarwala, A. Rajagopal, A. Hill, M. Joshi, S. Mullinnix, T. Anderson, R. Damodaran, L. Nardini, P. Wiley, P. Groves, J. Apostol, M. Gill, J. Flores, A. Chachad, A. Hales, K. Chirca, K. Panda, R. Venkatasubramanian, P. Eyres, R. Veiamuri, A. Rajaram, M. Krishnan, J. Nelson, J. Frade, M. Rahman, N. Mahmood, U. Narasimha, S. Sinha, S. Krishnan, W. Webster, Due Bui, S. Moharii, N. Common, R. Nair, R. Ramanujam, and M. Ryan. A 65 nm c64x+ multicore dsp platform for communications infrastructure. *Solid-State Circuits Conference, 2007. ISSCC 2007. Digest of Technical Papers. IEEE International*, pages 262–601, 11–15 Feb 2007.
2. J. H. Ahn. *Memory and Control Organizations of Stream Processors*. PhD thesis, Stanford University, 2007.
3. J. H. Ahn, W. J. Dally, and M. Erez. Tradeoff between data-, instruction-, and Thread-level parallelism in stream processors. In *proceedings of the 21st ACM International Conference on Supercomputing (ICS'07)*, June 2007.
4. J. H. Ahn, W. J. Dally, B. Khailany, U. J. Kapasi, and A. Das. Evaluating the imagine stream architecture. In *ISCA '04: Proceedings of the 31st Annual International Symposium on Computer Architecture*, page 14, Washington, DC, USA, 2004. IEEE Computer Society.
5. J. H. Ahn, M. Erez, and W. J. Dally. Scatter-add in data parallel architectures. In *Proceedings of the Symposium on High Performance Computer Architecture*, Feb. 2005.
6. J. H. Ahn, M. Erez, and W. J. Dally. The design space of data-parallel memory systems. In *SC'06*, Nov. 2006.
7. AMD. AMD ATI Radeon^TM HD 2900 Graphics Technology. http://ati.amd.com/products/Radeonhd2900/specs.html
8. AMD. Product brief: Quad-core AMD opteron^TM procsesor. http: http://www.amd.com/us-en/Processors/ProductInformation/0,,30_118_8796_152%23,00.html
9. AMD. AMD stream computing SDK, 2008. http://ati.amd.com/technology/streamcomputing/sdkdwnld.html
10. S. S. Bhattacharyya, P. K. Murthy, and E. A. Lee. *Software Synthesis from Dataflow Graphs*. Kluwer Academic Press, Norwell, MA, 1996.
11. I. Buck. Brook specification v0.2. Oct. 2003.

12. I. Buck. *Stream Computing on Graphics Hardware*. PhD thesis, Stanford University, Stanford, CA, USA, 2005. Adviser-Pat Hanrahan.
13. I. Buck, T. Foley, D. Horn, J. Sugerman, K. Fatahalian, M. Houston, and P. Hanrahan. Brook for GPUs: stream computing on graphics hardware. *ACM Transactions on Graphics*, 23(3):777–786, 2004.
14. J. Buck, S. Ha, E. A. Lee, and D. G. Messerschmitt. Ptolemy: a framework for simulating and prototyping heterogeneous systems. *Readings in Hardware/Software co-design*, pages 527–543, 2002.
15. J. B. Carter, W. C. Hsieh, L. B. Stoller, M. R. Swanson, L. Zhang, and S. A. McKee. Impulse: Memory system support for scientific applications. *Journal of Scientific Programming*, 7: 195–209, 1999.
16. C. H. Crawford, P. Henning, M. Kistler, and C. Wright. Accelerating computing with the cell broadband engine processor. In *CF '08: Proceedings of the 2008 Conference on Computing Frontiers*, pages 3–12. ACM, 2008.
17. W. J. Dally, P. Hanrahan, M. Erez, T. J. Knight, F. Labonté, J-H Ahn., N. Jayasena, U. J. Kapasi, A. Das, J. Gummaraju, and I. Buck. Merrimac: Supercomputing with streams. In *SC'03*, Phoenix, Arizona, Nov 2003.
18. W. J. Dally and W. Poulton. *Digital Systems Engineering*. Cambridge University Press, 1998.
19. A. Das, W. J. Dally, and P. Mattson. Compiling for stream processing. In *PACT '06: Proceedings of the 15th International Conference on Parallel Architectures and Compilation Techniques*, pages 33–42, 2006.
20. ELPIDA Memory Inc. 512 M bits XDR™ DRAM, 2005. http://www.elpida.com/pdfs/E0643E20.pdf
21. M. Erez. *Merrimac – High-Performance and Highly-Efficient Scientific Computing with Streams*. PhD thesis, Stanford University, Jan 2007.
22. M. Erez, J. H. Ahn, A. Garg, W. J. Dally, and E. Darve. Analysis and performance results of a molecular modeling application on Merrimac. In *SC'04*, Pittsburgh, Pennsylvaniva, Nov 2004.
23. M. Erez, J. H. Ahn, J. Gummaraju, M. Rosenblum, and W. J. Dally. Executing irregular scientific applications on stream architectures. In *Proceedings of the 21st ACM International Conference on Supercomputing (ICS'07)*, June 2007.
24. M. Erez, N. Jayasena, T. J. Knight, and W. J. Dally. Fault tolerance techniques for the Merrimac streaming supercomputer. In *SC'05*, Seattle, Washington, USA, Nov 2005.
25. K. Fatahalian, T. J. Knight, M. Houston, M. Erezand, D. R. Horn, L. Leem, J. Y. Park, M. Ren, A. Aiken, W. J. Dally, and P. Hanrahan. Sequoia: Programming the memory hierarchy. In *SC'06*, Nov 2006.
26. J. Gummaraju, J. Coburn, Y. Turner, and M. Rosenblum. Streamware: programming general-purpose multicore processors using streams. *SIGARCH Computer Architecture News*, 36(1):297–307, 2008.
27. J. Gummaraju, M. Erez, J. Coburn, M. Rosenblum, and W. J. Dally. Architectural support for the stream execution model on general-purpose processors. In *PACT '07: Proceedings of the 16th International Conference on Parallel Architecture and Compilation Techniques*, pages 3–12. IEEE Computer Society, 2007.
28. J. Gummaraju and M. Rosenblum. Stream programming on general-purpose processors. pages 343–354, 2005.
29. R. Ho, K. W. Mai, and M. A. Horowitz. The future of wires. *Proceedings of the IEEE*, 89(4):14–25, Apr 2001.
30. H. P. Hofstee. Power efficient processor architecture and the cell processor. In *Proceedings of the 11th International Symposium on High Performance Computer Architecture*, Feb 2005.
31. Intel® Corp. Pemtium®M processor datasheet. http://download.intel.com/design/mobile/datashts/25261203.pdf, April 2004.
32. T. Kanade, A. Yoshida, K. Oda, H. Kano, and M. Tanaka. A stereo machine for video-rate dense depth mapping and its new applications. *Proceedings CVPR*, 96:196–202, 1996.

33. U. J. Kapasi, W. J. Dally, S. Rixner, P. R. Mattson, J. D. Owens, and B. Khailany. Efficient conditional operations for data-parallel architectures. In *Proceedings of the 33rd Annual IEEE/ACM International Symposium on Microarchitecture*, pages 159–170, Dec 2000.

34. U. J. Kapasi, S. Rixner, W. J. Dally, B. Khailany, J. H. Ahn, P. Mattson, and J. D. Owens. Programmable stream processors. *IEEE Computer*, Aug 2003.

35. B. K. Khailany, T. Williams, J. Lin, E.P. Long, M. Rygh, D.W. Tovey, and W.J. Dally. A Programmable 512 GOPS stream processor for signal, image, and video processing. *Solid-State Circuits, IEEE Journal*, 43(1):202–213, 2008.

36. B. Khailany. *The VLSI Implementation and Evaluation of Area- and Energy-Efficient Streaming Media Processors*. PhD thesis, Stanford University, June 2003.

37. B. Khailany, W. J. Dally, A. Chang, U. J. Kapasi, J. Namkoong, and B. Towles. VLSI design and verification of the Imagine processor. In *Proceedings of the IEEE International Conference on Computer Design*, pages 289–294, Sep 2002.

38. B. Khailany, W. J. Dally, S. Rixner, U. J. Kapasi, P. Mattson, J. Namkoong, J. D. Owens, B. Towles, and A. Chang. Imagine: Media processing with streams. *IEEE Micro*, pages 35–46, Mar/Apr 2001.

39. B. Khailany, W. J. Dally, S. Rixner, U. J. Kapasi, J. D. Owen, and B. Towles. Exploring the VLSI scalability of stream processors. In *Proceedings of the Ninth Symposium on High Performance Computer Architecture*, pages 153–164, Anaheim, CA, USA, Feb 2003.

40. R. Kleihorst, A. Abbo, B. Schueler, and A. Danilin. Camera mote with a high-performance parallel processor for real-time frame-based video processing. *Distributed Smart Cameras, 2007. ICDSC '07. First ACM/IEEE International Conference*, pages 109–116, 25–28 Sept 2007.

41. E. A. Lee and D. G. Messerschmitt. Static scheduling of synchronous data flow programs for digital signal processing. *IEEE Transactions on Computers*, Jan 1987.

42. A. A. Liddicoat and M. J. Flynn. High-performance floating point divide. In *Proceedings of the Euromicro Symposium on Digital System Design*, pages 354–361, Sept 2001.

43. P. Mattson. *A Programming System for the Imagine Media Processor*. PhD thesis, Stanford University, 2002.

44. P. Mattson, W. J. Dally, S. Rixner, U. J. Kapasi, and J. D. Owens. Communication scheduling. In *Proceedings of the Ninth International Conference on Architectural Support for Programming Languages and Operating Systems*, pages 82–92, 2000.

45. MIPS Technologies. *MIPS64 20Kc Core*, 2004. http://www.mips.com/ProductCatalog/P_MIPS6420KcCore

46. NVIDIA®. NVIDIA's Unified Architecture GeForce® 8 Series GPUs. http://www.nvidia.com/page/geforce8.html

47. J. D. Owens, W. J. Dally, U. J. Kapasi, S. Rixner, P. Mattson, and B. Mowery. Polygon rendering on a stream architecture. In *HWWS '00: Proceedings of the ACM SIGGRAPH/EUROGRAPHICS Workshop on Graphics hardware*, pages 23–32, 2000.

48. J. D. Owens, B. Khailany, B. Towles, and W. J. Dally. Comparing reyes and OpenGL on a stream architecture. In *HWWS '02: Proceedings of the ACM SIGGRAPH/EUROGRAPHICS Conference on Graphics Hardware*, pages 47–56, 2002.

49. S. Rixner, W. J. Dally, U. J. Kapasi, B. Khailany, A. Lopez-Lagunas, P. R. Mattson, and J. D. Owens. A bandwidth-efficient architecture for media processing. In *Proceedings of the 31st Annual IEEE/ACM International Symposium on Microarchitecture*, Dallas, TX, November 1998.

50. S. Rixner, W. J. Dally, U. J. Kapasi, P. Mattson, and J. D. Owens. Memory access scheduling. In *Proceedings of the 27th Annual International Symposium on Computer Architecture*, June 2000.

51. S. Rixner, W. J. Dally, B. Khailany, P. Mattson, U. J. Kapasi, and J. D. Owens. Register organization for media processing. In *Proceedings of the 6th International Symposium on High Performance Computer Architecture*, Toulouse, France, Jan 2000.

52. Semiconductor Industry Association. *The International Technology Roadmap for Semiconductors*, 2005 Edition.
53. Texas Instruments. TMS320C6713 floating-point digital signal processor, datasheet SPRS186D, dec. 2001. http://focus.ti.com/lit/ds/symlink/tms320c6713.pdf, May 2003.
54. W. Thies, M. Karczmarek, and S. P. Amarasinghe. StreamIt: a language for streaming applications. In *Proceedings of the 11th International Conference on Compiler Construction*, pages 179–196, Apr 2002.
55. D. van der Spoel, A. R. van Buuren, E. Apol, P. J. Meulen -hoff, D. Peter Tieleman, A. L. T. M. Sij bers, B. Hess, K. Anton Feenstra, E. Lindahl, R. van Drunen, and H. J. C. Berendsen. *Gromacs User Manual version 3.1*. Nij enborgh 4, 9747 AG Groningen, The Netherlands. Internet: http://www.gromacs.org, 2001.

Chapter 9
Heterogeneous Multi-core Processors: The Cell Broadband Engine

H. Peter Hofstee

Abstract The Cell Broadband Engine[TM1] Architecture defines a heterogeneous chip multi-processor (HCMP). Heterogeneous processors can achieve higher degrees of efficiency and performance than homogeneous chip multi-processors (CMPs), but also place a larger burden on software. In this chapter, we describe the Cell Broadband Engine Architecture and implementations. We discuss how memory flow control and the synergistic processor unit architecture extend the Power Architecture[TM2], to allow the creation of heterogeneous implementations that attack the greatest sources of inefficiency in modern microprocessors. We discuss aspects of the micro-architecture and implementation of the Cell Broadband Engine and PowerXCell8i processors. Next we survey portable approaches to programming the Cell Broadband Engine and we discuss aspects of its performance.

9.1 Motivation

The job of a computer architect is to build a bridge between what can be effectively built and what can be programmed effectively so that in the end application performance is optimized. For several decades now it has been possible to maintain a high degree of consistency in how (uni-) processors are programmed, characterized by sequential programming languages with performance measured on sequentially expressed benchmarks. While microprocessor performance has increased exponentially over the last 25 years, efficiencies have declined. From 1985 (Intel 80386) to 2001 (Intel Pentium 4) microprocessor performance improved at a rate of about 52% per year, the number of transistors delivering this performance by about 59% per year, and operating frequencies by about 53% per year [1]. Thus architectural

H. Peter Hofstee (✉)
IBM Systems and Technology Group, 11500 Burnet Rd., Austin, TX 78758
e-mail: hofstee@us.ibm.com

[1]Cell Broadband Engine is a trademark of Sony Computer Entertainment, Inc.
[2]Power Architecture and PowerXCell are trademarks of IBM.

S.W. Keckler et al. (eds.), *Multicore Processors and Systems*, Integrated Circuits and
Systems, DOI 10.1007/978-1-4419-0263-4_9, © Springer Science+Business Media, LLC 2009

efficiency, performance divided by the product of the number of transistors and their frequency, declined in this period by about 60% per year, or almost a factor of 2000 decrease in efficiency in just 16 years. Thus dramatic decreases in the cost per transistor and in the energy per transistor switch have allowed us to deliver consistent performance increases while keeping the programming model the processors expose much the same at the expense of creating ever less-efficient microprocessor implementations.

The period of exponential performance growth without fundamental changes to the architecture of processors and to the way applications are programmed has now ended [2]. The easiest way to observe this change is by recognizing that microprocessor frequencies have barely changed in the last several years. While per-transistor cost continues to decrease, and therefore the number of transistors available to the microprocessor designer continues to increase, the available performance per transistor now remains constant because operating frequencies no longer increase. The market is not willing to accept further increases in processor power. Processor power is held roughly constant indicating that energy per switch improves from semiconductor generation to generation at roughly the same rate transistors are added if we assume transistor switching factors are roughly constant. The story with the wires connecting the transistors together is not much better, keeping the number of transistors that can communicate with one another in a cycle essentially the same [2].

While it is possible to continue to deliver modest performance increases on the historical benchmarks given more – but not faster – transistors, maintaining historical (50+%/year) performance growth has required a paradigm shift. The obvious way to increase performance given twice as many transistors at the same frequency is to instantiate two processors sharing the other system resources, including system memory. Doing so has a number of attractive characteristics:

(1) Applications do not degrade in performance when ported to this new processor. This is an important factor in markets where it is not possible to rewrite all applications for a new system, a common case.
(2) Because system memory is shared, applications will typically see more memory capacity and more memory bandwidth available to the application on a new system. Applications benefit from this even if they do not (optimally) use all the available cores.
(3) For systems that run many independent jobs, the new system delivers nearly twice the performance of the previous one without any application-level changes.
(4) Even when a single application must be accelerated, large portions of code can be reused.
(5) Design cost is reduced, at least relative to the scenario where all available transistors are used to build a single processor.

The major processor suppliers have therefore all switched to the shared-memory multi-core approach to building processors, and thus we have left the period of uniprocessors and entered a period dominated by (homogeneous) *chip multi-processors* (CMPs).

For server systems the paradigm shift is not so large, as many servers have been programmed using a shared memory abstraction since the 1960s. Most commercial server benchmarks measure throughput (e.g., transactions per minute or transactions per dollar, and probably soon transactions per Joule) and not single-thread performance. For PC processors and many embedded processors, however, the paradigm shift is much larger because few applications have been written to leverage concurrency effectively. Unless the performance of a majority of applications of interest to the consumer continues to improve at the historical rate, the size of the market is doomed to shrink due to longer product replacement cycles. Therefore, the PC industry is particularly eager to make it easier to program these CMP processors in order to encourage the development of applications that can leverage multi-core processors effectively.

It seems reasonable to assume that before CMOS technology stops improving altogether another fundamental transition will occur. This will happen when the price per transistor continues to drop, but the energy associated with switching a transistor no longer decreases significantly. Because most processors already operate at the power level the consumer is willing to tolerate this implies that we will no longer be able to increase the number of transistors that are simultaneously used. Once this occurs performance (on a fixed power budget) can only grow as efficiency is improved. Efficiency per computation can be improved in one of two fundamental ways:

Running each thread slower or
Specializing the core to the computation.

While running each thread slower is a viable option for highly concurrent applications, sequential sections and synchronization requirements limit the degree to which concurrency can be effective for a problem of a given size. This is a consequence of Gustafson's Law [3] (a generalization of Amdahl's Law) that recognizes that application performance on parallel systems is eventually limited by the sequential portion, but that this effect is mitigated by the fact that larger problems tend to allow more concurrency. Introducing processors that deliver less performance per thread makes it necessary to rewrite or retune all applications even to just maintain the level of performance on a previous system. This is hard to do, especially in the PC market with its myriad of applications. Thus, we predict that the multi-core era will be followed by a period where conventional cores are complemented by specialized cores of which only a subset will be activated at any one time. This is likely to be a period dominated by *heterogeneous chip multi-processors* (HCMPs). In this chapter, we discuss a processor that can be considered a forerunner for this period; the Cell Broadband Engine [4]. In order to make the HCMP approach viable, great care (even more so than for conventional CMP processors) will have to be taken to ensure that applications remain portable and can realize the expected performance gains. We will return to this issue in the section on programming.

Ultimately, neither the cost per transistor, nor transistor performance, nor the energy per switch will improve. When this happens it will become economically feasible to design processors with a narrowly defined purpose, as the design costs can be amortized over long periods of time. At this time we will (re)turn to building application-specific integrated circuits (ASICs) and single-function appliance-type devices and we are like to enter an era where ASICs dominate, the *appliance/ASIC era.*

Before this happens, however, multi-core processors and heterogeneous multi-core processors are likely to displace increasing numbers of ASICs because multi-core processors make much more efficient use of added transistor resources as they become available than was common during the uniprocessor period where the efficiency gap between processors and ASICS was always growing.

9.2 Cell Broadband Engine Architecture

The Cell Broadband Engine Architecture (CBEA) extends the Power architecture and aims to define a heterogeneous chip multi-processor with increased efficiency compared to homogeneous CMPs. The most significant sources of efficiency are

(1) Software-managed asynchronous transfer of data and code from system memory to a local store close to processor that allows efficient hiding of main memory latency: "memory flow control" (MFC) in the CBEA [5]. While there are significant implementation and architectural differences, the observation that software management of data movement can dramatically improve efficiencies follows the stream processors described in this book by Dally e.a.
(2) A SIMD-RISC execution core, the synergistic processor unit (SPU) [6] that co-optimizes efficiency and performance by:

   (A) Using a large unified register file that allows execution latencies to be efficiently hidden.
   (B) Branch hint and select instructions that allow an efficient means of enhancing branch performance.
   (C) Specifying a SIMD dataflow that improves performance and execution efficiency on applications that can be vectorized.

The synergistic processor element (SPE) implements both the MFC and SPU architecture and complements the Power architecture cores. A multi-faceted description of the Cell Broadband Engine architecture and micro-architecture can be found in [6, 5, 7, 8]. Here we focus on those aspects of the architecture and implementation that are particularly relevant to multi-core, and do not cover details of the instruction set architecture and implementation that have no direct bearing on multi-core.

9.2.1 Chip-Level Architecture

As explained in the previous section, our vision for heterogeneous chip multi-processors is one that extends a conventional homogeneous shared-memory multi-core architecture to include processor cores that are more efficient than conventional cores for a target set of applications (Fig. 9.1). In the Cell Broadband Engine Architecture (CBEA) [9], Power architecture [10] cores coherently share system resources with the synergistic processor element (SPE) that combines the memory flow control with the SPU. The Power architecture and the power processor element (PPE) that implements the Power ISA provide the foundation for this multi-core architecture. The CBEA shared memory organization follows the Power architecture, including the mechanisms for memory coherence, translation, and protection.

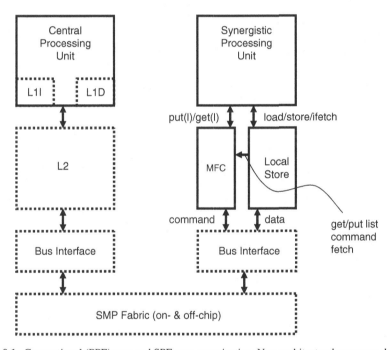

Fig. 9.1 Conventional (PPE) core and SPE core organization. Non-architectural resources dashed

9.2.2 Memory Flow Control

An access to system memory in a high-frequency processor can require close to a 1000 processor cycles, especially when one takes contention between the cores that share this memory into account. This causes substantial overhead: when one inspects a modern microprocessor it is immediately obvious that a dominant fraction of the silicon is devoted to the cache hierarchies required to shield the programmer from the fact that the latency to system memory is large. We concluded that as long as we would try to support a program semantics that is locality unaware we had no hope of delivering a substantially more efficient processor implementation with good single-thread performance on a broad set of applications.

The basic idea of the CBEA is that efficiencies can be improved by exercising program control to bring data close to the processor before program execution requires that data or code. Memory flow control therefore defines a set of commands (*put* and *get*) that transfer data between shared system memory and a "local store" memory. *Put* and *get* commands are executed by the memory flow control unit. As seen from the system, the semantics of *put* and *get* commands closely mirrors the semantics of the loads and stores in the Power architecture. *Put* and *get* commands use the same effective address that the Power cores use to access shared system memory, and access to shared system memory follows the Power architecture coherence rules. Note that the local store is a separate piece of memory and not a cache. A *get* command results in a copy in the local store of data or code elsewhere in shared system memory, much like a *load* instruction on the Power processor results in a copy of data in a general-purpose register. Unlike *load* and *store* instructions there is no implied sequential ordering between the completion of *get* and *put* commands and that of the other SPE instructions. This allows the memory flow control unit to process many commands concurrently and effectively hide main memory latency.

Each SPU has its own associated local store memory to store both code and data, and each SPU represents its own thread of control. Perhaps confusingly, transfers from the local store and the register file in the SPU are called (SPU) *load* and *store*. SPU *load* and *store* access the local store as private untranslated and unprotected memory. The SPU *load* and *store* use a private local store address for these accesses, and SPUs can only access shared system memory indirectly through memory flow control.

A local store can also be mapped into system memory. This allows the operating system to give it an effective address and enables memory flow control to access another SPE's local store with *put* and *get* commands, and also allows the Power cores to target any local store with conventional *load* and *store* instructions (though this is typically inefficient). The local store can be mapped as non-cacheable memory and is then trivially coherent. If a local store is cached (typically it is not), the cached copy is not coherent with respect to modifications by the SPE to which that local store belongs, but it does remain coherent with respect to accesses by all the other system resources.

MFC commands can be issued one at a time by the SPU through a set of synchronizing registers called *channels* or via system-memory-mapped registers by other processors in the system that have been granted the privileges to do so by the operating system. In addition, a program running on the SPU can prepare a "shopping list" in its local store of data to be fetched from system memory into the local store (or vice-versa) and issue a single command to the MFC to process this list. *Put*, *get*, and *put-list* and *get-list* commands contain a class identifier, and the SPU can interrogate the MFC to check if all commands in a class have completed. This allows the

SPU to amortize the memory latency penalty over all the accesses to shared memory in that class. In order to avoid "busy waiting," a blocking version of the *wait* command is also available. When an SPU is blocked on a channel command it enters a (low power) state in which no instructions are executed. A variety of interrupt mechanisms to enable conventional busy waiting are also supported.

There are also versions of *put* and *get* that mirror the atomic *load word and reserve* and *store word conditional* operations used in the Power architecture to create shared-memory-based semaphores. Through the atomic *get line and reserve* and *put line conditional* MFC instructions SPEs can participate in shared-memory-based synchronization with other SPEs and with the Power processors in the system. Because the Power architecture defines a minimum reservation granule, and thus prevents multiple locks in the same cache line, the two mechanisms fully interoperate as long as the SPEs atomic *get* line size in a particular implementation does not exceed the line size of the reservation granule. Having atomic access to a line rather than a word also allows small data structures to be accessed atomically without having to acquire locks. Memory-mapped *mailbox* and *signal notification* registers in the memory flow control can also be accessed through channel commands from the associated SPU and provide additional mechanisms for fast inter-core communication.

A conventional RISC processor accesses system memory with *load* and *store* operations that place a copy of the data in system memory in a private register file. The processor then performs operations on the data in the register file before placing data back in shared system memory, as opposed to CISC architectures where most operations are performed directly on data in memory with only a limited number of registers. RISC architectures promise greater execution efficiency than CISC architectures because software manages the transfer of data into and from fast registers, but they do so at the expense of reduced code efficiency. The CBEA's SPEs go a step further. *Get* operations issued to the memory flow control unit by the SPU transfer data from system memory to the private local store memory, asynchronous to the operation of the SPU. An explicit *wait* operation is required to ensure a class of get commands has completed. *Load* instructions, issued by the SPU, load data from the local store into the register file. As in a RISC processor, arithmetic operations occur only on data in the register file. *Store* and *put* operations transfer data from the register file to local store and, asynchronously, from the local store back to shared system memory. The decision to let the *put* and *get* operations operate asynchronously with the SPU and to require explicit synchronization between *put* and *get* commands and between *put* and *get* and SPU-*load* and -*store* allows the memory flow control unit to operate independently and issue as many put and get commands in parallel as are needed to effectively cover the latency to system memory. In contrast, the sequential semantics on conventional processors requires speculation to get multiple main memory accesses from a single thread to overlap [11]. Unlike speculative execution past a cache miss, or using speculative hardware or software pre-fetches into the caches of conventional cores, SPE put and get operations are non-speculative. When main memory bandwidth is a limiter to application performance, being able to saturate memory bandwidth with non-speculative operations is crucially important.

Because memory bandwidth can severely limit the performance of multi-core architectures and force the inclusion of large on-chip caches [12], this is an important aspect of the CBEA architecture.

Of course, the introduction of memory flow control places a new burden on the compiler writer or the application programmer in that it requires gathering data (and code) into a store of limited size (though much larger than a register file) before performing the operation. We will return to this issue at length in the section on programming, but for now let it suffice to say that time spent improving the locality and pre-fetch ability of data (and code) is an effort likely to benefit an application no matter the target platform. When the communication and computation phases of a program are of similar duration then multi-buffering, i.e., explicitly overlapping communication and computation, can further improve application performance. The performance effects of multi-buffering are usually significantly smaller than the benefits achieved by using the MFC to asynchronously access memory and pay the main memory latency penalty only once for multiple *get* commands in a class.

We believe that the memory flow control paradigm is more broadly appropriate for future heterogeneous chip multi-processors, because it fundamentally addresses the inefficiencies associated with accessing the shared memory resources. Memory flow control can be combined with a wide variety of specialized processors and single-function accelerators.

9.2.3 SIMD-RISC Core Architecture

Conventional multi-core processors can be expected to deliver performance growth at near historical rates as long as the number of transistors per dollar grows at historical rates and the energy per transistor switch continues to go down at near the same rate transistors become available. This may last for a decade and quite possibly longer [13]. Therefore, the SPE should not be too specialized a processor, or it would run the risk of not capturing a sufficient number of applications and a sufficient market size to allow it to be regularly re-implemented and thus benefit from the latest semiconductor technologies and maintain its competitiveness relative to conventional CMPs. The SPE is therefore a fairly generic processor. Power cores run the operating system and provide an initial target for porting applications to the Cell Broadband Engine. While in many cases the vast majority of the application cycles will run on the SPEs, and the PPEs can be considered a resource that provides system services for the SPEs, it can be equally valid to think of the SPEs as powerful accelerators to the PPEs.

While the major source of inefficiency is addressed by the memory flow control and local store architecture, the SPU ISA is defined to allow implementations that combine high single-thread performance with a high degree of efficiency. The SPU SIMD-RISC architecture differs from other RISC architectures in that a single shared register file supports all data-types and in that all instructions operate on 128-bit quantities. Doing so allows the SPU instruction set architecture to support 128 registers in an otherwise fairly traditional 32-bit instruction encoding [8]. This large number of general-purpose registers allows the SPUs to combine a high degree of application efficiency and a high operating frequency without having to resort to resource- and power-intensive micro-architectural mechanisms such as register renaming and out-of-order processing to hide pipeline latencies. *Branch hint* instructions are an efficient means to improve branch performance without the large history tables that tend to increase single-thread performance but have contributed significantly to the disproportionate growth of transistors per core. Finally, *select* instructions implement a form of conditional assignment and can be used to avoid branches altogether.

9.2.4 SPU Security Architecture

Because the SPU accesses its own local store with a private address, it was possible to define a robust security architecture [7]. An *isolate load* command issued to an SPU will cause it to clear its internal state and prevent further access to its internal state from the rest of the system. Next, the SPU will start to fetch, hardware-authenticate, and optionally decrypt, code and/or data from the address specified in the *isolate load* command. Decrypted and authenticated data is placed in the private and protected local store of the SPU and execution starts only if data and code was successfully authenticated. Thus *isolate load* creates a programmable secure processor on-the-fly within the Cell Broadband Engine. Neither the hypervisor, nor the operating system has the ability to interrogate the internal state of the SPU once it has initiated *isolate load* and is in isolate mode. An *isolate exit* command, issued either by the SPU, by the OS, by the hypervisor, or by an application with the right privileges, results in all internal state to be cleared prior to the SPU again becoming accessible to the rest of the system. Because a local store address is private and untranslated the integrity of its state is not dependent on the integrity of either the operating system or the hypervisor. This security architecture allows multiple independent secure processors to be created on the fly. With a co-operative approach where the isolated processors upon request encrypt and store their state prior to an isolate exit, isolated SPEs can be virtualized and the number of secure processors can exceed the number of physical SPEs in a system.

9.2.5 Optional Extensions to the Power Architecture

Besides the memory flow control and the synergistic processor unit architecture the CBEA defines a number of optional extensions to the Power architecture, mainly to improve real-time behavior [5]. One of these extensions, "replacement management tables," allows software to control the replacement policies in the data and translation caches by associating address ranges with subsets of the cache (ways). Another extension, "resource allocation" allows critical resources in the system such as memory and I/O bandwidth to be allocated on a per-partition basis. While resource allocation is perhaps not so difficult to conceptualize, it can be fiendishly difficult to implement. Strictly maintaining allocations results in resources that are always underutilized, or in significantly increased latencies, whereas oversubscription easily creates situations where the allocated bandwidth cannot be delivered. The solution to this problem is that the CBEA guarantees only steady-state behavior and does not attempt to provide instantaneous bandwidth guarantees.

A final set of optional extensions to the architecture defines a versatile set of power management states appropriate for a coherent multi-core system in which power management can occur on a per-core or per-unit basis. Defining these power management states for a coherent multi-processor required some special care in order to maintain system coherence as cores transition from one power management state to another in close co-operation with the operating system.

9.3 Cell Broadband Engine Micro-Architecture and Implementation

In this section, we discuss some of the key design decisions in the Cell Broadband Engine processors. We refer the reader to [6, 14, 15] for more detailed discussions including a more detailed description of the SPU. Here we focus on aspects of the micro-architecture that relate to aspects of multi-core design.

Perhaps the most significant design decision in the Cell Broadband Engine is the size of the local store memory in the SPE. The size of the local store represents a delicate balance between programmability, performance, and implementation efficiency. Our design goal was to have the processor operate at a nominal frequency of 4 GHz (3.2 GHz at near worst-case technology) so that it could compete with the fastest conventional cores. The initial concept defined a 64 KB local store, but this size was soon considered to present too large a hurdle for programmability. A first SPE prototype in silicon had a 128 KB local store, but a study revealed that the performance benefits of growing the local store to 256 KB outweighed the increases in SPE size and power. Thus the local store in CBE is 256 KB and a load instruction has a latency of six cycles.

The local store occupies about half the silicon area of the SPU and about a third of the area of the SPE. The six-cycle latency can be effectively hidden for many applications by using the 128 general-purpose registers to enable unrolling and interleaving of the inner loop. At eight cycles (the latency we anticipate for a 512 KB local store) performance would start to degrade due to an insufficient number of registers to hide the latency and at 10 or more cycles (1 MB local store or more) performance would degrade significantly. The decrease in efficiency implies that the average application on CBE should benefit by significantly more than 30% (taking the increased load-latency into account) to justify growing the local store to 512 KB. For the majority of applications we studied, we did not find this to be the case and thus the local store size was set at 256 KB. It is important to note that at 256 KB the SPE often outperforms conventional cores of similar frequency that have 1 or 2 MB caches. We believe this is the result of the efficiencies gained by placing the management of this local memory under program control and because a memory, unlike a cache, does not have a minimum storage granule (cache line).

Because the local store is accessed both by the SPU and the MFC and must support *load*, *store*, *put*, *get*, and SPU and MFC (list) instruction fetch, it could easily become a performance bottleneck. In order to address these demands the local store supports both 16-byte (16-byte aligned) and 128-byte (128-byte aligned) accesses. Each of these operations is executed in a single cycle on the SRAM array, and is fully pipelined. *Put* and *get* operations have the highest priority. Data associated with *put* and *get* operations are accumulated to 128 bytes and instruction fetch also occurs in chunks of 128 bytes. These design decisions enable the SPE to achieve a high percentage of peak performance on numerous applications with a single-ported efficient local store design. It is not uncommon to see the local store utilized 90% of the available cycles.

Another key design decision is the number of outstanding *put* and *get* requests the MFC can handle in parallel. The MFC for each SPE in the Cell Broadband Engine breaks commands into transfers of 128 bytes or less, and is able to process up to 16 of these commands in parallel. With eight SPEs this means that up to 128 *put/get* operations can be in flight concurrently, far more than are needed to cover the latency to main memory. It should be noted that if future implementations would expose the SPEs to much larger memory latencies it is not difficult to grow the number of outstanding transactions per SPE.

Because the Cell Broadband Engine was first implemented in 90 nm, a technology where the industry at large was transitioning from single- to dual-core designs, there was limited chip resource available for the Power processor. As a result, the Power processor in the first generation of Cell/B.E. processors does not deliver the same amount of performance of high-end Power cores (or high-end competing cores) in the same technology. This means that the current Cell/B.E. is biased towards applications that can run effectively on the SPEs.

In the Cell Broadband Engine, the PPE, the SPEs, the memory controller, and the I/O and coherence controllers are all connected through an on-chip coherent fabric dubbed the "element interconnect bus" (EIB) [15] (Figs. 9.2 and 9.3). The EIB control fabric is organized as a tree, but its data paths are constructed as rings, two of which run clockwise around the chip and two counterclockwise. The EIB delivers about a third of a Terabyte per second peak bandwidth, an order of magnitude more than the peak off-chip memory bandwidth of 25.6 GB/s. Clearly, applications that can leverage on-chip communication can derive substantial benefit, but thus far it has proven difficult to leverage this on-chip bandwidth resource in a portable programming framework.

Figure 9.4 shows a high-performance dual-core design (IBM Power5) and the Cell Broadband Engine side by side. It is clear that the Cell Broadband Engine supports a (lightweight) Power core and eight SPEs in less area than is required for two conventional cores. Comparisons to x86 dual-core designs present a similar picture. Each SPE requires about 1/6th the area of a leading-performance conventional core

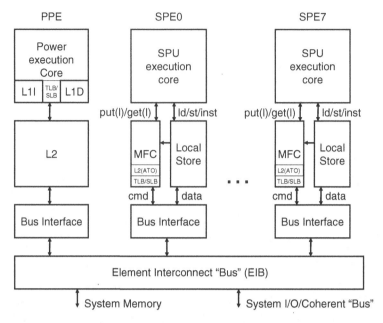

Fig. 9.2 High-level organization of the Cell Broadband Engine

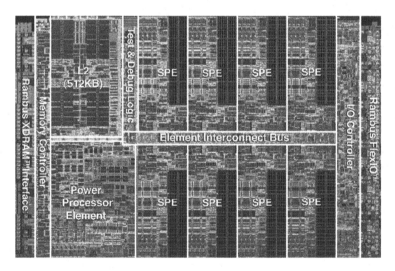

Fig. 9.3 90 nm Cell Broadband Engine. Photo courtesy: Thomas Way, IBM Microelectronics

in the same technology. We will see in Sect. 5 that in many cases each SPE can out-perform each of these conventional cores. At the risk of stating the obvious, it should be noted that this resource ratio is not technology dependent. In every technology generation, one can expect a CBEA processor with about 4× the number of SPEs

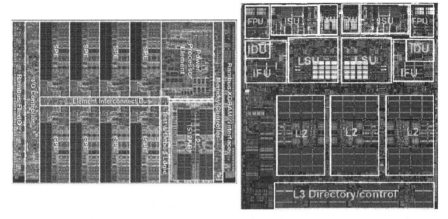

Fig. 9.4 Cell Broadband Engine and IBM Power 5 processor on a similar scale. Micrographs courtesy: Thomas Way, IBM Microelectronics

compared to a CMP built out of high single-thread performance conventional cores.

In the Roadrunner supercomputer at Los Alamos National Laboratory, a variant (PowerXCell8i) of the Cell Broadband Engine is used. This processor has added double-precision floating-point performance (100+ DP GFlops per chip) and supports increased main memory capacity (16 GB DDR2 per chip) at the same peak bandwidth (25.6 GB/s) as the original Cell/B.E. In Roadrunner the PowerXCell8i processors are used as PCI-express attached accelerators to each of the nodes of a cluster of conventional (AMD Opteron) processors. This arrangement limits the amount of application code that must be ported to the SPEs for base performance reasons. Organized as a cluster of clusters interconnected by a fat-tree infiniband network, the Roadrunner supercomputer was the first to deliver Linpack performance in excess of 1 Petaflop [16]. As more applications run natively on the SPEs and as the programming tools discussed in the next section mature, it will become more often beneficial to directly cluster Cell/B.E. processors together as the PPE provides more than adequate performance to provide OS services to the SPEs in most cases.

9.4 Programming the Cell Broadband Engine

9.4.1 The Cell Broadband Engine as a CMP

There are significant similarities between the programming model for the Cell Broadband Engine and that of a conventional chip multi-processor. In either case the basic abstraction is that of threads running on shared system memory. While the Cell Broadband Engine certainly allows other programming abstractions, several of which have been implemented on top of the thread abstraction, this is the one supported by the Linux operating system and used (directly or indirectly) by most applications. Also, within each thread, the use of SIMD data types is essentially the same as that on the media units of conventional processors. While scalar data types are not supported directly on the hardware, they are supported by the various compilers that target the synergistic processor unit.

The CBEA is an extension of the 64-bit Power architecture and includes hardware-based hypervisor support. Address translation in the memory flow control unit in the SPE mirrors that of the Power processor. Thus, access privileges of threads that run on the SPEs can be controlled in the same manner (page and segment tables) that threads are restricted or protected on a multi-core Power processor.

There is a fundamental difference between a thread running on an SPU and a thread on the Power processor because the SPU's access to its own local store is untranslated and unprotected. The SPU does not support a supervisor or hypervisor mode for accessing its local store and thus local store is part of the state. This does not imply that an SPE is user-mode only. It is entirely reasonable to map either a supervisor or hypervisor process on the SPE and have an SPE take on a part of the system management burden.

In order to context switch an SPE process its entire state must be stored in system memory and a new process state must be loaded before the SPE is restarted. Because the 256 KB local store is a part of the state of an SPE, a (pre-emptive) context switch is an expensive operation. Thus operating system calls are preferably serviced by a processor other than the SPE. In Linux for Cell/B.E. this issue is addressed by starting SPE threads as Power threads [17]. The initial code on the Power processor loads code to the SPE and starts the SPE. The Power thread remains to service OS calls from the associated SPE thread. While this arrangement can put a heavy burden on the Power core in programs with frequent OS calls it has the benefit that the code running on the SPE is only minimally disturbed.

It should be noted that while a context switch to a task that is going to do only a small amount of work presents a large overhead on the SPE, context switches between tasks that do a substantial amount of work can be significantly faster on an SPE than on a conventional processor. The reason is that while on a conventional processor the context switch itself may be fast, the caches tend to be "cold" after this switch, and all required thread context is eventually demand-fetched line by line into the cache. In the SPE on the other hand, local store process state is restored with a few large *get* transfers into the local store. This can be considerably more efficient than the demand misses because the main memory latency is seen only once. A "substantial amount of work" above can therefore be quantified as a process that would require in the order of 256 KB of code and data before the next thread switch.

From the application programmer's perspective a thread on an SPE differs from a thread on the Power core principally in that the thread operates on shared system memory only indirectly. According to today's APIs for the SPEs data structures reside in the (finite) local store by default. It is the responsibility of the application programmer to move code and data between the local store and shared system memory. If the entire application fits in the local store, then application code by and large does not have to be modified, but if the code and/or data do not fit, which is often the case; the burden on the programmer can be substantial.

Because code tends to have significantly better temporal and spatial locality than data, it is reasonable to ask if placing the burden of code locality management on the application programmer is warranted. In the open-source software development kit for Cell/B.E. we have made steady progress towards relieving the programmer of this burden. The most recent development kit introduces automated code overlays and mechanisms in the compiler to segment the code into partitions that fit in part of the local store. Prototype compiler/loader/runtimes manage part of the local store as software-managed instruction cache with a very modest impact (usually less than 5%) on application performance [18, 19]. Thus, it is our expectation that except for applications with the most stringent real-time requirements or those that are tuned to the last few percent of application performance, programmers of Cell/B.E. and PowerXCell in the future will not be concerned with code partitioning. While memory latencies can be expected to grow as the number of cores increases, microprocessor frequencies have now by and large stagnated, and thus we do not expect that memory latency will increase to the point where a software instruction cache would fail us any time soon.

For data the situation appears to be substantially different. Our experience with Cell/B.E. shows us that effort to improve data locality and pre-fetching data prior to use results in dramatic performance improvements (on the Cell/B.E.). We discuss several examples in the next section. While compiler/runtime-based data pre-fetching schemes show promising results on a number of codes, it is not reasonable to expect a compiler to "solve" this problem in the general case if an application is not structured to be locality aware. Thus we must look for ways to allow programmers to express data locality and to provide hints on what data will be needed next. Our goal should be to do this in a way that is not specific to the Cell Broadband Engine Architecture. We next discuss a number of programming paradigms that can help in this regard. This is still an area of active investigation and as yet no definitive standard has emerged though OpenMP and the emerging OpenCL open standards appear to be headed for broad use.

9.4.2 Multi-core Programming Abstractions

A programming concept that appears to be highly effective for programming the Cell/B.E., but that also shows significant promise for programming homogeneous CMPs, is the abstraction of tasks and task queues. Especially if the tasks access shared memory only to obtain their parameters and return their results, much of the burden of the difficult semantics of shared memory multi-programming can be taken off the application programmer. Load balancing within the node is automatic; this seems appropriate within a shared-memory node as performance is not highly sensitive to the placement of tasks on the cores within the node. Task queues can be found at the application level, in the runtime system, or can be a part of a library API. Task queues can be maintained by a single authority that issues work to processors that are ready or the processors themselves can interrogate and update a data structure in shared system memory to load the next piece of work. Organizing the computation in the second way avoids the performance bottlenecks associated with a central controlling authority that takes an interrupt every time a task is completed. A task-based programming framework is effective for conventional CMPs, but also allows acceleration of an application with I/O connected (e.g., PCI-express) accelerators. Finally, task-based APIs allow a future transition to accelerators more specialized and more efficient than the Cell/B.E. Tasks play an important role in many of the programming frameworks for the Cell/B.E. and other multi-core processors.

We now discuss a number of programming frameworks that have shown good results on the Cell Broadband Engine, and that present abstractions to the programmer that are not Cell/B.E. specific.

The most abstract programming frameworks provide a tailored development environment specific to a class of applications. Two examples of such environments that have been optimized for the Cell Broadband Engine are the Gedae [20] and Sourcery VSIPL++ [21] development environments. These development environments have thus far been targeted predominantly at signal processing applications. The benefits of these environments is that the programmer is unaware of the details of the Cell/B.E., that code remains portable, and that within the application domain performance can be compelling, but these types of development environments are fairly application domain specific, and depend on a high-quality implementation of a pre-defined set of (library) functions. Some game development environments similarly provide physics engines or AI engines that hide the multi-core implementation from the user.

The Rapidmind framework [22] is one of a series of programming languages that present a sequential programming model. This framework also abstracts away both concurrency and heterogeneity, but introduces opportunity for parallelization with an API that defines a computation as a sequence of transformations on arrays. This framework is commercially available for conventional multi-core processors, for the Cell Broadband Engine, and for GPUs allowing code to be ported across a wide variety of multi-core platforms. Perhaps the biggest downside to this approach is that thus far the variations on C++ that Rapidmind introduces have not been adopted in a standard. Combining "optimistic" execution with runtime detection that maintains a simple semantics is a promising approach. Especially when the application is structured so that the optimistic execution is valid this approach results in surprisingly good performance.

In CellSuperScalar (CellSs) [23] the programmer annotates functions (tasks) to be executed on the SPEs. As in OpenMP, the language semantics declares parallel sections to be interference-free (if they are not it is considered a programmer error) and thus opens up new opportunities for leveraging concurrency. As long as this condition holds, sequential execution and concurrent execution of the code will yield the same result. In the current runtime for CellSs, the main thread and the SPE task scheduler run on the PPE processor. In most task-based frameworks, arguments to a task have to be fully resolved when the task is placed in the queue, meaning that every task is ready to run. In the CellSs runtime, the tasks are placed in a task-dependency graph before all arguments are resolved and thus a task may have dependencies on the results of other tasks. The current implementation of CellSs mitigates the performance implications of having a single scheduler at run-time by scheduling multiple tasks to an SPE at once. CellSs presents a portable framework and relieves the programmer from the many difficult multi-programming issues.

In the Codeplay language, a task-like concept is introduced by defining "sieves." Within a sieve block or function all modifications of variables declared outside the sieve are delayed until the end of the sieve block. This allows the runtime to execute these blocks by reading the external variables, executing the block, and then writing the results. This semantics introduces significant opportunities for parallelization. While the introduction of a sieve block can modify the meaning of the original (sequential) program, the effects of introducing sieves can be found and debugged by executing the code sequentially.

The starting point for the IBM single-source compiler for Cell/B.E. [24] is sequential C/C++, but the language contains explicitly parallel constructs. The IBM prototype single-source compiler for Cell/B.E. is a C/C++ compiler that allows the user to specify concurrency using the OpenMP 2.5 standard pragmas [25]. The compiler also incorporates technology for automated SIMD vectorization. This is a portable approach to programming the Cell/B.E., but the parallel semantics of OpenMP 2.5 may still be too sophisticated for the average programmer. OpenMP 3.0 is a new draft standard incorporating the task concept [25] and may provide a path towards incorporating notions of tasks in a standards-based programming language.

Charm++ for Cell [26] is another portable variant of C++. Similar to the task concept, Charm++ introduces objects called "shares" and methods invoked on these objects or arrays of objects result in asynchronous, task-like execution. A number of applications, including NAMD have been demonstrated to achieve good performance in this framework.

The "Sequoia" language from Stanford University [27] allows the programmer to express both concurrency and data locality in a non-architecture-specific way. Programs are annotated to express movement of data across a more abstract memory hierarchy. Significant performance improvements can be achieved when applications are (re)written to account for data locality. The transformations not only benefit processors such as the Cell/B.E., but also conventional multi-core processors. Applications written in Sequoia have demonstrated excellent performance, but as with many of the proposed languages the path towards standardization for this language is not yet clear.

In the "accelerator library framework" (ALF) [28] that is supported on the Cell/B.E. processors, the concept of tasks and task management is handled more explicitly. Instead of having a function or task executed synchronously on an SPE, tasks to be accelerated are placed in task queues, and task queues can be interrogated for task completion. One can think of this framework as a generalization of a remote procedure call. The task queues and placement of tasks are handled automatically by the runtime system. Of the approaches described here the ALF runtime perhaps allows the greatest degree of scalability as task queue management is distributed to the processors.

Some of the most efficient applications on Cell/B.E. use direct local store to local store communication, and program the Cell/B.E. as a collection of processors that communicate through message passing. While this has the benefit that it allows as much data as possible to be kept on-chip where there is an order of magnitude more bandwidth available, it places the burden of data partitioning and load balancing the application on the programmer. A "micro-MPI" library supports the most common MPI calls on the SPEs themselves, while leveraging the Power cores for connection setup, teardown, and error handling [29]. For users who are willing to manage placement of the computation and data partitioning fully, message passing can be a very effective way to program the Cell/B.E. processor. When message passing is combined with one of the approaches that allow hiding the local store size, the result is code that is less Cell/B.E. specific.

In summary, there is a very large variety of portable programming models and languages available for Cell/B.E. and also for other shared memory multi-core processors. A large number of these introduce task-like concepts to C++ in order to introduce locality and allow for concurrent execution. Several of these approaches have demonstrated good application performance. What is perhaps needed most at this point is a sustained effort to bring the most effective approaches into the standards process, so that programmers can write code that can be handled by multiple competing compilers for multiple competing multi-core processors.

9.5 Performance of the Cell Broadband Engine Processor

This section contains a survey of some of the kernels and applications that have been ported to the Cell/B.E. processor. We refer to the original papers for detailed discussions. The purpose of this section is to highlight the breadth of applications that perform well on the Cell Broadband Engine and to shed some light on why they perform as they do.

Perhaps counter to expectations, Cell performs best in comparison to conventional processors on applications with parallel but unstructured memory access. Less surprisingly, Cell also performs well on computationally bound problems and on streaming applications.

We look at samples from the following application categories:

(1) Scientific math kernels: DGEMM, Linpack, Sparse Matrices, and FFTs.
(2) Basic computer science kernels: sorting, searching, and graph algorithms.

(3) Media kernels: (de)compression, encryption/decryption, graphics.
(4) Sample applications.

9.5.1 Scientific Math Kernel Performance

An early paper by Williams et al. [30] highlighted the promise of the Cell/B.E. for scientific computation.

As is the case with most processors, the SPEs achieve a high fraction of peak performance (over 90%) on dense-matrix multiplication [31]. Achieving this performance is made easier by the fact that the SPU can dual-issue a *load* or *store* with a floating-point operation. Some of the address arithmetic takes away floating-point instruction issue slots, but the impact can be minimized by unrolling the inner loop.

Linpack (LU decomposition) is more interesting. On the Roadrunner supercomputer PowerXCell8i processors are used to accelerate a cluster of multi-core Opteron nodes. The majority of the code remains on the Opteron processors, but the most compute-intensive components are offloaded to the PowerXCell processors. Overall performance is about 75% of the peak flops of the Opteron–PowerXCell combination, even in very large systems [16]. This percentage of peak is considerably higher than the percentage of peak quoted on other supercomputer architectures that use PCI-attached accelerators and even higher than that of several homogeneous clusters of conventional processors. The main reason for this is the balance between local store size and main memory bandwidth within the core, and main memory size and PCIe bandwidth within the PowerXCell8i-based QS22 accelerator blade.

An even more challenging application is sparse matrix math. Williams et al. [32] performed an in-depth study comparing the Cell/B.E. to a variety of other multi-core architectures. Sparse-matrix operations are generally bandwidth bound, and the performance achieved on the various processors that were studied is an indication of how effectively these processors access main memory. The Cell/B.E. processor achieved a substantially higher fraction of peak memory bandwidth (over 90%) than the other leading multi-core systems Williams studied, and, because this application is bandwidth bound, significantly outperformed conventional processors of similar size. This is a consequence of the many outstanding requests and the non-speculative nature of these requests in the Cell processors.

While dense-matrix multiply is typically flop bound, and sparse-matrix multiply tends to be memory bandwidth bound, fast Fourier transform (FFT) typically operate closer to a balance point. The best FFTs on Cell/B.E. operate at a higher fraction of peak than on most conventional architectures [33], and thus FFT performance on Cell/B.E. is again significantly faster than that on a conventional multi-core processor of similar size. FFT performance is the basis for the competitiveness of the Cell/B.E. in many signal processing applications. While most applications on Cell/B.E. are written to the model of threads on shared memory, the highest performing FFTs make extensive use of direct local store to local store communication to mitigate the off-chip memory bandwidth limitations.

9.5.2 Sorting, Searching, and Graph Algorithms

Sorting a large data set is all about orchestrating memory accesses. The best sorting algorithms on Cell/B.E. outperform the best sorting algorithms on conventional processors by a significant margin [34, 35]. Interestingly, the best algorithm on conventional multi-core processors is not the best performing algorithm on Cell/B.E. Within the core a SIMD bitonic merge is preferable over a quicksort (which is not easily SIMD-ized) for the SPEs due to absence of scalar data types. The most efficient published algorithms for the Cell/B.E. also leverage the ability to communicate directly between the SPE local stores, rather than communicate between the cores through shared memory. This allows a bitonic merge sort to be very effective for data sets that fit within the Cell/B.E.

Searching through large data sets appears to be a straightforward streaming-type problem and one would expect the Cell/B.E. processor to do well. However, in practice detecting matches in the data stream can be highly non-trivial, especially if the data stream needs to be matched against a large library that does not fit within the on-chip storage [36]. The results of this application on the Cell/B.E. indicate that the ability to manage data locality can be a major boost to application performance.

"List rank" [37] is a graph ordering problem and another significant example. In the most interesting case, the initial graph is unordered and very little locality of access can be found. Programmers faced with this type of problem often favor programming languages that allow the programmer to express high degrees of concurrency in the application but ignore data locality (because there is none or little to be had). It may come as a surprise that the Cell/B.E. performs exceptionally well on this application. Historically the best-performing processors on these types of applications were the Cray vector machines. The SPE *get-list* operation can be considered the equivalent of the critical Cray "vector-load" instruction that gathers a list of data given a list (vector) of random addresses. One can think of the local store as a giant vector register file in this case. Each SPE supports 16 outstanding *put* or *get* operations, thus Cell/B.E. can be thought of as a 128-threaded processor. Each SPU has sufficient instruction execution bandwidth to perform the needed operations on the data brought to the processor by the independent memory fetch operations. One might think that with 128 general-purpose registers even a user-level thread switch on the SPE is prohibitively expensive, but for applications that are purely memory access bound the computational requirements per thread can easily be met by using just a few general-purpose registers, thus dramatically reducing the overhead of a (software) thread switch on the SPE. Partitioning the register file is also a possibility, but while this reduces thread switch overheads to the minimum, it prevents code from being shared between the threads.

9.5.3 Codecs and Media

As one might expect, the performance of each SPE on problems that are computationally bound such as compression/decompression or encoding/decoding is comparable to that of a conventional processor with 128-bit media extensions. In many cases CBE does better, primarily as a result of having 128 general-purpose registers available to hide execution latencies with loop unrolling. Because an SPE requires less than a quarter of the chip resources of a conventional core, the two effects often combine to almost an order of magnitude performance advantage over a conventional (dual-core) processor of similar size and power in the same technology node [31].

Initial implementations of the media codecs (specifically H.264) placed the three major computational phases on separate SPEs and organized the computation by streaming data from one SPEs local store to another. This type of approach, while it minimizes main memory bandwidth requirements, causes significant load balancing issues whenever one phase of the computation is re-optimized or if the problem size changes and the phases do not scale the same with problem size. Current implementations of the media codecs either perform the entire computation on a single SPE or communicate between the phases of the computation through shared memory [38].

9.5.4 Sample Applications

Performance on application kernels does not always carry over to significant performance enhancements at the application level. Some of the major reasons this can occur are

(1) The accelerator core is too specialized and the remaining part of the computation takes too much time.
(2) Synchronization and communication overhead between the general-purpose core and the accelerator core is too large.

In the case of the Cell Broadband Engine, while there is a significant effort required to port code, it has been our experience that the processor is generic enough to accelerate all or nearly all of the application. Task-based frameworks allow the synchronization and communication overheads to be reduced. As a result, there are now numerous examples of applications that have been successfully ported to or written for the Cell Broadband Engine. We name but a few.

a) The largest single distributed computation in the world is Stanford's "folding@home." This distributed computation uses time donated by PC processors, graphics processors, and the Cell Broadband Engine on Playstation®3. More than 40,000 Playstations contribute in excess of 1 PFlop (single precision) application performance and each Playstation contributes more than 25 times as much to the computation than the average PC [39].

b) One of the earliest commercial uses of the Cell Broadband Engine outside the game space is optical proximity correction (OPC). Cell processors accelerate a majority of this computation essential to prepare the masks required to make chips in advanced semiconductor technologies [40].

c) As a part of the evaluation of the Cell Broadband Engine for the Roadrunner supercomputer, Los Alamos National Laboratory ported three full applications (and an application kernel). In each case the Cell processors performed significantly better than the dual-core processors they accelerate [41].

d) Williams et al. ported (LBMHD) Lattice Boltzman Magneto Hydro Dynamics to a number of processors including the Cell Broadband Engine [42] with performance and power-efficiency results comparable to those in their sparse-matrix study.

Further indications as to the breadth of applications where Cell/B.E.-based systems can be quite competitive can be found in a recent paper on the applicability of Cell/B.E. to data mining [43] and to the "mapreduce" programming framework [44].

9.6 Summary and Conclusions

The Cell Broadband Engine Architecture (CBEA) is a heterogeneous chip multi-processor and is an early example of an architecture where a heterogeneous collection of processor cores operates on a shared system memory. We believe that these types of architectures will become commonplace, just like homogeneous shared memory multi-core processors are commonplace today. We predict the transition to heterogeneous multi-core will become necessary when transistors continue to become more affordable, but stop becoming more energy efficient. When this occurs efficiency must be improved to further improve performance. The Cell Broadband Engine Architecture defines a method for shared memory access; "memory flow control," that addresses the major source of energy inefficiency in modern processors; the cache hierarchies needed to maintain the illusion that system memory is near to the processing elements. Memory flow control, and the associated programming models, can be readily extended to processors with much more specialized cores than the synergistic processor elements of the Cell Broadband Engine. Because the CBEA builds on conventional shared-memory multi-core architecture it allows programming models that will work effectively on CMPs as well as highly heterogeneous and highly specialized ASIC-like architectures. By accessing memory more efficiently, and by re-architecting the core to use a single shared register file, the Cell Broadband Engine can accommodate about four to six times the number of cores with the same resources that are required to support a conventional core of similar performance. The Cell Broadband Engine delivers comparable and sometimes significantly better performance per core than conventional CMP cores on a wide variety of applications. Perhaps counter to expectations, the Cell Broadband Engine performs best on problems that are unstructured, but allow high degrees of concurrent access to main memory.

References

1. J. L. Hennessy and D. A. Patterson. *Computer Architecture, A Quantitative Approach, 4th edition*. Morgan Kaufmann, 2006.
2. V. Agarwal, M.S. Hrishikesh, S.W. Keckler, D. Burger. Clock rate vs. IPC: The end of the road of conventional microarchitectures. In *Proc. 27th Annual International Symposium on Computer Architecture (ISCA). ACM Sigarch Computer Architecture News*, 28(2), May 2000.
3. J. L. Gustafson. Reevaluating Amdahl's law. *Communications of the ACM* 31(5):532–533, 1988.
4. J. A. Kahle, M. N. Day, H. P. Hofstee, C. R. Johns, T. R. Maeurer, and D. Shippy. Introduction to the Cell multiprocessor. *IBM Journal of Research and Development*, 49(4/5), 2005.
5. C. R. Johns and D. A. Brokenshire. Introduction to the Cell Broadband Engine Architecture. *IBM Journal of Research and Development*, 51(5):503–520, Oct 2007.
6. B. Flachs, S. Asano, S. H. Dhong, H. P. Hofstee, G. Gervais, R. Kim, T. Le, P. Liu, J. Leenstra, J. S. Liberty, B. Michael, H.-J. Oh, S. M. Mueller, O. Takahashi, K. Hirairi, A. Kawasumi, H. Murakami, H. Noro, S. Onishi, J. Pille, J. Silberman, S. Yong, A. Hatakeyama, Y. Watanabe, N. Yano, D. A. Brokenshire, M. Peyravian, V. To, and E. Iwata. Microarchitecture and implementation of the synergistic processor in 65-nm and 90-nm SOI. *IBM Journal of Research and Development*, 51(5):529–554, Oct 2007.
7. K. Shimizu, H. P. Hofstee, and J. S. Liberty. Cell Broadband Engine processor vault security architecture. *IBM Journal of Research and Development*, 51(5):521–528, Oct 2007.
8. M. Gschwind, P. Hofstee, B. Flachs, M. Hopkins, Y. Watanabe, and T. Yamazaki. Synergistic processing in cell's multicore architecture. *IEEE Micro*, pp.10–24, Mar 2006.
9. Cell Broadband Engine Architecture Version 1.02, Oct 2007. http://www.ibm.com/chips/techlib/techlib.nsf/products/Cell_Broadband_Engine
10. http://www.ibm.com/chips/techlib/techlib.nsf/products/PowerPC
11. D. Burger, J. R. Goodman, and A. Kagi. Memory bandwidth limitations of future microprocessors. In *Proceedings of the 23rd Annual International Symposium on Computer Architecture*, pp. 79–90, May 1996.
12. J. Huh, D. Burger, and S. Keckler. Exploring the design space of future cmps. In PACT'01: In *Proceedings of the 10th International Conference on Parallel Architectures and Compilation Techniques*, pp. 199–210, Washington, DC, USA, 2001. IEEE Computer Society.
13. International Technology Roadmap for Semiconductors, www.itrs.net
14. D. Pham, T. Aipperspach, D. Boerstler, M. Bolliger, R. Chaudhry, D. Cox, P. Harvey, P. Harvey, H. Hofstee, C. Johns, J. Kahle, A. Kameyama, J. Keaty, Y. Masubuchi, M. Pham, J. Pille, S. Posluszny, M. Riley, D. Stasiak, M. Suzuoki, O. Takahashi, J. Warnock, S. Weitzel, D. Wendel, and K. Yazawa. Overview of the architecture, circuit design, and physical implementation of a first-generation cell processor. *IEEE Journal of Solid-State Circuits*, 41(1):179–196, Jan 2006.
15. S. Clark, K. Haselhorst, K. Imming, J. Irish, D. Krolak, and T. Ozguner. Cell Broadband Engine interconnect and memory interface. In *Hot Chips 17*, Palo Alto, CA, Aug 2005.
16. Top 500 list of supercomputers (www.top500.org)
17. A. Bergmann. Linux on Cell Broadband Engine Status Update. *Proceedings of the Linux Symposium*, June 27–30, 2007 Ottawa. http://ols.108.redhat.com/2007/Reprints/bergmann-Reprint.pdf
18. B. Flachs and M. Nutter. Private communication.

19. T. Zhang and J. K. O' Brien. Private communication.
20. W. Lundgren, K. Barnes, and J. Steed. "Gedae Portability: From Simulation to DSPs to Cell Broadband Engine", In *HPEC 2007*, Sep 2007. (poster) (http://www.ll.mit.edu/HPEC/agendas/proc07/Day3/10_Steed_Posters.pdf).
21. J. Bergmann, M. Mitchell, D. McCoy, S. Seefeld, A. Salama, F. Christensen, T. Steck. Sourcery VSIPL++ for the Cell/B.E. *HPEC 2007*, Sep 2007.
22. M. McCool. A unified development platform for Cell, GPU, and Multi-core CPUs. *SC'07*, (http://www.rapidmind.com/pdfs/SC07-TechTalk.pdf).
23. J. M. Perez, P. Bellens, R. M. Badia, and J. Labarta. CellSs: Making it easier to program the Cell Broadband Engine processor. In *IBM Journal of Research and Development*, 51(5):593–604, Oct 2007.
24. A. E. Eichenberger, J. K. O'Brien, K. M. O'Brien, P. Wu, T. Chen, P. H. Oden, D. A. Prener, J. C. Shepherd, B. So, Z. Sura, A. Wang, T. Zhang, P. Zhao, M. K. Gschwind, R. Archambault, Y. Gao, and R. Koo. Using advanced compiler technology to exploit the performance of the Cell Broadband Engine^{TM} architecture. In *IBM Systems Journal*, 45(1):59–84, 2006.
25. http://openmp.org
26. D. Kunzman, G. Zheng, E. Bohm, and L. Kale. Charm++, offload api, and the cell processor. In *Proceedings of the Workshop on Programming Models for Ubiquitous Parallelism* (at PACT 2006).
27. K. Fatahalian, T. J. Knight, M. Houston, M. Erez, D. R. Horn, L. Leem, J. Y. Park, M. Ren, A. Aiken, W. J. Dally, and P. Hanrahan. Sequoia: Programming the memory hierarchy. In *Proceedings of the 2006 ACM/IEEE Conference on Supercomputing*, 2006.
28. ALF for the Cell/B.E. Programmer's guide and API reference. http://www.ibm.com/chips/techlib/techlib.nsf/products/IBM_SDK_for_Multicore_Acceleration
29. M. Ohara, H. Inoue, Y. Sohda, H. Komatsu, and T. Nakatani. MPI microtask for programming the Cell Broadband Engine^{TM} processor. In *IBM Systems Journal*, 45(1):85–102, 2006.
30. S. Williams, J. Shalf, L. Oliker, S. Kamil, P. Husbands, and K. Yelick. The potential of the cell processor for scientific computing. In *ACM International Conference on Computing Frontiers*, 2006.
31. T. Chen, R. Raghavan, J. N. Dale, and E. Iwata. Cell Broadband Engine Architecture and its first implementation-A performance view. In *IBM Journal of Research and Development*, 51(5):559–572, Oct 2007.
32. S. Williams, L. Oliker, R. Vuduc, J. Shalf, K. Yelick, and J. Demmel. Optimization of sparse matrix-vector multiplication on emerging multicore platforms. In *Supercomputing (SC07)*, 2007
33. L. Cico, R. Cooper, and J. Greene. Performance and Programmability of the IBM/Sony/Toshiba Cell Broadband Engine Processor. *White Paper*, 2006. http://www.mc.com/uploadedFiles/CellPerfAndProg-3Nov06.pdf
34. B. Gedik, R. R. Bordawekar, and P. S. Yu. CellSort: High performance sorting on the cell processor. In *Proceedings of 33rd International Conference on Very Large Databases*, pp. 1286–1297, 2007.
35. H. Inoue, T. Moriyama, H. Komatsu, and T. Nakatani. AA-Sort: A New Parallel Sorting Algorithm for Multi-Core SIMD Processors. In *Proceedings 16th International Conference on Parallel Architecture and Compilation Techniques*, pages 189–198. PACT 2007, 15–19 Sept 2007.
36. O. Villa, D. P. Scarpazza, and F. Petrini. Accelerating Real-Time String Searching with Multicore Processors. *IEEE Computer*, 41(4), Apr 2008.
37. D. A. Bader, V. Agarwal, K. Madduri, and S. Kang. high performance combinatorial algorithm design on the cell broadband engine processor. In *Parallel Computing*, 33(10–11):720–740, 2007.
38. L.-K. Liu, S. Kesavarapu, J. Connell, A. Jagmohan, L. Leem, B. Paulovicks, V. Sheinin, L. Tang, and H. Yeo. Video Analysis and Compression on the STI Cell Broadband Engine Processor. In *Proceedings of IEEE International Conference on Multimedia and Expo*, pages 29–32, 9–12 July 2006.
39. Folding@home (http://folding.stanford.edu/).

40. F. M. Schellenberg, T. Kingsley, N. Cobb, D. Dudau, R. Chalisani, J. McKibben, and S. McPherson. Accelerating DFM Electronic Data Process using the Cell BE Microprocessor Architecture. In *Electronic Data Process (EDP) Workshop*, Monterey CA, April 12, 2007.

41. J. A. Turner et al. Roadrunner Applications Team: Cell and Heterogeneous Results to date. Los Alamos Unclassified Report LA-UR-07-7573. http://www.lanl.gov/orgs/hpc/roadrunner/rrperfassess.shtml

42. S. Williams, J. Carter, L. Oliker, J. Shalf, and K. Yelick. Lattice Boltzmann simulation optimization on leading multicore platforms. In *International Parallel & Distributed Processing Symposium (IPDPS)*, 2008.

43. The Potential of the Cell Broadband Engine for Data Mining (ftp://ftp.cse.ohio-state.edu/pub/tech-report/2007/TR22.pdf)

44. MapReduce for the Cell B.E. Architecture. University of Wisconsin Computer Sciences Technical Report CS-TR-2007-1625, Oct 2007.

Index

A

Address interleaving
    data address, 94, 95
    instruction identifier, 98, 99
    register name, 95
Address resolution buffer, 122, 134, 135
Amdahl's law, 181, 273
Atomicity
    failure, 150, 151, 159
    strong, 151, 155
    weak, 151, 155

B

Bandwidth hierarchy, 233, 237, 239, 240–241, 248, 251
Block-atomic execution, 79, 81, 97
Brook, 234, 243, 244–245, 246
Bulk data transfer
    DMA, 243
    gather, 237
    scatter, 237

C

Caches
    coherence, 37, 101, 133, 135, 164, 165, 166, 182, 184, 191, 194–195, 217, 257
    interleaved, 94, 95, 100
    multi-level, 6, 175, 182
    private, 96, 101, 121, 135, 136, 157, 163, 165, 182
    shared, 4, 68, 101, 104, 121, 122, 134, 165, 182, 183, 186, 198, 211, 216, 218, 220
Cell Broadband Engine, 4, 234, 245, 266, 271–293
Cell Broadband Engine Architecture (CBEA), 274–280, 282, 284, 285, 292
Charm++, 287
Clusters, 241–242
Complexity wall, 174

Composable processors, 73–107
Conflict detection
    optimistic, 152, 156, 157, 166
    pessimistic, 152, 156, 166
Control-flow graph, 115, 125, 127
Cryptography coprocessor, 211, 223

D

Data-level parallelism (DLP), 2, 5, 27, 29, 235, 237, 241, 242, 245, 250, 257, 260, 261, 262, 263, 264, 265
Data locality, 237, 285, 287, 290
Data versioning
    eager, 151, 152, 155, 157, 161, 165
    lazy, 152, 155, 161, 165, 167
Deadlock, 9, 41, 43, 52, 60, 112, 141, 146, 150
Distributed protocol
    commit, 88, 90, 91, 95, 96
    execution, 88, 89–90, 91, 92, 93, 95, 106
    instruction fetch, 88, 89, 91, 95
Dynamic instruction stream, 113, 115, 116

E

Explicit Data Graph Execution (EDGE), 77, 78, 79, 80, 81, 82, 95, 96, 98, 99, 102
Exposed communication, 28, 233, 237, 240, 266

F

Fine-grained locking, 146, 149
Flow control
    flow control digit (flit), 37, 38, 42, 43, 44, 45, 46, 47, 48, 52, 57, 59, 60, 62
    packet, 42, 43, 44, 45, 46, 47, 52, 53, 55, 58, 59, 60, 61, 62, 83, 89, 194, 199
    store-and-forward, 42
    virtual cut-through, 42, 50
    wormhole, 42, 43
Frequency wall, 174

G

Garbage collector, 155, 163
Gedae, 286
Gustafson's law, 273

H

Hardware-accelerated software transactional
    memory (HASTM), 162–165, 170
Hardware transactional memory (HTM), 141,
    146, 165, 166, 167, 168, 170
Heterogeneous chip multi-processor (HCMP),
    273, 275, 278, 292
Heterogeneous processors, 75–76, 202,
    271–293
HyperTransport, 188, 198, 199, 200
Hypervisor, 179, 181, 183, 196, 201, 279, 284

I

Imagine, 11, 25, 27, 29, 151, 234, 243, 244,
    246–251, 253, 255, 256, 257, 258, 261,
    266
Instruction-level parallelism (ILP), 1, 2, 4, 5,
    6, 7, 8, 12, 15–16, 19, 24, 25, 26, 27,
    28, 29, 30, 76, 77, 78, 92, 93, 102, 103,
    104, 111, 112, 115, 139, 142, 174, 205,
    207, 208, 209, 227, 235, 237, 241, 244,
    250, 253, 260, 261, 262, 263, 264, 265
Isolation, 76, 125, 147, 149, 151, 168, 197, 279

J

Just in time (JIT) compiler, 154, 155, 161

K

Kernel, 18, 20, 101, 178, 232, 233, 234, 235,
    236, 237, 238, 239, 240, 241, 243–244,
    245, 246, 247, 249, 250, 251, 253,
    256, 260, 263, 264, 288, 289–290,
    291, 292

L

Latency, 6, 7, 8, 9, 13, 14, 23, 24, 27, 28, 29,
    36, 37, 38, 39, 42, 44, 49, 50, 52, 56,
    57, 59, 60, 63, 64, 66, 79, 85, 88, 89,
    98, 101, 121, 162, 169, 175, 177, 182,
    184, 185, 189, 190, 193, 195, 196, 197,
    199, 200, 206, 207, 208, 209, 212, 213,
    216, 218, 219, 223, 224, 231, 232, 233,
    234, 235, 237, 239, 240, 243, 250, 257,
    258, 264, 266, 274, 276, 277, 278, 280,
    281, 284, 285

M

Mailbox synchronization, 277
Managed runtime environment, 180, 181

Memory
    coherence, 37, 275
    consistency, 104, 154, 155, 156, 272
    disambiguation, 87, 100, 123
Memory flow control (MFC), 274, 275–278,
    279, 281, 282, 284, 292
Memory-level parallelism (MLP), 185, 186,
    187
Merrimac, 234, 240, 243, 244, 245, 246, 251,
    253–259, 261, 263, 264, 265, 266, 267
Micronet, 83, 88
Moore's law, 1, 2, 3, 35, 173, 174, 201
Multiple instruction multiple data (MIMD),
    260, 261, 264
Multiscalar, 28, 113, 123
Multithreading
    coarse-grained, 74, 146, 148, 149, 150, 209
    fine-grained, 146, 149, 150, 209, 211
    simultaneous, 117, 209, 211
    thread selection, 213, 214

N

Network clock timing
    mesochronous, 49, 59, 63–64
    synchronous, 63, 66
Network interface, 9, 18, 35, 37, 49, 50, 52, 57,
    88, 206, 248
Niagara 2 processor, 220, 226
Niagara processor, 4, 36, 205, 210–227, 228

O

Operand network, 7, 8, 27, 28, 29, 48, 49,
    50–57, 83, 84, 93, 102, 103
Opteron processor, 175, 188, 189, 190, 191,
    192, 193, 195, 197, 201, 283, 289

P

Pipeline, 2, 3, 6, 7, 8, 12, 23, 24, 27, 28, 29,
    37, 44, 45, 46, 47, 49, 50, 53, 54, 57,
    58, 59, 60, 61, 62, 66, 85, 87, 88, 89,
    90–91, 93, 95, 96, 122, 126, 182, 185,
    189–191, 193, 206, 207, 209, 211, 212,
    213, 214, 215, 216, 218, 219, 220, 221,
    222, 223, 224, 237, 239, 240, 241, 244,
    248, 253, 256, 258, 261, 279, 281
Power
    dynamic, 197, 202, 207, 213, 214, 225
    efficiency, 75, 77, 93, 106, 142, 174, 197,
        248, 249, 253, 292
    management, 64–65, 66, 93, 177, 178,
        197–198, 202, 280
    static, 209, 213, 214
    wall, 174

Power processor element (PPE), 275, 278, 281, 282, 283, 287
PowerXCell8i, 283, 289
Prefetching, 175, 177, 285
Processor granularity, 73, 75, 76, 82, 93, 103, 104, 105, 106
Program dependencies
    control-flow, 111, 113, 117, 118, 119, 125, 127, 139
    data, 111, 113, 117, 118, 119, 120, 123, 124, 125, 126, 127, 128

R
Rapidmind, 286
Raw processor, 1, 2, 5, 7, 10, 11, 12, 27, 29
Read
    barrier, 154, 155, 157, 159, 161, 163, 164
    set filtering, 158
Register
    mask, 130, 131, 132
    renaming, 76, 87, 96, 121, 122, 130, 279
Reliability, 37, 38, 41, 44, 67, 174, 194, 257
Rock CMP, 168
Rollback, 118, 119, 122, 123, 124, 126, 127, 131, 134, 138, 151, 159
Router design
    arbitration, 48, 52, 53, 55, 56, 60, 62, 216, 223
    crossbar, 9, 29, 36, 45, 46, 47, 48, 53, 60, 62–63, 65
    pipeline, 44, 45, 46, 47, 49, 50, 53, 54, 57, 58, 59, 60, 61, 62, 66
    router buffers, 41, 47
Routing algorithm
    dimension order, 9, 28, 37, 38, 40, 41, 45, 47, 50, 52, 59, 74, 77, 83
    dynamic, 8, 9, 10, 15, 28, 47, 49, 50, 51, 58, 62, 104
    static, 8, 9, 49, 59, 60, 62

S
Scalar unit, 238, 239, 252
Scheduling
    kernel, 237, 239
    memory access, 243, 252
    stream, 237, 243
Security, 169, 223, 279
Sequoia, 234, 243, 245–246, 266, 287
Serializability, 147
Server systems, 26, 273
Signature-accelerated software transactional memory (SigTM), 163

Single instruction multiple data (SIMD), 225, 239, 241, 242, 243, 244, 250, 258, 260, 274, 278–279, 284, 287, 290
Software managed cache, 8, 9, 37, 49, 50, 57, 274, 285
Software transactional memory (STM), 153–156, 157, 158, 159, 160, 161, 162–165, 167, 168, 169
Specialization, 5, 7, 9, 12, 25, 76, 239
Speculative
    multithreading, 113, 135, 139, 140, 141
    thread, 212
    versioning cache, 122, 135, 136
Storm I, 74, 234, 240, 251–253, 254, 261, 266
Stream, 2, 5, 6, 7, 10, 12, 13, 15, 17–21, 23, 24, 25, 26, 27, 29, 30, 75, 88, 96, 113, 114, 115, 116, 117, 121, 222, 223, 231–267, 274, 288, 290, 291
StreamC, 234, 243–244
Stream controller, 238, 239, 247
Stream flow graph (SFG), 235, 236, 237, 239
StreamIt, 18–19
Stream processing, 75, 222, 231–266, 267, 274
Stream register file (SRF), 29, 238, 240, 241, 242, 243, 244, 245, 248, 249, 250, 252, 255, 256, 257, 258, 261, 263, 264, 265
    lane, 240, 265
Stream unit, 238, 251, 256
Symmetric multiprocessor, 162, 175, 179, 188, 206, 228
Synchronization, 49, 58, 59, 60, 64, 79, 112, 119, 120, 124, 127, 138, 141, 145, 146, 151, 169, 181, 182, 218, 239, 261, 264, 265, 273, 277, 291
Synchronous dataflow (SDF), 232, 235, 237, 243
Synergistic processor unit (SPU), 274, 275, 276, 277, 278, 279, 280, 281, 282, 284, 289, 290

T
Task
    commit, 117, 123
    selection, 123, 124, 125, 126, 127, 128, 132, 139
TeraFLOPS processor, 35, 49, 58–66
Thread-level parallelism (TLP), 2, 5, 6, 12, 27, 29, 50, 73, 74, 76, 77, 106, 113, 141, 235, 260, 262, 263, 264, 265, 267
Throughput, 3, 13, 22, 23, 24, 36, 38, 39, 41, 42, 43, 44, 47, 48, 66, 67, 75, 77, 78, 101, 103, 104, 105, 177, 178, 179, 181, 184, 185, 186, 188, 193, 195, 201,

Throughput (*cont.*)
        205–228, 232, 233, 237, 239, 240, 241,
        243, 256, 257, 258, 261, 263, 266, 273
Tile, 1–30, 36, 49, 50, 51, 52, 55, 56, 57, 58,
        59, 60, 62, 63, 64, 65, 66, 82–88, 89,
        91, 92, 93, 98, 102, 105
Tiled microarchitecture, 20, 82–85
Tilera processor, 3, 29, 36
Topology
        mesh, 28, 29, 38, 39, 40, 45, 47, 48, 50, 52,
        58, 59, 65, 66, 83, 84, 85, 101, 256, 265
        ring, 4, 38, 119, 121, 122, 129, 132
        torus, 38, 39, 50
Transactional memory (TM), 113, 141, 145,
        146, 147, 148, 149, 150, 151–153, 154,
        155, 156, 157, 159, 163, 165, 166, 167,
        168, 169, 170
Transaction descriptor, 155, 159, 160, 161
Transaction records, 155, 163, 165
Transactions, 68, 80, 141, 145–170, 177, 185,
        194, 216, 217, 273, 281
Translation lookaside buffer (TLB), 85, 86,
        87, 88, 183, 184, 191, 193, 211, 222,
        224, 282

TRIPS processor, 3, 27, 28, 35, 48, 50–57, 78,
        79–80, 81, 82, 83, 84, 85–88, 90, 91,
        92, 93, 95, 97, 98, 101, 102, 103, 105,
        106

U
Undo log filtering, 158

V
Versatility, 4, 7, 24, 26, 28, 30, 280
Very long instruction word (VLIW), 5, 15, 29,
        58, 241, 242, 243, 250, 251, 252, 253,
        256, 260, 262, 264
Virtual channels, 43, 44, 45, 46, 47, 48, 50, 52,
        59, 60, 84, 201
Virtual machine, 180, 196, 197
VSIPL++, 286

W
Wire delay, 2, 5, 6, 8, 11, 12, 29, 51, 57
Write barrier, 153, 155, 156, 157, 158, 159,
        160, 161, 162, 163, 165